Y0-BTD-407

Systematic Materials Analysis

VOLUME III

MATERIALS SCIENCE AND TECHNOLOGY

A. S. Nowick and B. S. Berry, Anelastic Relaxation in Crystalline Solids, 1972

E. A. Nesbitt and J. H. Wernick, Rare Earth Permanent Magnets, 1973

W. E. Wallace, Rare Earth Intermetallics, 1973

J. C. Phillips, Bonds and Bands in Semiconductors, 1973

H. Schmalzried, Solid State Reactions, 1974

J. H. Richardson and R. V. Peterson (editors), Systematic Materials Analysis, Volumes I, II, and III, 1974

A. J. Freeman and J. B. Darby, Jr. (editors), The Actinides: Electronic Structure and Related Properties, Volumes I and II, 1974

In preparation

A. S. Nowick and J. J. Burton (editors), Diffusion in Solids: Recent Developments

J. W. Matthews (editor), Epitaxial Growth, Parts A and B

Systematic Materials Analysis

VOLUME III

Edited by

J. H. RICHARDSON

Materials Sciences Laboratory
The Aerospace Corporation
El Segundo, California

R. V. PETERSON

Materials Sciences Laboratory
The Aerospace Corporation
El Segundo, California

ACADEMIC PRESS New York San Francisco London 1974

A Subsidiary of Harcourt Brace Jovanovich, Publishers

ACADEMIC PRESS, INC.
111 Fifth Avenue, New York, New York 10003

United Kingdom Edition published by
ACADEMIC PRESS, INC. (LONDON) LTD.
24/28 Oval Road, London NW1

Library of Congress Cataloging in Publication Data

Richardson, James H
Systematic materials analysis.

(Materials science series)
Includes bibliographies.
1. Materials–Analysis. 2. Instrumental analysis.
I. Peterson, Ronald V., joint author. II. Title.
QD131.R48 543 72-12203
ISBN 0–12–587803–6 (v.3)

PRINTED IN THE UNITED STATES OF AMERICA

Dedicated to the One Perfect Instrument:

HEBREWS 1:2
JOHN 3:17

J.H.R.
R.V.P.

Contents

Chapter 23 Gas Analysis Techniques and Combustion Methods

R. V. Peterson

Chapter 24 Gas Chromatography

Gerald R. Shoemake

Chapter 25 Ion-Scattering Spectrometry for Surface Analysis

Robert S. Carbonara

Chapter 26 Mössbauer Spectrometry

P. A. Pella

Chapter 27 Optical Microscopy

J. H. Richardson

Chapter 28 X-Ray Diffraction

G. M. Wolten

Chapter 29 X-Ray Fluorescence and Absorption Spectrometry

N. Spielberg

List of Contributors

Numbers in parentheses indicate the pages on which the authors' contributions begin.

RAMON M. BARNES (23), Department of Chemistry, University of Massachusetts, Amherst, Massachusetts

ROBERT S. CARBONARA (229), Battelle Memorial Institute, Columbus Laboratories, Columbus, Ohio

CHARLES E. KLOPFENSTEIN (1), Departments of Chemistry and Computer Science, University of Oregon, Eugene, Oregon

P. A. PELLA (241), Institute for Materials Research, Analytical Chemistry Division, National Bureau of Standards, Washington, D. C.

R. V. PETERSON (125), The Aerospace Corporation, El Segundo, California

JUAN RAMÍREZ-MUÑOZ (85), Beckman Instruments, Inc., Irvine, California

J. H. RICHARDSON (269), The Aerospace Corporation, El Segundo, California

GERALD R. SHOEMAKE (159), Texas Engineering and Science Consultants, Inc., Houston, Texas

N. SPIELBERG (333), Department of Physics, Kent State University, Kent, Ohio

CHARLES L. WILKINS (1), Department of Chemistry, University of Nebraska, Lincoln, Nebraska

G. M. WOLTEN (299), The Aerospace Corporation, El Segundo, California

Preface

It is both exciting and dismaying to observe the parade of new and refined instrumental methods available for the analysis of materials—exciting because these instruments provide opportunities for faster and more reliable answers to material analysis problems, dismaying because one is hard pressed to evaluate these various instruments for a given task. Materials analysis often involves the complete characterization of a material, including structural and textural analyses in addition to chemical analysis.

It has been the aim of the editors of *Systematic Materials Analysis* to satisfy the needs of the materials analyst in these areas by presenting brief discussions on a broad range of instrumental methods and bringing to their selection new approaches that will yield the desired information about a given material. These volumes not only comprise a brief, comprehensive reference for the materials analyst but also provide a source of information for the engineer or researcher who must select the appropriate instrument for his immediate needs. Although the volumes are directed toward the physical sciences, they can also be of value for the biological scientist with materials problems and of use to the laboratory administrator as both convenient reference and guide for the purchase of new instrumentation.

Chapter 1 focuses on the selection of analytical methods on the bases of specimen limitations and information desired. The selection is made by use of flow charts encompassing the various instruments outlined in the succeeding chapters. The unique character and utility of this work lie in the use of these charts, since they present a complete listing of analytical instrumentation arranged so as to permit selection of the best method(s) for a given analytical task. The student may thus gain insights into thought processes that are usually acquired only after years of experience in this field. Thus, these volumes can appropriately serve as a college text (third year to graduate level) as well as a reference work.

The chapters on specific instruments briefly outline the theories of operation, with detailed discussions of theory fully referenced, and describe the capability of the methods for qualitative and quantitative measurements of chemical composition, structure, and texture (as applicable).

Topics such as the sensitivity and selectivity of each method are emphasized. References illustrating the operation of the instrument, as well as references to user-constructed accessories that extend and improve the instrument's capabilities, are included when applicable.

The wide variety of commercial instruments available precludes the inclusion of instructions for the operation of instruments and, consequently, the inclusion for the student of experiments based on these instructions. For the same reason, comprehensive descriptions and the inevitable comparisons of commercial instruments are beyond the scope of this work.

Acknowledgments

We want to thank all the authors of this work for their willing participation in this endeavor, and we gratefully acknowledge their corrections and comments on the flow charts in Chapter 1.

We also want to thank our many colleagues at The Aerospace Corporation who gave support in various ways, especially Mrs. Genevieve Denault, Camille Gaulin, Dr. Wendell Graven, Henry Judeikis, Dr. Gary Stupian, and Dr. Hideyo Takimoto, who rendered specific suggestions and reviewed chapters. We remember with special affection the late Dr. Thomas Lee, whose remarks and comments were very valuable in the development of the concept of this work.

We also wish to thank Miss Debra Levy and Mrs. Myra Peterson for help in the critical review of the work and Miss Rosalie Hernandez, Mrs. Jean Hill, Mrs. Carolyn Thompson, and Mrs. Marsha Graven for typing assistance.

To Ann and Myra Ann, our wives, we are grateful for their love and their spiritual challenge to us.

Contents of Other Volumes

CHAPTER 20

Computer–Instrument Interfacing

Charles L. Wilkins

Department of Chemistry
University of Nebraska
Lincoln, Nebraska

and

Charles E. Klopfenstein

Departments of Chemistry and Computer Science
University of Oregon
Eugene, Oregon

Introduction

It is clear that in the space allotted here we cannot expect to present an in-depth analysis of all the details of computer–instrument interfacing. Rather, we hope to provide a basic introduction to the considerations involved, some guidelines regarding useful approaches to the interface problem, and finally, some discussion of ways to implement interfaces once they have been adequately defined. In order that there be no misunderstanding, we simply define the term "interface" as that which describes the total hardware and software (or program) package required to allow for all necessary communication between an experiment or an instrument and a digital computer. It should be recognized that, in general, the optimal interface for any particular application will be very much a function of the requirements and limitations imposed on it by the specific details of the analytical instrument. Our subsequent comments should therefore be interpreted in this light and should be understood to represent general, although not necessarily optimum solutions, to the particular sorts of problems we discuss. It is more important, we believe, to outline a methodology for attacking computer–instrument interface problems rather than to present specific solutions for particular problems. It is our hope that this chapter will successfully reflect that philosophy.

1 Theory

An excellent approach to any interface problem is to define precisely what one wishes to accomplish. This can be done systematically and the requirements conveniently categorized for ease of later use. It is absolutely essential that, before any construction of a hardware interface be initiated, the analyst have a rather complete definition of experimental requirements for the interface and an awareness of exactly what measurements and control functions will be required. Therefore, a qualitative overview of the problem should be the first item of business.

If measurements are to be made, monitored, or logged by a digital computer and possibly used for control purposes, it is necessary that the electronic and mechanical characteristics of these measurements be completely described (or at least completely enough so that the necessary information about the interface may be deduced). Therefore, before any consideration at all is given to the details of the interface hardware or the computer system, a careful consideration of the experiment itself is in order. This consideration should include, at a minimum, delineation of the conditions of the operation of the instrument or experiment, some description of the transfer functions required, and a brief description of the instrument's function as a transducer. It is probably obvious that a description of the operating conditions of an instrument should also include consideration of the environmental conditions to which it is normally subjected, the degree of training and knowledge of its operators, and reliability requirements. It is, perhaps, not so obvious what is meant by the terms "transfer functions" and "transducer." A transfer function is a description of the steps required to change one type of chemical or physical information to another. For example, the statement "shaft rotation motion must be converted to a digital number proportional to wavelength" describes a transfer function explaining that a mechanical motion must somehow be converted to wavelength. This very common problem in optical spectrometry may be solved in a number of ways (Willis *et al.*, 1970; Klauminzer, 1970; Johnson *et al.*, 1969). It is important to note that statement of the requirements of a transfer function does not necessarily imply the knowledge of how this transfer function is to be carried out. In other words, it is possible to define such functions without understanding whether the required transfer functions are realizable in terms of existing hardware technology. Many times, a transducer may be required to implement the transfer functions. It is therefore clear that if a particular transfer function requiring a transducer is to be carried out, a suitable transducer must exist. The transducer can thus be considered a hardware device used to carry out a part of a given transfer function. In this sense, the instrument itself is a transducer and can be analyzed in precisely the same terms as any other kind of transducer. Its only difference is that it is a somewhat more complex transducer than those engineers are generally accustomed to deal with.

After preliminary consideration of the technical aspects of a particular computer–interface problem is complete, the objectives and possible benefits from use of the computer in an analytical application remain to be defined. Although it is often assumed that for almost any particular application the addition of a computer to the procedure will result in the realization of improvements, this may only be the case in a certain narrow sense. For example, if the expense of such a procedure is far in excess of the value of the improved results, use of a computer may be contraindicated. One of the greatest sources of dissatisfaction with computerized laboratory systems has been the failure to clearly define the objectives to be achieved and the benefits expected before proceeding with the project. If a clear statement about these matters in not available at the outset, it is unlikely that all of the objectives will be satisfied, at least in the first computer–instrument interface produced. Although it may seem so obvious as to be unworthy of mention, it is usually unsatisfactory to attempt to modify the characteristics of an interface after it has been fabricated in order to satisfy new objectives. For one thing, as mentioned above, the interface consists of more than an electronic or electromechanical hardware package; it also includes the programs necessary to allow operation of these elements. Changes in design objectives will almost invariably require changes in the software or programs used. It is for this reason that we stress the importance of a fairly careful definition of the results and benefits expected before detailed interface designs are made.

Of course, if a qualitative overview of the interface problem is to be completed in a useful form, it is necessary that the analyst have some awareness of the broad general functions and capabilities of the digital computer in this context. Furthermore, he should be aware that for some applications simple logging of the experimental data may be satisfactory and that a computer, at least an on-line system, may not be required at all. Briefly, we can arrange possible tasks in order of complexity and difficulty in the following way. Data logging, as mentioned above, may be satisfactory for the purpose. All that need be used in this circumstance is a simple device which will permit monitoring of an experiment or instrument, digitization of the data, and recording of such data on a medium compatible with later computer analysis. Sometimes, of course, computer analysis is not required at all, since logging the data for record-keeping purposes may be sufficient. In that case, some kind of printout will serve quite satisfactorily. Examples of mediums used for recording and storing data are punched paper tape (Oberholtzer, 1967), punched cards (Brown *et al.,* 1966), magnetic tape storage (Hites and Biemann, 1967; Venkataraghavan *et al.,* 1967), or a simple keyboard printer such as a teletype. In all of these cases, the data would be converted to digital form as a preliminary step. Of course, it is also possible to record an instrument's output in analog form using a strip chart recorder or an analog magnetic tape logging device. For a great many cases it will, however, be necessary to perform analog-to-digital conversion before recording the data. Sometimes, the expense of rapid logging devices will dictate the use of the small dedicated computer rather than a relatively unintelligent, although fast, logging device. The reasons for this are twofold. First, a small dedicated computer has a great many capabilities in addition to its ability to be used as a rapid data-logging device. All other things being equal, a more general-purpose device is often more desirable than one with fewer capabilities. Second, for the most rapid data-logging devices, the expense may well exceed that which would be involved if a judicious use of a small computer were implemented. For implementing special memories there are, certainly, data-logging devices which provide the highest speed capabilities at costs comparable to those of minicomputers. For certain applications these are the reasonable solution, since in some cases the ability of these special devices to *log* data at high speeds exceeds that of general-purpose computers. Rather frequently, it is found that data acquisition, as the logging process is often called, is only the first step in a sequence of events which leads to full automation of a particular experiment. Many times the requirements are such that a simple logging device will no longer serve. An example of this is the very common requirement that monitoring be carried out for long periods of time and data only be stored when certain conditions are detected. The rest of the time, in this situation, the primary function of the monitoring device is to determine the absence of events of chemical interest and to reject incoming data. An example of such an experiment is gas–liquid partition chromatography where the appearance of a peak, the time required for the peak to appear, and the area of the peak are major items of interest. Generally, the experimenter is not the least bit interested in watching, or recording, data indicating the absence of peaks. In fact, it is desirable to dispense with this entirely. Additionally, it is very rare that raw data are actually the objects of interest. More often, computations must be performed and the data reduced to useful form such as identities of compounds or relative ratios of components of an unknown mixture. In these applications, the ability to log the data automatically, preferably in digital form, is a very great advantage. It is far more advantageous to be able to log only the data of interest and to eliminate the data which are not pertinent to the solution of a particular problem. This transfer function can be accomplished in two ways. One way is through the use of suitable electronic discriminating circuitry (Gill, 1969), or alternatively, through a digital computer which examines the data as they are collected and makes decisions as to their value. For slow experiments, the data reduction may be performed "simultaneously" with this process so that at the moment the experiment is complete, the reduced information is available (Westerberg, 1969a; Hancock *et al.*, 1970). This can then be printed by the computer in a reasonable report form. In this application, although it is a simple one, the computer performs the very useful function of filtering and condensing a large amount of information and presenting the data in a format which requires the scientist to consider only pertinent information. Of course, if this application is to be possible, the parameters of the experiment must be well defined and the distinction between vital data and useless data must be clear. It is also essential that algorithms (or detailed methods for computational treatment of the required control, discrimination, and data reduction functions) be available.

The next possible step is adding to the capability of the computer so that, while examining the data, it can provide some control information to the experiment. In the simplest cases, control decisions can be completely preprogrammed, and the computer need only activate a particular control sequence when a predefined condition is detected in the incoming data. To use a simple example, the computer can be so programmed that if it finds the temperature of a heating bath has exceeded an allowable limit, a relay which removes power from the heat source will be activated and thus prevent a catastrophic explosion. This sort of control and decision-making capability is often called "open-loop control" to distinguish it from a more sophisticated type of control which allows the computer more discretion. In the open-loop control situation, no computer feedback is generally used in reaching the decision. In other words, the programmer has anticipated all possible situations and programmed the computer to carry out a certain preset sequence of operations depending on the conditions found. A more sophisticated type of control, "closed-loop control," puts the computer in a feedback loop with the experiment. As before, when data are received by the computer, they are analyzed, and actions which affect the continuation of the experiment are carried out. This in turn, may modify the incoming data, which are further analyzed, and possibly may initiate further control sequences from the computer. One example of this is the use of a computer to monitor a nuclear magnetic resonance spectrometer and, if it finds the magnetic field to be drifting, to adjust the magnetic field so as to optimize its homogeneity and thus, presumably, to minimize the drift. In this case, the computer might examine a particular reference peak, corresponding to the absorption of some nucleus in a reference sample in the spectrometer, and determine whether that peak is decreasing or increasing in intensity. If it finds that an increase in intensity has occurred as a result of a particular control operation, the computer may execute that operation again. If, on the other hand, a decrease is found, the computer may execute a different control decision. By continual reexamination of the signal in question, followed by appropriate responses, an iterative procedure resulting in continuous optimization of the homogeneity may be carried out. This is feedback or closed-loop control in its truest sense. As can be seen from the foregoing discussion, there is a very wide range of possibilities for computer–instrument interaction. The level of such interaction required for any particular problem will be the result of a number of factors including the objectives expected, the economic restraints, the experimental hardware, and the benefits required. Factors influencing these things will be the speed and cost of available peripheral devices, the data rates and experiment lengths for various types of potential applications, and the cost of suitable computer equipment. Rough estimates of some of these quantities are included in Tables 1–3. Table 1 simply lists some of the standard peripheral computer equipment with which most readers may be familiar. A comparison of both costs and speeds shows that the variety is large, but usually, the higher the speed of a device, the more expensive it is. For convenience we have included devices which may be used as either logging or computer input–output devices. In this category are the various tape punches, printers, and magnetic tape units. Table 2 provides some very rough estimates of the prices for a minimal minicomputer at the present time. It should be understood that these prices are still changing so rapidly that the table ought only be used as the roughest qualitative guideline of what maximum prices for common computers should be. Even now, a trend

TABLE 1

STANDARD PERIPHERAL EQUIPMENT

Device	Speed	Cost ($)	Use
ASR-33	10 char/sec	≃1,000	Most common hard copy I/O device
IBM 2741 or ASR-37	15 char/sec	≃4,000	Hard copy I/O
Cathode-ray-tube units	10–10,000 char/sec	≃1,500–10,000	Graphic display
X–Y plotters (analog)	5–10 in./sec	≃1,000–5,000	Graphic hard copy
Incremental plotters	300 step/sec (0.005–0.01 in./step)	≃2,800–10,000	Graphic hard copy
Nonimpact printers	10–11,000 char/sec	≃2,000–25,000	Printing and/or plotting
Paper tape readers	10–1000 char/sec	≃500–3,000	Input
Paper tape punches	10–150 char/sec	≃800–3,500	Output and data logging
Cassette tape	50–1600 char/sec	≃1,000–8,000	I/O auxiliary storage
Magnetic tape	2000–30,000 16-bit words/sec	≃7,500–15,000	I/O bulk storage; program storage
Punched card reader	50–1200 cards/min	≃4,000–30,000	Input
Card punch	100–600 cards/min	≃10,000–30,000	Output
Magnetic disk and drum	10,000–300,000 16-bit words/sec	≃6,000–31,000	Bulk storage systems base

toward more and more inexpensive computer hardware, some of it employing new technology, continues to reduce these prices. In any case, it is clear that general-purpose computer hardware is becoming far less expensive than many of the laboratory instruments commonly employed. Costs of computer equipment may be further minimized by avoidance of the very common tendency to "over specify" (i.e., purchase equipment in excess of that actually required for the purpose).

Table 3 contains a summary of some of the data rates for common analytical instrumentation. A point to be noted here is that for any particular experiment, the data-acquisition requirements may be characterized both by the rate at which data must be collected (and digitized) and the total length of the experiment, which we have called the epoch. For experiments requiring relatively low data-acquisition rates (20–100 points/sec), the total

TABLE 2

MINICOMPUTER ECONOMICS—1972
PRICE ESTIMATES[a]

Manufacturer	Model	Price ($)
Varian Data Machines	620/L	9,500
Varian Data Machines	620/F	14,800
Digital Equipment Corp.	PDP-11	19,150
Data General Corp.	Nova 1220	9,500
Data General Corp.	Nova 820	11,000
Hewlett-Packard	2100A	12,750
Texas Instruments	960A	7,000

[a] These are estimates of current prices for an 8K 16-bit word computer equipped with an ASR-33 teletype and multiply–divide hardware.

TABLE 3

DATA RATES AND EPOCH LENGTHS FOR CHEMICAL INSTRUMENTATION

Device	Data rate (points/sec)	Length of epoch (sec)
Nuclear magnetic resonance	20–100	10–100,000
Electron spin resonance		
Infrared spectrometry		
Ultraviolet spectrometry		
Gas chromatography		
X-ray diffractometer		
Slow kinetics		
Low-resolution mass spectrometer	500–4000	150–5000
Potentiometer		
Medium speed kinetics		
Mass spectrometry	10,000–100,000	10–100
Chronopotentiometry		
Stopped-flow (fast) kinetics		
Flash photolysis		

length of the experiment may be relatively long. In some of these cases experiments as long as 100,000 sec are not uncommon. On the other hand, as the rate at which data must be collected increases, typically, the total length of the epoch decreases. The net result of this trend is that total quantities of data to be acquired are somewhat constant. Generally on the order of 10^3 to 10^5 individual samples (digitizations) may be required, independent of the rate at which acquisition takes place. To use an example, again from the field of nuclear magnetic resonance, the typical continuous-wave proton nuclear magnetic resonance spectrum of an organic material may span 1000 Hz in 250–500 sec. If resolution approaching that of the instrument is desired (0.2 Hz), this will require a storage of 5000 individual samples of the detector output. Alternately, the same data may be collected in time domain as the result of a high-power rf transmitter pulse which causes a transient signal which is then observed for 3–4 sec. During this time as many as 5000 samples of the output pulse decay may be collected and stored. Then a Fourier transformation (transfer function) may be carried out and the resulting frequency domain spectrum displayed (Farrar and Becker, 1971). The total time (including computer time) for the latter experiment would be on the order of 20 sec. So it is seen that even though the experiment is accomplished approximately 12 times faster, the total quantity of data required remains the same. This is fortunate, for if this were not the case, it would be necessary to provide a storage medium that could hold a very large amount of data. An example of such a situation is found when experiments of widely differing epoch lengths are combined as in gas chromatography–mass spectrometry experiments (Bonnelli, 1971). Depending upon the resolution required of its mass spectrometer, storage of up to 10^8 data points may be required. If the speed at which an experiment produces data is fast enough, it will be impossible to do any processing between the acquisition of individual points. The net result will be that the data must be stored and treated at a later time. Clearly, if it were not for the fact that total quantities of data remain somewhat constant, it would be impossible to design or purchase a computer system that would be capable of handling both the low and the high data-rate experiments. Table 3 summarizes these facts for some common analytical methods.

2 Quantitative Interface Characteristics

Once the interface problem has been surveyed in the manner outlined above, it is then necessary to come to grips with the details of the hardware part of the interface. It is at this stage that attention to the actual hardware which will be used to implement a particular interface is in order. In this section, we will assume that some digital computer is available and that an interface of this computer to a particular instrument or experiment is in order. We can conveniently divide the required hardware into three separate functional categories: first, the required acquisition functions; second, the control functions; and finally, the display functions. We will consider each of these in order.

2.1 Acquisition Functions

In order for the computer to satisfactorily communicate with the experiment or instrument involved, it is necessary that, if the instrument produces analog data (which is the most common case), there be some means of translating these data into the digital information the computer understands. Furthermore, if the instrument needs analog control, it will be necessary to provide the computer with some means of translating its digital information into the analog information that the instrument understands. Analog information is, of course, simply continuously variable data of the kind exemplified by the recording of a voltage output on a strip chart recorder. Digital data can be characterized as a discrete and finite number of individual samples. It can be seen from Fig. 1 that analog data can be very closely approximated by a sufficient number of digital samples. Of course, if an infinite number of digital samples were taken, a digital representation of an analog signal would be indistinguishable from the original signal. Since the digital computer has only the capability of storing and using discrete numbers, and since the laboratory world is an analog world, if an interface is to be successfully designed, it is necessary to understand the available means of analog-to-digital conversion (for instrument to computer communication) and digital-to-analog conversion (for computer to instrument communication). Of course, there are instruments which produce digital information and can accept digital information for control. When this is the case, it is not necessary to provide digital-to-analog conversion or its converse. It will, of course, be necessary to consider the implications of digital control logic. Under the category of acquisition functions, only two of the above-mentioned elements need be considered. These are the means of analog-to-digital conversion and the way digital in-

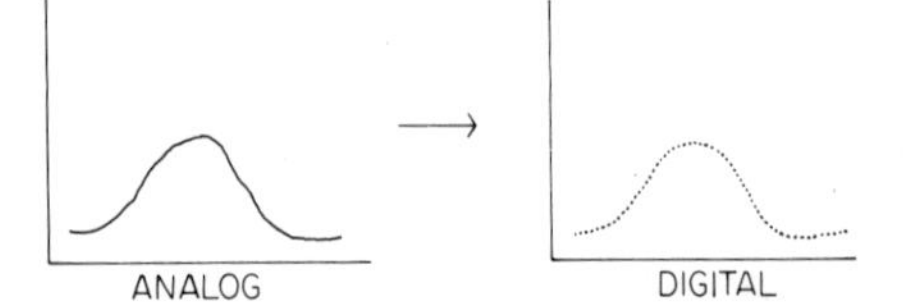

Fig. 1. Analog-to-digital conversion.

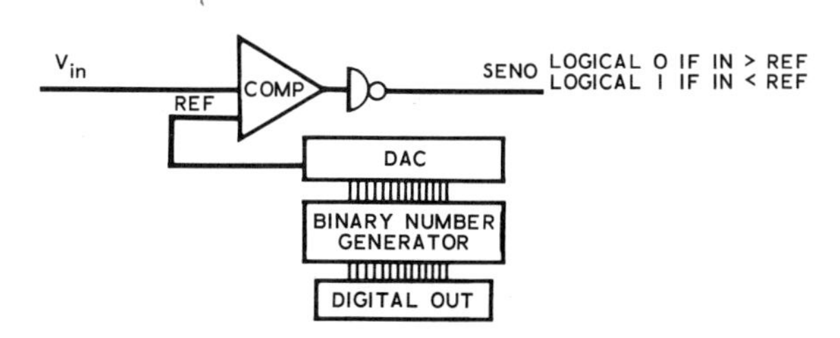

Fig. 2. Simple counter converter (block diagram).

formation may be passed to the computer, if the instrument provides such information. We will consider each of these in turn. Nearly all electronic analog-to-digital converters (ADCs) include a digital-to-analog converter (DAC) and a comparator. The various techniques for conversion differ mainly in how the binary number is generated (Fig. 2). By far the simplest method of number generation is the use of a counter. The counter starts at zero and is incremented until the digital-to-analog voltage output is greater than the input voltage. The binary number is then transferred to the computer and the process is repeated. It can be seen from the representation in Fig. 3 that for this procedure to provide accurate conversions, the clock rate of the counter must be very much higher than the rate of change of the signal. A minor modification of the circuit provides for much more rapid conversion rates and for more accurate tracking. The trick is to replace the counter with one capable of counting both up and down and to use the output of the comparator to control the direction of counting. Such an ADC is illustrated in Fig. 4, and a typical oscilloscope display is shown in Fig. 5. The speed of this converter is much increased over that of the first, but still, the rate of change of the input signal can be no greater than the weight of the least significant bit count rate.

Successive approximation techniques yield even faster conversion rates than the continuous converter (the previous one). Here, the binary word is generated one bit at a time starting with the most significant bit. The possible pathways for a 3-bit converter are shown in Fig. 6, and a typical

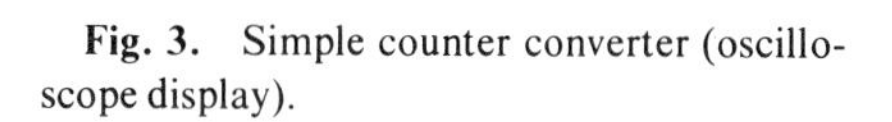
Fig. 3. Simple counter converter (oscilloscope display).

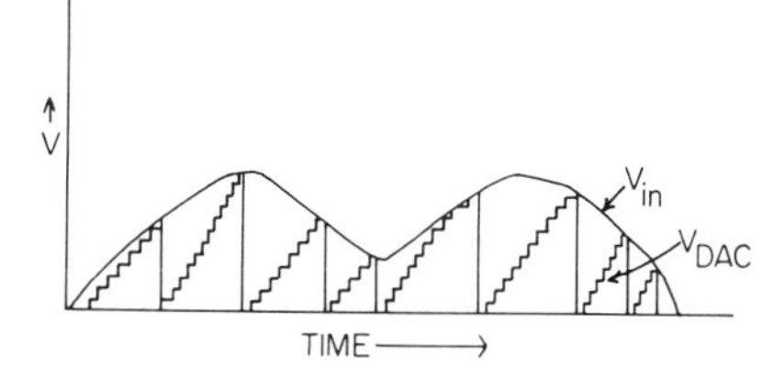

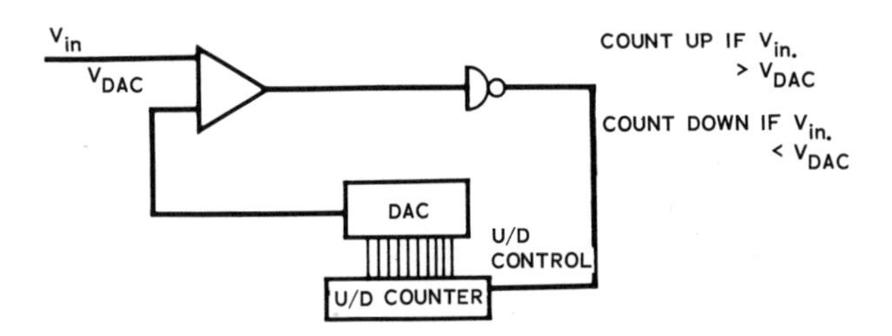

Fig. 4. Continuous converter (block diagram).

oscilloscope display is shown in Fig. 7. The same length of time is required to convert each bit, so in contrast with the first example, where $T \cdot 2^n$ sec were required, here, only $T \cdot n$ sec are required (where T is the time per bit and n the number of bits). Each of these means of conversions has its particular applications, and it is important to understand when one or the other method is indicated. Usually the counter converters are used in situations where relatively high precision is desired, but data conversion rates are slow. The advantage of this procedure is that it is possible to build integrating ADCs of this type which will integrate the sample over the aperture time (the time the sample is actually being observed) and, in this way, reduce the noise and yield an average value of the signal level. This is made possible by suitable design of the sampling part of the ADC. For the more rapid converters, this advantage is of necessity sacrificed, and instead somewhat higher sample rates are possible, although analog noise filtering is no longer so easy to carry out. In this situation, noise, if present, may be removed by digital techniques. The economics of electronic ADCs are straightforward. For a given speed, the more bits of precision required (i.e., the greater the number of binary digits available at the digital output), the more expensive is the converter. This is certainly logical since it simply requires more electronic components to provide higher numbers of bits.

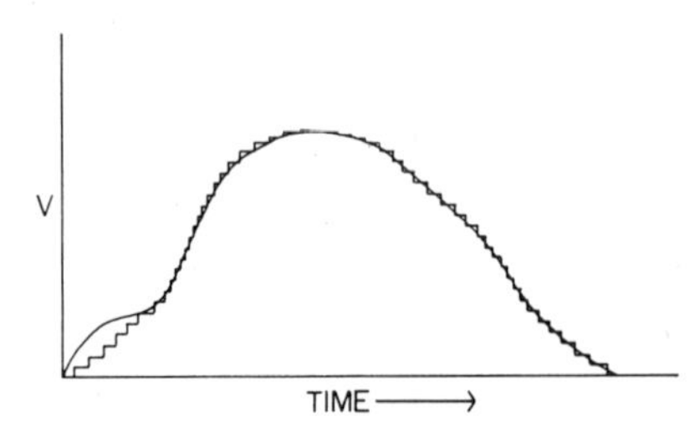

Fig. 5. Continuous converter (oscilloscope display).

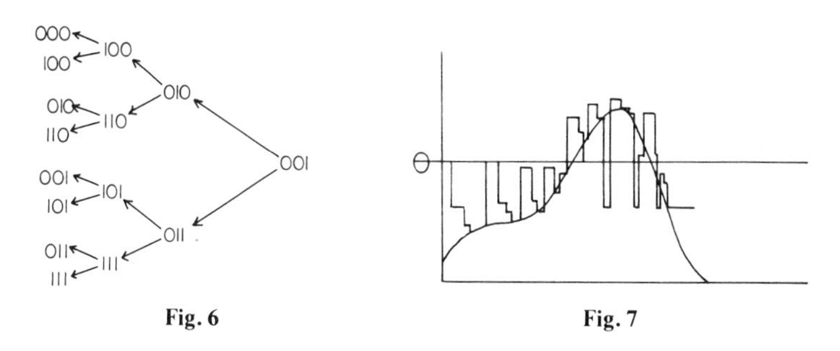

Fig. 6. Successive approximation pathways.
Fig. 7. Successive approximation converter (oscilloscope display).

There are also converters which carry out the conversion of analog mechanical motion into the required digital form. The single most common device of this type is the optical shaft encoder. This device, when attached to a shaft which can rotate, can produce digital information in the form of square wave pulses as a result of the mechanical motion of the shaft. The way this is accomplished is to provide optical masks through which a suitable light source shines, impinging, under certain conditions, on a photoelectric detector. The design of the mask determines when the light will be allowed to impinge on the detector. As the shaft rotates, so does the mask, and at appropriate points, elements of the mask align in such a way as to permit light to penetrate and hit the detector which, in turn, generates the required pulses. If these pulses are monitored by the computer through suitable circuitry, it is then possible to detect not only the direction of the rotation but its magnitude. Usually the pulses are simply counted, since the number of pulses produced is a known function of the number of degrees of rotation of the shaft. A very great variety of such encoders are available, and their uses are many. The main applications of shaft encoders are to convert instruments to allow computer monitoring of such things as the monochromator drive (for optical spectrometers) or other similar parameters for the other instruments. Another type of electromechanical converter is one which uses, instead of an optical mask and a photoelectric detection scheme, actual electrical contacts which are either made or broken depending on the mechanical motion. Perhaps one of the most well known of these devices is that employed in many punched-card readers. Here the presence of a hole in a card allows an electrical contact to be made, the absence of a hole prevents contact. This is analogous to what occurs in the encoders of the type just mentioned.

Of course, many instruments today provide digital output in one form or another, and for a computer to acquire data from this source, timing control

TABLE 4

BCD CODE

Decimal	BCD
0	0000
1	0001
2	0010
3	0011
4	0100
5	0101
6	0110
7	0111
8	1000
9	1001

and voltage level conversions may be required. In a great many instruments binary coded decimal (BCD) output is provided as a low-cost option. The BCD is one of several special data codes developed in an attempt to provide an easy means of handling decimal data. It is a code requiring four specified sequential bits of a binary number in order to represent decimal numbers. The high-order bit is given the value 8, the next 4, the next 2, and the low-order bit 1. The BCD code for a particular decimal number is simply derived by placing a 1 in the corresponding bit locations required to total the value of the decimal number. This is summarized in Table 4. In any case, it is relatively easy to change BCD code to a suitable code for computational use in the computer. A computer can either monitor this code directly after voltage level conversions (if any are required) and translation to the binary number system it requires, or it may make use of simple readily available digital electronic devices called decoders which can convert codes of this sort to other codes.

2.2 CONTROL TECHNIQUES

If the computer is to control instruments which accept analog information for control (voltages, for example), it is necessary that the computer possess digital-to-analog conversion capabilities. The principles of digital-to-analog conversion are simple to understand (Karp, 1972). One type of converter transforms the digital number into a current which is proportional to the magnitude of the number. A schematic of a 3-bit converter is shown in Fig. 8. The precision of such a converter is gauged by the number of bits in the resistor weighting network, the stability of the power

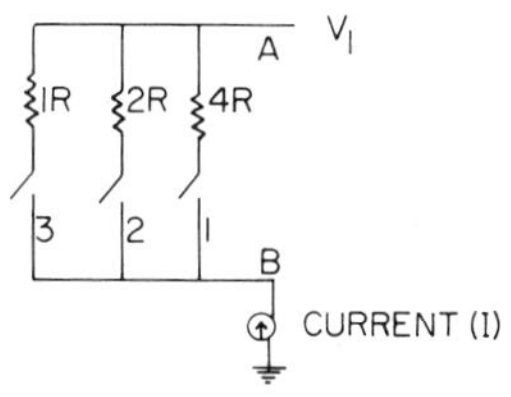

Fig. 8. 3-Bit digital-to-analog converter (schematic).

supply, and the accuracy of the resistors. The switches in Fig. 8 could be replaced by solid-state devices so that the binary weighting network could be placed under computer control. Table 5 lists the current output of this DAC for the 8 possible switch settings. Commercial converters are commonly available in 8- to 16-bit versions costing from about $20 to approximately $3000 depending upon the precision and speed of the conversion. For most applications (oscilloscope display, recorder control, etc.) 10 bits will provide sufficient precision. In some cases, 15 bits or more are required. The example discussed here illustrates a current mode converter. Resistor ladders for voltage mode conversion may also be constructed, or the current output can be translated to voltage by proper buffering and amplification.

In addition to digital-to-analog conversion, other control techniques utilizing digital logic circuitry may be divided into several categories. There is single-bit or logical control which is generally used to communicate binary control information. One of the two possible stable voltage levels for binary logic devices is defined as true and the other false.

TABLE 5

CURRENT OUTPUT FOR 3-BIT RESISTANCE NETWORK

Switches 3 2 1	Octal	Resistance AB	Current (I)	Current if $E = 4$ V, $R = 1$ kΩ (MA)
0 0 0	0	∞	0	
0 0 1	1	$4R$	$E/4R$	1
0 1 0	2	$2R$	$2E/4R$	2
0 1 1	3	$4R/3$	$3E/4R$	3
1 0 0	4	R	E/R	4
1 0 1	5	$4R/5$	$5E/4R$	5
1 1 0	6	$4R/6$	$6E/4R$	6
1 1 1	7	$4R/7$	$7E/4R$	7

It is a simple matter to provide one of these 1-bit control lines for an external device and another for the computer. Control lines from the computer can provide peripheral devices and experiments with logical signals for the control of various timing functions. These can range from turning on lamps and relays to synchronization of stimuli and time domain experiments. On the other hand, the sense lines can be examined by the computer and used as inputs from the instrument to control conditional branching instructions in programs monitoring the instruments. In some cases, where several logical lines need to be set simultaneously, or several sense lines need to be sampled simultaneously, often the simplest solution is to make use of each bit of an entire word of data in input or output. Simple and low-cost interfaces providing this sort of capability are available. All numerical data are passed to and from the computer over digital input and output channels. The analog-to-digital and digital-to-analog conversion equipment is interfaced through such channels. In those instruments which provide digital information directly, there is generally some kind of ADC built into the instrument itself. In any case, the principles of communication and control remain the same, whether these elements reside in the instrument, in the interface, or at the computer.

2.3 Display Functions

Finally, we come to the question of display characteristics that may be required. As we have hinted, it is possible to use control elements to control display devices as well as instruments or experiments. After all, a display device is simply a special kind of instrument designed to serve the display function. One of the most useful of such devices is the oscilloscope. As we are all well aware, tables of numerical information are generally far less meaningful to people than are graphic displays. Such displays often provide the most rapid way of assessing the status of an experiment or the outcome of data reduction. It is also possible to display the outcome of simulations and to make ready comparison of such experiments to the actual data in this way. The most common devices of this kind are included in the list of Table 1. When selecting display devices, the exact place of the operator in the experiment must be carefully considered, for, very often, the selection of devices made available will dictate the degree of operation interaction that is possible. It is often found that a human operator can make decisions and evaluate data in ways which would be very difficult or impossible to program (within the restraints of limited memory often present in a small computer). It is for this reason that careful choice of peripheral devices must be made in order to maximize the opportunity to use the advantages that can be gained by placing a human in the computer-experiment feedback loop (Perone, 1971).

3 The Interface Specification

Once the preliminary steps of obtaining a qualitative overview of the problem, defining the transfer functions and conditions of operation, and outlining the objectives and benefits to be realized have been accomplished, it is then possible to consider the computer–instrument interface problem in terms of functional control and interaction requirements. The person armed with the knowledge of the functional requirements of an interface can often specify, in general terms, the kinds of transducers, ADCs, digital control capabilities, and operator interaction required. It is not necessary that this person be capable of either designing or implementing the detailed electronic components of the interface. Rather, he must understand the functions which are required. In the following section, where we discuss the problem of implementation of an interface, it will also be noted that sometimes the decision of whether to provide software (or programs) to fulfill a certain function or to perform the same task by hardware must be made. This decision will very often depend upon the economics of the situation and, as we have emphasized earlier, is very much a function of the specific appplication and the individual circumstances. It is impossible to state a general rule which will always provide an optimum distribution of required interface functions between hardware and software. All that can be done is to note the capabilities and elements of each. When these are kept in mind by the individual specifying the interface, he will be able to evaluate the possibilities, based upon his knowledge of his own circumstances.

4 Interface Implementation

There are a number of approaches available to an experimenter who has completed the preliminary steps of specifying the functional capabilities required in the particular instrument interface. In the discussion which follows, it is presupposed that the steps outlined earlier have been followed and a particular desired task has been outlined in sufficient detail so that hardware and software to implement the interface may now be considered. There are a variety of routes open to the experimenter at this point. He may, if the application is common, find that a predesigned and packaged interface system including both the necessary hardware and programs is available. If this is the case, it will very often prove to be the most economical solution to a particular problem. For example, it is now possible to buy complete interface packages for such diverse applications as gas–liquid partition chromatography for multiple chromatograph systems, mass spectrometry systems, Fourier nuclear magnetic resonance spectrometers, optical spectrometers, x-ray diffraction apparatus, and a wide variety of other experiments (Perone, 1971). It would be foolish, in all but the most unusual circumstances, for an experimenter to attempt to duplicate these sophisticated hardware and software packages. Furthermore, the end result could very well be disastrous not only in terms of time delay until implementation, but also in financial terms. Thousands of dollars have gone into the development of existing commercial interface packages. Unless there is a demanding reason to do otherwise, the most intelligent move on the part of an experimenter who wishes to interface a particular common experiment to a computer is to evaluate existing prepackaged systems and to buy the one which most nearly fits the requirements of the situation. Of course, it is sometimes the case that funds are simply not available to allow this approach. In laboratories where programmers and electronics technicians are already on the payroll, it may be possible to make tradeoffs that permit development of less sophisticated and elaborate interface packages for a fraction of the cost of their commercial counterparts. One of the ways this may be done is through the use of general-purpose input and output devices. At present, there are available a variety of modular systems which include all of the important elements to permit computer instrument interfacing, even by a novice. This approach presupposes an understanding of the functional requirements of the interface and a reasonably sophisticated knowledge of programming. It is now possible to buy such modular general-purpose interface devices complete with interfaces to the appropriate input and output buses of common laboratory computers. These may be configured in a variety of ways, depending upon the needs of the individual laboratory. Since, in these cases, the primary cost is for electronic hardware components, these packages are often, in spite of their general-purpose nature, less expensive than interface systems designed for a particular specific application. On the other hand, the intangible cost of software development is often underestimated, particularly by the novice. It can easily cost an amount of money equivalent to the entire hardware cost to complete the programming for all but the simplest computer–experiment interface. Accordingly, consideration of this fact should not be neglected. On a more fundamental level, provided that personnel knowledgeable in digital electronics are available, it is possible to build complicated interfaces from rather inexpensive integrated circuits and blank computer cards. Such interfaces are generally custom designed for a specific application and are often "one-of-a-kind" interfaces. It is entirely possible that the parts for an interface of this type for some simple laboratory purposes (e.g., a shaft-encoder pulse counter) may cost less than $300. The amount of time required for design and fabrication may well result in a cost equal or even double that of the components. It is difficult to say what the cost of the time involved in designing, fabricating, and testing such an interface may be. It may range from a very low sum in a university, where students often work for no pay and the work is part of their training, to a considerable sum (in an industrial laboratory where time is quite literally equivalent to money). In any case, this approach does require a level of expertise not necessary with either of the two previous methods.

5 Software Techniques

An equally important part of the interface is the software or program packages which are necessary to drive the interface hardware. Here, just as in the selection of the hardware, there are a variety of approaches available to the prospective interfacer. These range from the use of low-level programming techniques (using the assembly language of the computer) all the way to the highest level programming techniques (using languages such as FORTRAN, BASIC, or other, nonstandard languages). In order to decide which of these is most appropriate for a particular computer–experiment interface, it is necessary once more to consider the objectives, how often the programs will be required to change, and the background level of those who will implement the interface. In general, if relatively frequent changes of the software are anticipated, the assembly language approach is the least successful and the most time consuming. Here, even with adequate documentation, it is difficult for anyone but the original programmer to make changes. Since projects often change hands, the loss of a key member of the laboratory staff may very well render a software package of this kind rather expensive to change. Accordingly, it is to be expected that such programming techniques will be used only when no other means is possible, or when there are sufficient personnel available to provide on-going support and documentation.

A somewhat more flexible approach, which maintains all of the capabilities of assembly language programming but lessens the logistic problems, is the use of so-called macrolanguages. These languages generally consist of packages of assembly language subroutines which, under the direction of a suitable executive program, may be called upon whenever required. The macrolanguage itself very often simply serves to process a string of these package (or subroutine) calls and can be very flexible in the ways in which it carries out these tasks. This provides, in software, essentially an analogous approach to the modular interface design discussed previously for the hardware. Here, experimenters may put together different combinations of subroutine packages to yield different results, depending upon the experimental requirements. Packages for acquiring data from an ADC, providing necessary control and sensing functions, or initiating digital-to-analog conversion may be included. The problem of documentation is therefore much reduced, since one need only document how he has linked the standard elements for a particular experiment. The elements themselves need not be changed often, and new personnel can rapidly become familiar with the use of such a system. This type of programming is, however, still at a more fundamental level and requires a greater knowledge of the working of the computer hardware and interface than does the use of a language such as FORTRAN or BASIC which, once the language is understood, requires very little knowledge of how the computer hardware works. We have successfully used the high-level language approach in courses in which students with no previous background in computer programming are required to construct hardware and software interfaces to chemistry experiments (Klopfenstein and Wilkins, 1971). In these cases, of course, we use general-purpose interfaces which provide certain functional capabilities, and simply instruct the students in the details of implementing these capabilities by inclusion of appropriate statements in their BASIC or FORTRAN programs (Wilkins and Klopfenstein, 1971; 1972a,b).

The latter approach certainly requires that assembly language programming be done, but once the appropriate sets of assembly language subroutines (analogous to those used in the macroprogramming approach) are developed, it is a relatively easy matter to allow large numbers of experimenters to make use of the subroutines, without requiring that the user either learn a computer assembly language or how to build interfaces. The required knowledge is thus reduced to the necessity for understanding the elements of information processing necessary for the particular experiment, the elements of control, the timing requirements, and the analytical method itself. The laboratory scientist is therefore able to concentrate, as he should, on the experiments he is performing and the data which he will obtain. It is through an approach such as this that we have been highly successful in making general use of laboratory computers available to a variety of experimenters, even for nonstandard experiments. Such an approach seems to offer the best compromise between flexibility and expense of program development. Many of the minicomputer manufacturers aiming at the experimental and analytical laboratory market have begun to provide so-called real time FORTRAN and/or BASIC compilers for just this reason. It seems likely that, with increased availability of support of this kind from manufacturers, an increasing amount of interface programming will be done using this technique. The availability of high-level languages will make it even more feasible for the scientist to consider the laboratory computer as just another general-purpose laboratory instrument and to use it accordingly.

6 Conclusion and Trends

It is already apparent that the rapid rate of computer technology development will continue for the foreseeable future. The implications this has for the individual laboratory are encouraging. Price trends seem to be continuing toward lower levels, and the investment in computer hardware will make the need to understand computer–instrument interfacing even more important than it has been in the past. There is at present no good reason why any laboratory should lack at least a minimal laboratory computer system. The availability of general-purpose interface hardware has reduced the difficulty of the instrument interfacing problem to the point that even the relatively untrained experimenter can begin to consider taking advantage of the capabilities laboratory computer–instrument interaction offers him. We expect that availability of a variety of high-level programming languages, designed to take advantage of the general-purpose interface options now available, will make possible increased implementation of computer systems in the analytical laboratory. It is even more clear that experimenters in the physical sciences will have to inform themselves on the methods and techniques of such laboratory computer use if they are to stay abreast of the significant developments in the analytical instrumentation area. There is no question that there will be an increased number of microprogrammed computers built directly into analytical instruments. The use of most of these controllers will undoubtedly be directed toward sophisticated control of the instruments of which they are a part. The interface problem will, of course, become easier in these cases because of the ease with which digital computers can be made to communicate with one another. For, although the sophisticated microprogrammable controller is now and will continue to be a reality, its capabilities will be sufficiently limited for the general-purpose laboratory computer still to play an important role in laboratories in the future. More sophisticated data-processing, display, and storage capabilities will require the continued use of general-purpose minicomputers. An understanding of the functional requirements of a computer–instrument interface will be important if maximum use of such interfaces is to be made. It is certainly true that the analytical techniques of the future will, in many ways, be much different from those we have used, even in the recent past. Instrument design is undergoing a rapid change toward providing outputs and modes of operation more compatible with direct digital computer interaction. This will, of course, simplify the problem of interfacing such instruments. Nevertheless, there is a continuing need to understand the basic principles involved in order to make maximum use of the exciting capabilities offered by such intereaction.

Acknowledgments

We must give proper credit to Jack Frazer, Sam P. Perone, and our other colleagues in ASTM Committee E-31 (Computerized Laboratory Systems) for their valuable contributions to the development of many of the ideas presented here. We will, however, take the credit for any shortcomings of the chapter. We would also like to acknowledge the support of the National Science Foundation for our work in laboratory computing through grants GJ-393, GJ-441, GY-8018, and GP-18383.

References

Bonnelli, E. J. (1971). *Amer. Lab.* **3** (2) 27.

Brown, E. R., Smith, D. E., and Ford, D. D. (1966). *Anal. Chem.* **38,** 1130.

Farrar, T. C., and Becker, E. D. (1971). "Pulse and Fourier Transform NMR." Academic Press, New York.

Gill, J. M. (1969). *J. Chromatogr. Sci.* **7,** 731.

Hancock, H. A., Jr., Dahm, L. A., and Muldoon, J. F. (1970). *J. Chromatogr. Sci.* **8,** 57.

Hites, R. A., and Biemann, K. (1967). *Anal. Chem.* **39,** 965.

Johnson, B., Kuga, T., and Gladney, H. M. (1969). *IBM J. Res. Develop.* **13,** 36.

Karp, H. R. (1972). *Electronics* **45,** (6), 84.

Klauminzer, G. K. (1970). *Appl. Opt.* **9,** 2183.

Klopfenstein, C. E., and Wilkins, C. L. (1971). *AFIPS Proc.* **39,** 435.

Oberholtzer, J. E. (1967). *Anal. Chem.* **39,** 959.

Perone, S. P. (1971). *Anal. Chem.* **43,** 1288.

Venkataraghavan, R., McLafferty, F. W., and Amy, J. W. (1967). *Anal. Chem.* **39,** 178.

Westerberg, A. W. (1969a). *Anal. Chem.* **41,** 1595.

Westerberg, A. W. (1969b). *Anal. Chem.* **41,** 1770.

Wilkins, C. L., and Klopfenstein, C. E. (1971). *Proc. Conf. Comput. Undergrad. Curriculum, Dartmouth College,* p. 269.

Wilkins, C. L., and Klopfenstein, C. E. (1972a). *Chem. Tech.* **2,** 560.

Wilkins, C. L., and Klopfenstein, C. E. (1972b). *Chem. Tech.* **2,** 681.

Willis, B. G., Bittikofer, J. A., Pardue, H. L., and Margerum, D. W. (1970). *Anal. Chem.* **42,** 1340.

CHAPTER 21

Emission Spectrometry: Arc, Spark, Laser, and Plasmas

Ramon M. Barnes

Department of Chemistry
University of Massachusetts
Amherst, Massachusetts

Introduction

Emission spectrometry encompasses a number of diverse techniques for the qualitative and quantitative determination of elemental chemical compositions of materials having widely varied forms. With appropriate facilities, qualitative analyses for most elements, especially metals and metalloids, can be accomplished in less than half an hour. Quantitative analysis on a routine basis, such as in production control, provides multiple-element results in a few minutes. For the occasional, special, or unique sample, quantitative analyses may involve considerable method development unless a semiquantitative or universal method is used. Recent developments have made possible rapid, multielemental quantitative analysis of miscellaneous samples with high precision and good accuracy.

Emission spectrometry is characterized by simultaneous, multielement analysis over wide concentration ranges from major constituent to ultratrace. Although the sample is destroyed during analysis, the sample weight necessary is small. The sample need not be conducting for analysis, and applications range from agriculture and biochemistry to oceanography and zoology.

A variety of excitation sources are available, and significant research in emission spectrometry is concentrated on the characterization, improvement, and development of excitation sources for analysis. Gratings are internationally accepted for dispersion, and selection of photoelectric or photographic recording systems depends upon the particular needs. Photographic detection is used most often for qualitative and quantitative analyses, and photoelectric spectrometers are used exclusively for quantitative work. Data recording and computation are becoming more computer oriented, and many of the manual operations have been completely eliminated.

1 Theory

1.1 Atomic Spectra

Emission spectrometry measures the energy and intensity of radiation as photons emitted from excited atoms during radiative transitions of valence electrons from upper to lower atomic energy states. The energy of each transition is directly related to the frequency ν, as expressed by the relationship in Eq. (1). The symbols and units of Eq. (1) include energy E in wave numbers or electron volts (eV); Planck's constant h; frequency ν in waves per second; velocity of light, c; and wavelength λ in angstrom units. One angstrom unit (Å) equals 10^{-10} m.

$$E_2 - E_1 = h\nu = hc/\lambda = hc\lambda^{-1} \tag{1}$$

Qualitative identification of elements is accomplished through the location of the unique emission frequencies for each element, whereas the quantitative determination of an element is achieved by measuring the number of photons at the unique frequencies. The concentration of the atoms in the source is proportional to the spectral line intensity.

Because each element is characterized by its unique arrangement of electrons surrounding the nucleus, a representation of the energy states occupied by these electrons permits the complete definition of the possible atomic emission spectrum. The atomic energy levels for elements have been determined experimentally from painstakingly careful spectral measurements (Moore, 1971). For spectrochemical analysis, reference to atlases of spectral lines or tabulation of energy levels is sufficient. Table 1 provides a listing of some of the major spectral atlases used in spectrochemical analysis. A discussion of the theory of atomic spectra can be found in Devlin (1971).

TABLE 1

MAJOR SPECTRAL ATLASES

M. I. T. Wavelength Tables (Harrison, 1969)
Tables of Spectral Lines (Zaidel' *et al.*, 1970)
Tables of Spectral-Lines Intensities (Meggers *et al.*, 1961)
Coincidence Tables for Atomic Spectroscopy (Kuba *et al.*, 1965)
Atomic Emission Lines below 2000 Angstroms (Kelly, 1968)
An Ultraviolet Multiplet Table (Moore, 1950, 1952a, 1962)
A Multiplet Table of Astrophysical Interest (Moore, 1972)
Identification of Molecular Spectra (Pearse and Gaydon, 1963)
Tabelle der Hauptlinien der Linienspektren aller Elemente (Kayser and Ritschl, 1939)

Elements with one valence electron beyond closed shells or subshells display similar spectra, which are fairly simple and systematic. Comparison shows that the spectra from alkali metals, the boron family (earth metals), copper, silver, and gold are very similar. In the same way the alkaline earths, helium, the carbon family, zinc, cadmium, and mercury have two electrons. Their spectra have close similarities with roughly twice as many lines as the alkali-type elements (Candler, 1964).

As the complexity of the atom increases, especially in the transition, lanthanide, and actinide series, most elements emit extremely line-rich spectra, and some elements have thousands of lines. In the analysis of these extremely line-rich spectra elements, special techniques, such as carrier distillations, have been developed by means of which the emission of the matrix material, e.g., thorium, is suppressed while the spectrum of trace elements is obtained.

The spectra of nitrogen, oxygen, and halogen families display six to seven times more lines than spectra from the alkaline-type elements. As a consequence of their energy arrangement, nitrogen, oxygen, and halogen families are substantially more difficult to analyze than metallic elements. Except for the high atomic weight members they require special excitation conditions, sampling methods, and spectrometric readout systems.

Superimposed on the atomic emission spectrum are two other types of spectra which can increase the spectral complexity or can be used for indirect analysis. In addition to atomic line spectra, continuous and molecular band spectra are observed. A continuum appears as broad wavelength background. Continua arise from incandescence in the discharge due to particles or electrodes in the optical path or from ion–electron recombination.

Molecular band spectra arise from electronic, vibrational, and rotational energy transitions in molecules. Compendiums of band spectra have been prepared by Pearse and Gaydon (1963), and Gaydon (1968). Molecular spectra from species such as CN and OH may interfere in spectrochemical analysis, although indirect determination of nitrogen, fluorine, sulfur, hydrogen, oxygen, and chlorine have been possible using the CN, CF, CS, CH, OH, and CCl molecular bandheads (Heemstra, 1970).

1.2 Spectral Intensity

Spectrochemical sources emit as radiating volumes, although they are commonly viewed by an optical system as plane sources with observed intensities integrated along the line of sight. The emitted volume radiation for n_2 atoms per unit volume is given as the spectral emittance in Eq. (2) as watts per steradian per cubic centimeters. In every direction from the unit volume $A_{2\cdot 1}n_2$ photons/sec are emitted at frequency $\nu_{2\cdot 1}$. The spectral emission per unit volume per unit solid angle results when the product $A_{2\cdot 1}\, n_2\, h\, \nu_{2\cdot 1}$ is divided by 4π.

$$J_{2\cdot 1} = A_{2\cdot 1}\, n_2\, h\, \nu_2/4\pi \tag{2}$$

Since n_2 varies in the source volume, $J_{2\cdot 1}$ is a function of the spatial coordinate. The spectral emittance of a source defined by the effective aperture of the spectroscopic observation device gives the spectral line intensity used in spectrochemical analysis. In spectrochemical analysis the spectral line profile as viewed by the detector is assumed to be nearly rectangular, so that peak intensity instead of integrated line profile measurements can be made. Thus the spectral intensity is related to the number of excited atoms by the transition probability $A_{2\cdot 1}$, the Einstein coefficient of spontaneous emission.

Under equilibrium conditions the excited atom population n_2 is related to the total number of atoms per unit volume by the Boltzmann equation

$$N_2 = N_1\, g_2/g_1 \exp(-E/kT) \tag{3}$$

where E is the energy difference between upper and lower transition levels, T the absolute source temperature, K the Boltzmann constant, and g_2 and g_1 the statistical weights for the excited and ground states.

Clearly the line intensity measured for spectrochemical analysis depends upon numerous parameters other than the distribution of energy levels of the element. The source excitation energy, the spatial and temperal distribution of emitting species, the transfer optics between the discharge and the dispersing systems, the dispersing system, the detector, and the readout systems all influence the final spectrochemical measurement.

1.3 Spectra Excitation Processes

Common to the production of emission spectra from an electrical discharge in spectrochemical sources is the competition among a limited number of physical excitation and deexcitation processes. Emission spectra result from excited levels in ions, atoms, and molecules produced from both discharge atmosphere and sample materials.

Since emission results from the radiative transition from high- to low-energy levels within the electronic configuration of an atom, the processes which populate and depopulate the upper levels must be considered. A number of excitation and deexcitation phenomena compete simultaneously, and the dominance of one type relative to another depends upon the energy balance of the discharge, the distribution of energy among species of the discharge, and the relative concentration of source species. Three major types of physical processes account for excitation and deexcitation of atoms (Table 2).

TABLE 2

Excitation and Deexcitation Processes

1a. Excitation of atoms by absorption of light quanta
 b. Spontaneous or induced emission
2a. Collisions with particles in which atomic species are excited from a low level to a high level (collisions of the first kind)
 b. Collisions in which excited levels are destroyed and no radiation is emitted (collisions of the second kind)
3a. Collisions with electrons which excite upper levels (collisions of the first kind)
 b. Collisions with electrons which destroy excited levels without radiation (collisions of the second kind)

The *principle of microscopic reversibility* specifies that for every excitation process, a deexcitation process occurs. The *principle of detailed balance* requires that the rates of excitation and deexcitation be equal.

Therefore, under thermodynamic equilibrium the relative population of an excited state compared to the nonexcited state remains unchanged regardless of the prevailing (de-) excitation process (Boumans, 1966, 1972a).

1.3.1. Absorption Excitation

a. LIGHT ABSORPTION. The conditions under which efficient excitation occurs by absorption of light quanta (Table 2, 1a) include matching of photon energy with energy differences between two allowed transition states. In calcium, for example, the Ca (I) 4426.73-Å line requires 23,652.324 cm^{-1}. If a photon of that energy interacts with the Ca in the $^1S_0$ ground state, absorption takes place. In atomic absorption spectrometry, the necessary photon is provided by an external radiation source, such as a hollow cathode lamp, emitting the Ca (I) 4226.73-Å line.

If the photon energy is other than 23,652 cm^{-1}, no absorption occurs until the photon energy coincides with another transition. In addition to the coincidence of photon and transition energy, the probability of the excitation from specific low- to high-energy levels must be great. This requirement is most easily confirmed when the atomic absorption spectrochemical analysis of calcium is performed. The differences in sensitivity reflect the relative values of the absorption rate constant, commonly called the transition probability, for each of the calcium lines.

Although the rate constant for (de-) excitation of selected lines is high, excitation by absorption is a minor route in most flames and electrical discharge sources because of the low radiation efficiencies (i.e., spectral volume density of radiant energy). Excited states are chiefly produced and destroyed in thermal equilibrium discharges by collisions rather than by radiation (Boumans, 1972a).

b. LASER RADIATION. The production and absorption of laser radiation illustrate the predominance of absorption compared to collisional excitation. A complete detailed balancing of radiation includes spontaneous emission, induced emission, and absorption.

Laser action requires a population inversion of an upper energy level with relatively long lifetime. This condition is produced by irradiation with a high-intensity flashlamp and absorption by the lasing system. Pumping of an excited state is precursory to induced transition to a low state by interaction with a light quantum of appropriate frequency in an optical resonant cavity. The intense, monochromatic, coherent, parallel laser radiation provides a unique light source in spectrochemical analysis. In electrical discharge sources induced emission is negligible.

1.3.2 Collisional (De-)Excitation

Collisions resulting in excitation are called "collisions of the first kind," and collisions which destroy excited states are known as "collisions of the second kind." In these classes of inelastic collisions, some kinetic or translational energy of the colliding species (i.e., electron, atom, molecule) is converted to potential (internal) energy and results in excitation, ionization, or attachment. A distinction between types 2 and 3 in Table 2 is made for atom–atom (ion–atom, ion–ion) collisions and electron–atom (molecule–ion) collisions, each of which has a different collision probability.

The (de-)excitation efficiency of free electrons (type 3) differs significantly from that of free atoms and molecules (type 2) as the result of the much smaller mass and larger mean velocity of electrons under thermal conditions. The cross section for electrons is high, but in some sources, notably flames, the concentration of electrons is very low compared to the concentration of molecular constituents (Alkemade and Zeegers, 1971; Boumans, 1972a).

The majority of (de-)excitation processes in spectrochemical electrical sources are by electron–atom collisons.

1.3.3 (De-) Ionization

The processes given in Table 2 account for ionization as well as excitation. In the majority of atmospheric pressure electrical discharges, ionization takes place mainly by electron collision and rarely by photon absorption, because the volume radiation density is low, and electron collisions need not be matched to excitation or ionization levels.

Electron collisions with molecules to produce molecular ions may result in internal energy in excess of the molecular or molecular-ion dissociation limit, which leads to fragmentation, often producing an atomic ion. The formation of nitrogen ions in an atmospheric spark discharge during breakdown apparently follows this process (Walters, 1968a, 1969; Walters and Malmstadt, 1965).

Collision of atoms and ions produce charge-transfer reactions when a minimum kinetic energy is required or released, momentum changes are minimized, and population density is high.

The reverse processes (deionization) can take place through recombination attachment or charge-transfer reactions. Positive ions and electrons recombine to form neutral atoms, and the excess energy may be transferred to a third body as kinetic, excitation, or dissociation energy; emitted as radiation; or contained by the withdrawing partners. The relative probabilities for two- or three-body ion–electron recombination depend upon the energy levels and electron velocity involved. The third body may be an electron, ion, or molecule in an excited or ground state. In low-pressure discharges the container walls act as the third body.

The inverse of stepwise excitation and ionization can populate upper atomic levels through recombination collisions and radiation decay. The populating of excited levels through photon emission from upper levels is called radiative cascading. Stepwise excitation and radiative cascading have been proposed to account for excitation and population processes in spark discharges (Walters, 1968a, 1969).

1.3.4 Nonthermal Discharges

In some spectrochemical discharges (de-)excitation and radiation processes can be completely characterized by a thermal model. In a thermal discharge the population of various energy levels satisfy the Boltzmann's distribution law; the dissociation equilibria are described by the general equations for chemical equilibria, and the ionization equilibria are represented by Saha's relationship. The inelastic collisional processes are balanced between excitation and deexcitation. Electron velocity distribution is given by Maxwell's equation (Boumans, 1972a).

Departures from thermal mechanisms may occur at reduced pressures, in pure atomic gases, in the cathode fall region of a normal carbon arc, and in arcs between metal electrodes (Boumans, 1972a).

In emission spectrochemical analysis the thermal character of the emission is not critical in analysis, for the dependence of intensities upon sample concentrations is sufficient. However, knowledge of the departure of the discharge from a thermal model can provide useful information to influence detection limits, calibration curve shape, and interelement effects (Alkemade and Zeegers, 1971).

1.3.5 Excitation in Arcs

The dc arc operating in air at atmospheric pressure can be described adequately by a thermal model (Boumans, 1966, 1972a), whereas an arc in a pure atomic gas is expected to be substantially different. However, as little as 1% impurity of molecular gas in the atomic gas will convert the (de-) excitation processes to those characteristic of a molecular gas. In the discharge in a molecular gas, energy exchange between electrons and heavy particles takes place mainly through inelastic collisions and vibrational excitation. The competition among deexcitation processes provides alternative pathways to radiation by which excited atoms can lose energy. In an atmospheric air arc, excited atoms of substances volatilized from the electrode crater undergo collisions of the second kind with molecules (e.g., N_2, CO, NO) and radicals (e.g., CN) that can accept widely differing amounts of energy through rotational and vibrational energy changes. This highly efficient deexcitation process removes most of the excited-state energy at atmospheric pressure and leaves a small fraction for emission. Excitation in arcs for spectrochemical analyses has been treated in unique detail by Boumans (1966, 1972a).

1.3.6 Excitation in Sparks

a. PLASMA SPECIES. In contrast to arc discharges, excitation processes in sparks are directly or indirectly dependent upon the time changes of the discharge current. Spark events are rapid, many on the order of less than a microsecond. The excitation processes outlined in Table 2 must be considered on a time basis. The high-voltage spark discharge in air at atmospheric pressure is a special environment for excitation, characterized by times of high vapor and current density, low direct voltage drops, and short mean free paths. Those physical processes which favor energy transfer without high kinetic energy predominate, and stepwise excitation and radiative cascading appear dominant (Walters, 1968a, 1969).

This high-frequency, low-energy collisional process provides the rapid production of highly excited and often simultaneously ionized plasma species which are the precursers to electrode sampling and electrode vapor excitation for analysis, while minimizing the need for formation of relatively high-energy electrons in short times and in short mean free paths at atmospheric pressure (Walters, 1968a). Details of spark formation in argon are presented by Walters (1972b).

b. SPARK SAMPLE EXCITATION. Once the electrode is sampled and electrode vapor propagates into the spark gap, two independent systems must be considered in sample excitation processes: (a) the expanding cloud of electrode material vapor, and (b) the current-modulated supporting plasma species defining the spark channel. The interactions of these two systems must be viewed in two spatially separated regions of the discharge: (a) the central channel region and (b) the fringe region (Sacks and Walters, 1970; Walters, 1972a).

The time dependence and current sensitivity of emission from a spark discharge, particularly a rapidly changing current-modulated discharge, results from the interaction of the support plasma species and expanding electrode vapor. The interaction and specific excitation levels critically depend upon the energy state and dimensional extent of the conducting plasma, which, in turn, depend directly upon the magnitude of the spark current.

Sequential stepwise excitation in the first microsecond of the spark discharge has been clearly illustrated by Walters (1968a). Electrode material is maintained in a state of high internal energy by undergoing continual charge-transfer reactions while traversing the spark gap. The greater the internal energy, the more restricted the distribution of excited species is to the region near the plasma axis, where electrode material interacts strongly with the supporting plasma species. Volume relaxation is prevented in the channel during periods of high current density. In the fringe regions electrode material interacts indirectly and weakly with the supporting plasma species (Sacks and Walters, 1970).

1.4 Sampling Processes

The transport mechanisms by which the sample is removed from the specimen and carried into the discharge plasma are critical in the determination of the number of atoms and ions formed in states which emit useful radiation. Since the arc and spark are the most commonly used discharge sources in emission spectrometry, a brief discussion of the sampling and transport phenomena for each is included in the following sections.

1.4.1 Arc Sampling

The schematic representation of an ideal arc configuration shown in Fig. 1 illustrates the location of the sample relative to the core. The core carries

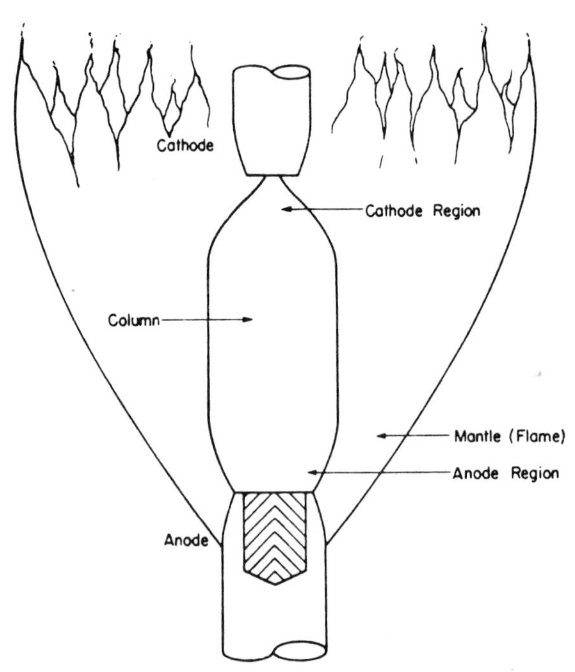

Fig. 1. Schematic representation of ideal vertical arc. The electrodes are connected by the core, which carries the electrical current and has the highest temperature. The core contracts toward the electrodes. If the arc length exceeds a few millimeters, the central part of the core is more or less cylindrical and is designated the *uniform column*. The arc core is surrounded by the *mantle*, which is not heated immediately by the electrical current but receives heat by thermal conduction from the core. The mantle has a more or less parabolic profile. [From Boumans, 1972a, by courtesy Marcel Dekker, Inc.]

most of the arc current, and its temperatures are maximum. The sample-containing anode heats to temperature of 4000°K compared to the arc column temperatures in air of 5000 to 7000°K. The mantle region is heated to 3000 and 4000°K owing to thermal conduction and convection from the arc column. The temperature of the arc gas is completely determined if the composition of the gas, the diameter of the column, the current strength, and the pressure of the surrounding atmosphere are fixed (Boumans, 1972a).

The addition of metal vapors to the arc column introduces significant changes in the electrical conductivity of the arc, resulting in a decrease in temperature and field strength, and an increase in electron concentration. Thus the concentration and vaporization rates of elements with relatively low ionization potential can exert a substantial effect on the characteristics of the arc. If an arc is burning in air, the ionization potential of components of air is about 14 eV. The addition of a metal vapor with lower ionization potentials will supply sufficient electrons to maintain the energy balance at a lower arc temperature. Since the evaporation of metals into the arc depends upon their relative boiling points, the arc temperature changes as different fractions of elements enter the arc column. To reduce the effects of individual elements on the arc temperature, one spectrochemical technique is to supply to the arc a large quantity of low-ionization-potential material such as lithium carbonate to set the arc temperature during the entire spectral exposure. These spectrochemical "buffers" promote regular volatilization, give a smooth burn, and stabilize the arc temperature. The evaporation rate of sample and buffer is further controlled by the composition of the arc discharge atmosphere. Chemical, physical, and mechanical processes in the electrode cavity alter the pathways, vaporization rates, and distribution of elements as they leave the anode crater. Thermochemical reactions which occur in the electrode include oxidation, reduction, carbide formation, halidation, and sulfidation. Sample volatilization and transport in the arc have been studied by Boumans (1966, 1972a), who concedes that volatilization is the least understood arc process.

For the development of arc spectrochemical analysis, element vaporization is determined empirically (see Fig. 4 below).

1.4.2 Spark Sampling

In contrast to the arc discharge, the high-voltage spark is characterized by a high-frequency current flow between electrodes which may change polarity every few microseconds. The transient current requires that sampling, excitation, and emission in the spark gap occur rapidly, as diagrammed schematically in Fig. 2 (Walters, 1968a).

The sampling mechanism of the cathode electrode requires that the sample material be ejected from the electrode at velocities near 10^5 cm/sec,

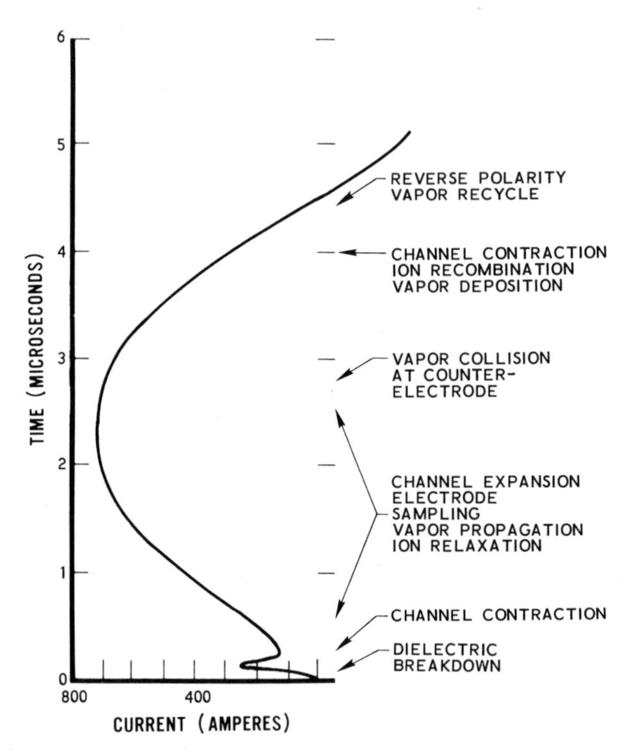

Fig. 2. Critical time events controlling radiation patterns during a single-current half-cycle in a high-voltage spark discharge. Material sampled from the spark electrode must propagate through high-density space charges at the electrode surface. Ejected electrode material velocity is not controlled by the spark current. Time dependence of emission spectra depends upon interaction of discharge plasma species and expanding electrode vapor. Events reoccur for each current half-cycle. [From Walters, 1968a.]

and ionized and excited during the first microsecond of the discharge. To account for the phenomena, two independent systems are active in the spark discharge. One controls the plasma channel, and one controls the movement of the sample electrode vapor. The spark discharge current controls the conducting discharge channel shape and the amount of material eroded from the cathode. However, the current does not control the axial or radial velocity of ejected electrode vapor. The residence time is therefore independent of the spark current. The electrode vapor movement from the cathode results from a thermal explosion. The amount of material removed is proportional to the current at any time, but the velocity depends upon the atomic mass and boiling point of the material—not on the current amplitude (Walters, 1972a). The interaction of the conducting channel with the electrode vapor in air and nitrogen is described by Sacks and Walters (1970).

When the cathode spot is located by direct formation of ionization, as occurs in argon (Walters, 1972b) and the polarity of the electrodes is unchanged during current conduction, controlled and stable electrode sampling becomes realistic. The quarter-wave resonant cavity spark source described by Walters and Bruhns (1969) has provided sampling stability exceeding any previously reported.

Spark discharges other than high-voltage sources have been documented in some detail, and a good recapitulation of publications through mid-1960 can provide details to observed electrode–spark interactions (Boumans, 1972a; Rozsa, 1972).

2 Instrumentation

2.1 Introduction

The steps and instrumental devices required in performing an emission spectrochemical analysis are summarized in Table 3. The critical steps are sample excitation, radiation dispersion, and data recording. The capability of emission spectrometry to obtain qualitative and quantitative analyses of every element in many sample forms is provided through a variety of excitation sources.

Radiation dispersion provides selectivity and sensitivity in emission methods. Dispersion with gratings mounted in a number of popular arrangements predominates. A complete description of the theory of prisms and gratings, their mounting arrangements, and commercial instruments can be found in recent chapters by Faust (1971), and Barnes and Jarrell

TABLE 3

Performing an Emission Analysis

Steps	Devices
(a) Sample preparation	
(b) Sample excitation	Excitation source
(c) Radiation dispersion	Spectrometer/spectrograph
(d) Radiation detection	Photomultiplier/photographic emulsion
(e) Data recording	/developing
(f) Data processing	/photometry
(g) Information analysis, reporting	Computer processing

(1971). Detection systems include both photographic and photoelectric methods, each requiring its particular data recording and processing. The digital computer has made much of the information-analysis and reporting stages in emission laboratories rapid and routine.

Thus uniqueness in emission spectrometry instrumentation results from the need for simultaneous, multielement data recording and analysis for emission from samples in many different forms containing many elements in concentration ranges from ultratrace to major components.

2.2 Excitation Sources

A systematic classification of excitation sources for emission spectrometry has been developed by the American Society of Testing and Materials (ASTM, 1971). Although the classification does not include all excitation sources used, most conventional laboratory sources fall into the definitions. Five types of excitation sources are defined. Many commercial excitation sources units contain two or more of the five types of excitation in a single cabinet, so that common components may be rapidly interconnected by switches and relay networks to provide the appropriate configuration desired. Typical electrical characteristics are summarized in Table 4.

The five categories are defined as follows (ASTM, 1971):

(1) *Continuous dc arc:* "A self-maintaining d-c discharge." Most dc arc sources use graphite electrodes and are commonly called "carbon arcs." An extensive study of the spectrophysics of arcs has been compiled by Boumans (1966, 1972a).

(2) *Noncapacitive ac arc:* "A series of electrical discharges in which the individual discharges are either self-initiating or in which they are initiated by separate means. These discharges are extinguished when the potential across the analytical gap falls to a value that no longer is sufficient to maintain it. Each current pulse is in the reverse direction from the previous one."

The current (5–10 A) and the output potential of the transformer (2200–4400-V rms) are typical. If a separate low-power igniting circuit is used, ac arcs may be operated at voltages above 200 V. Similar to the dc arc, the ac arc sacrifices sensitivity to provide better sampling and improved precision as the result of relocation of the anode spot at each current half-cycle.

(3) *Noncapacitive, intermittent dc arc:* "A series of electrical discharges in which d-c pulses are initiated by separate means, either mechanically or electrically. Each current pulse has the same polarity as the previous one and lasts for less than 0.1 second."

TABLE 4

TYPICAL ELECTRICAL CHARACTERISTICS OF EXCITATION SOURCES[a]

Classification	Duration (μsec)	Peak current (A)	Repetition rate (pulses/sec)	Frequency (kHz)
High-voltage spark				
Oscillatory	0.2–9	100–1000	600–31,200	1000–100
Medium-voltage spark				
Oscillatory	30–2000	75–400	300–780	40–0.2
Unidirectional (overdamped)	25–6000	10–250	30, 60, 240, or 400	
Low-voltage arc				
Oscillatory	2000–7000	1.4–42	120	0.4–0.03
Unidirectional (polarized)	3000–6000	1.4–11	30, 60, or 120	
High-voltage arc				
Oscillatory	8000	0.6–7 or 14–42	120	0.120
Low-voltage arc				
Unidirectional (dc arc)	1,000,000+	3–56		
High-voltage arc				
Unidirectional (hollow cathode, high-frequency discharge)	1,000,000+	0.5–5		

[a] Adapted from Rozsa (1972, p. 477), by courtesy of Marcel Dekker, Inc.

Electrically characterized by the primary power supply and the radio frequency, the spark-ignited low-voltage dc and ac arcs have replaced the high-voltage ac arcs.

(4) *Triggered capacitor discharge:* "A series of electrical discharges from capacitors, the energy of which is obtained from either an a-c or d-c electrical supply. Each discharge may have either an oscillatory, critically damped, or overdamped character. It is initiated by separate means and is extinguished when the potential across the analytical gap falls to a value that no longer is sufficient to maintain it.

"The spark source is differentiated from the triggered capacitor discharge mainly because of the range of electrical parameters available to each and because the triggered capacitor discharge needs separate means for initiating each discharge. In general, a spark source produces potentials of the order of 10,000 to 40,000 V using capacitances between 0.001 and

0.02 μF. A triggered capacitor discharge produces potentials of 200 to 2000 V and uses capacitances of 1 to 250 μF."

In recent developments, solid-state control circuits rather than auxiliary spark gaps are being used for switching at high repetition rates.

(5) *Spark*: "A series of electrical discharges, each of which is oscillatory and has a comparatively high maximum instantaneous current resulting from the breakdown of the analytical gap or the auxiliary gap, or both, by electrical energy stored at high voltage in capacitors. Each discharge is self-initiated and is extinguished when the potential across the gap, or gaps, is no longer sufficient to maintain it.

"Some spark sources use direct current to charge the capacitors. This means that the capacitors are always charged with the same polarity. Since peak amperage is reached only during the first half-cycle of an oscillating discharge, this type of circuit may give a discharge with predominantly un-directional behavior."

Three notable developments by Walters and co-workers have expanded the capabilities of spark discharges. These developments include (a) the semiconductor diode shunt across the analytical spark gap, (b) an adjustable waveform spark using a diode-shunted tank circuit, and (c) a 162-MHz quarter-wave, resonance cavity spark discharge. By installation of the semiconductor diode shunt (Walters and Malmstadt, 1965) parallel to the analytical gap in an electronically triggered discharge (Walters and Malmstadt, 1966), the alternative half-current cycles shown in Fig. 2 are passed through the diode rather than through the analytical gap. This significantly reduces background and allows complete recombination of ions to excited atoms. Line-to-background ratios are obtained in excess of those obtained from a typical dc arc. By combining the diode in a tank circuit, Walters designed a high-voltage spark source capable of producing high-current discharges ranging in waveform from fully oscillatory at kilohertz frequencies to fully unidirectional in the conventional overdamped mode previously restricted to low-voltage triggered capacitor discharges (Walters, 1968b).

The quarter-wave, current-injection spark source is unique among atmospheric pressure spark sources in its ability to provide temporally precise, low-power variable frequency sparking with completely electronic control (Walters, 1972b; Walters and Bruhns, 1969). Through simple current injection and the absence of interference, the spark source may be used with state-of-the-art electronic components and systems.

Boumans has recapitulated in a recent chapter much of the conventional work on excitation sources through the mid-1960s (Boumans, 1972a). A comparison of the performance of some of the conventional excitation sources is given in Table 5. These general comparisons can be misleading if

TABLE 5

RELATIVE SPECTROSCOPIC PERFORMANCE OF MAJOR EXCITATION MODES[a]

Excitation mode	Precision	Obtainable detection level	Susceptibility to matrix effect
AC spark	Excellent	Fair	Low
Triggered capacitor—discharge underdamped	Good	Good	Low to moderate
Triggered capacitor—discharge mildly damped ("critical")	Good	Fair	Moderate to low
Unidirectional arc—pulsed	Good to fair	Good	Moderate
Ignited ac arc	Fair	Good	Moderate
Triggered capacitor—discharge overdamped	Fair	Good	Moderate to high
DC arc	Poor to fair	Excellent	High
Sustaining ac arc	Fair	Excellent	High to moderate

[a] Adapted from Rozsa (1972, p. 479), by courtesy of Marcel Dekker, Inc.

not considered closely in the light of recent advances. For example, the Gordon arc (Gordon and Chapman, 1970) has significantly improved the precision for dc arc operation, although major modifications in the arc arrangements are required. On the other hand, the development of new spark sources have improved the sensitivity and thus the limit of detection remarkably, so that for some analyses the sensitivity can be made to surpass that obtained from the conventional arc. The research in induction-coupled plasmas indicates that precision and sensitivity may both be incorporated into a usable source (Dickinson and Fassel, 1969).

2.3 SPECTRAL DISPERSION: RECORDING PROCESSING

The spectral results from excitation are recorded either photoelectrically with a single- or multichannel spectrometer or photographically with emulsion-coated film or plates.

2.3.1 Spectrometric Readout

In the spectrometer arrangement narrow slit bodies are positioned at locations along the focal curve of the spectrometer corresponding to the wavelengths of the lines to be studied. The light passing through the slits is directed and sometimes focused onto the photocathode of individual photomultiplier tubes. Whereas a monochromator uses a single slit and photomultiplier and requires that the grating be rotated to present different wavelengths to the exit slit, many multichannel (direct reading) spectrometers have fixed grating mountings. Thus the slit location and alignment are critical, and a series of adjustments are required to locate, position, and finely peak the slits in the correct location.

The multiplier phototubes provide a linear response to radiation over many decades, and a variety of readout systems have been used. The appropriate photocathode material is selected for the wavelengths detected, and photomultipliers are pretested and selected for sensitivity and stability. In a monochromator, the wavelength range of the photocathode material is also important. Commonly, the phototube current is collected during the exposure by means of various approaches. An electrical signal for each channel phototube is obtained and processed by analog systems in older readouts and by digital systems in modern readouts. The signal for each channel and for each concentration level is used to construct a calibration curve. In computer processing, a curve-fitting program may provide the parameters of the fitted function in place of the actual curve. If background corrections or internal standard ratios are required, the calibration curve may read the corrected signal as a function of concentration. Finally, the signals from an unknown sample are collected and processed in the same way, and the calibration curve equation is solved for the sample signal to obtain the unknown concentration. The data system used by Gordon and Chapman (1970), for example, contains the background and intensity corrections for all elements for every readout channel. Thus in modern spectrometer systems, the readout, data collections, and data computation have been reduced to a minimum. Pictorial data display systems provide rapid concentration readouts. This approach yields improved accuracy, precision, and speed compared to previous analog and manual systems.

One major limitation of the multichannel spectrometer is the relative difficulty of changing the slit position once a number of slits have been located on the focal curve. Although a few spectrometers can accommodate up to 50 slits, others require replaceable slit holders to change types of analyses. In production applications a limited number of different materials are analyzed, so that major changes in slit location (or "program") do not occur.

2.3.2 Spectrographic Readout

The spectrograph with either film or glass plate photographic detectors is commonly used for qualitative analysis, special work, unique samples, multiwavelength detection, or the development of a wavelength position for the multichannel spectrometers. Unfortunately the wavelength and the exposure response of the photographic emulsion are nonlinear. For qualitative analysis, emulsions are selected to respond in the spectral wavelength of interest. Since most sensitive lines of the common elements appear between 2500 and 5000 Å, selection of the Kodak Spectrum Analysis No. 1 (2300–4400 Å) and No. 3 (2300–5200 Å) emulsions with their relatively high contrast and low speed provides good coverage for qualitative and quantitative analyses.

Standardized processing and calibrating procedures are specified by ASTM E 115-71 and ASTM E 116-70a (ASTM, 1971). Much of the data reduction is carried out commonly by straightforward computer programs (Margoshes and Rasberry, 1969). The plate provides a complete spectral display over the wavelength coverage. This permanent record permits measurement of multiple lines and background corrections at will. For qualitative analysis and some semiquantitative analyses, visual inspection will provide rapid identification of elements and estimation of concentrations. For more complete description the plate can be positioned in a comparator–microphotometer. The plate image is usually projected at a magnification of 10 to 20 ×, and a split screen permits simultaneous projection of a "master" or comparison plate. The master plate may contain calibration wavelength scales and major line identification, or it might be a series of exposures from a variety of elements to be determined (Wang *et al.*, 1972). The blackening or transmittance of the photographic spectral line image can be quantified by spectrophotometry. Recent microphotometers have digital as well as analog readout systems, so that line profiles as well as minimum transmissions can be measured. The recording and processing of microphotometer readings for quantitative analyses are readily computer automated (Margoshes and Rasberry, 1969).

2.4 Laboratory Records

Bookkeeping operations in an emission spectroscopy laboratory are essential to insure that each sample analysis request is defined, sample preparation and excitation conditions are recorded, and results are reported in a meaningful arrangement. Although methods vary with laboratories, two or three forms fill most needs.

Each photographic emulsion must be cataloged with a unique number, commonly located in the same relative position. Plates and films are stored in individual envelopes, which may be preprinted or stamped to include further identifying and operating information.

Analysis reports may be prepared by hand or may be generated in computer readout stages. Duplicate copies of the report form provide a record for the laboratory and a receipt for the analysis. "Hard copy" reproductions from visual display computer terminals should speed data recording in future systems.

3 Applications of Emission Spectrometry

The techniques of emission spectrometry provide the practicing instrumental analyst with specimen in hand a diversity of approaches in the characterization of the chemical composition, particularly the elemental analysis, of materials in all forms, as outlined by the flow charts in Chapter 1.

The selection of a particular emission spectrometric technique depends upon a number of criteria, which must be evaluated by the analyst when informed that an emission spectrometric analysis might be necessary. Table 6 outlines a number of the common considerations used in choosing an emission spectrometric technique.

TABLE 6

INFORMATION NECESSARY FOR TECHNIQUE SELECTION

(a) Element(s) desired	(f) Sample size
(b) Concentration range	(g) Time available
(c) Concentration distribution	(h) Result accuracy
(d) Sample form	(i) Result precision
(e) Sample matrix (environment)	

3.1 ELEMENTS DESIRED

Since each element is characterized by its unique emission spectrum, in theory every element can be analyzed by emission spectrometry. Mixtures of elements, as well, can be determined simultaneously because of the discriminatory ability of the spectrograph. In practice, however, the simultaneous, multiëlement analysis of materials is limited by instrumental conditions to approximately 20–30 elements.

For example, standard methods published by the American Society for Testing and Materials (ASTM, 1971) include emission analyses of aluminum for 23 elements; indium for 18 elements; gold, platinum, and palladium for 26, 27, and 28 elements, respectively; and uranium oxide for 32 elements. The scope of one aluminum analysis is illustrated by Table 7.

Within the wavelength region between 2000 to 8500 Å, emission spectra for all elements can be obtained. Some groups of elements, however, may require specific attention. The permanent and noble gases along with the halogens and metalloids are less easily excited than the vast majority of other elements. To obtain useful spectra from these elements, either the resonance lines in the vacuum ultraviolet or the visible lines resulting from

TABLE 7

ANALYSIS OF ALUMINUM AND ALLOYS[a]

Element	Concentration range (%)	Element	Concentration range (%)
Copper	0.001–30.0	Beryllium	0.001–1.2
Silicon	0.001–14.0	Zirconium	0.001–1.0
Magnesium	0.001–11.0	Lead	0.002–0.7
Zinc	0.001–10.0	Bismuth	0.001–0.7
Nickel	0.001–10.0	Titanium	0.001–0.5
Manganese	0.001–8.0	Calcium	0.001–0.2
Tin	0.001–7.5	Sodium	0.001–0.05
Silver	0.001–5.0	Vanadium	0.001–0.05
Iron	0.001–4.0	Boron	0.001–0.05
Chromium	0.001–4.0	Gallium	0.001–0.05
Cadmium	0.001–2.0	Barium	0.001–0.05
Cobalt	0.001–2.0		

[a] From American Society for Testing and Materials (1971): ASTM E 227–67.

transitions among high-energy states must be employed. Many of these elements, like the metalloids, are routinely determined using vacuum spectrometers and flushed spark chambers (ASTM, 1971). The developments in vacuum spectrometry as demonstrated by the work of Berneron and Romand (1964), Malmand (1970), and Zaidel' and Shreider (1970) may well make these analyses routine in the future. Alternative procedures using nonthermal excitation discharges provide the excitation energies required for emission in the visible region.

The emission analysis of rare elements can be performed on a routine basis. For example, lighter actinides (e.g., Th, U, Pu) are commonly analyzed for 30–35 trace elements at 1 to 200 ppm using the dc arc carrier distillation technique (ASTM, 1971).

3.2 Sample and Concentration Conditions

The selection of an emission spectrometric technique is often determined by the sample and concentration conditions (Table 6, b–f) regardless of the elements present. Concentration ranges can be arbitrarily divided into major (>5%), minor (0.1–5%), trace (<0.1–0.0001%), and ultratrace (<0.0001%) levels. Sample size also varies, and emission spectrometric techniques can accommodate a range from bulk to minute specimens.

The application of emission spectrometry for high concentrations, especially major constituents, is usually limited by the precision of analysis. Precision not better than 1% of the amount present (often considerably poorer) favors the selection of other analysis methods.

3.2.1 Concentration Range

Concentrations from ultratrace to major components can be analyzed by suitable emission spectrometric methods. Not all elements can be determined simultaneously over this range, and the spectroscopist needs some type of comparison of merit for each technique. One approach is to list elements by their limits of detection, expressed either as absolute (micrograms) or relative concentrations (ppm). In favorable cases, absolute quantities of 10^{-5} to 10^{-9} g can be detected, corresponding to detections of ppm to ppb. The ability to detect an element varies considerably with technique and the sample conditions.

A comparison of relative concentration detection limits for a limited number of elements by emission methods is given in Table 8. Detection limit tabulations are useful as provisional guidelines. Practical analyses, especially simultaneous, multielement ones, are generally performed at concentration ranges at least an order of magnitude higher than the limits of detection. Limits of detection are often collected for single elements in relative matrix-free environments and under optimum experimental conditions. For maximum certainty, the analysts should attempt to compare the range of actual calibration curves in the matrix of interest. Table 9 illustrates a complete report for the 23 elements in aluminum by high-voltage spark (ASTM, 1971). A number of analysis lines may be used for each element to extend the range of the calibration curves. The background equivalent is the concentration which gives a line intensity equal to the background equivalents. The detection limit is reported as the standard deviation at the 10-ppm concentration level. Unfortunately few literature reports are as detailed and documented as the methods specified by ASTM.

3.2.2 Sample Size

Sample size requirements for most emission spectrometry are modest by routine chemical standards. However, not all emission methods can utilize samples in all sizes, especially when the sample size is small. For example, small pieces of materials for which major constituents are to be determined may be readily analyzed by dc arc methods; powders may be analyzed by spark methods through the use of briquettes or solutions. Powder samples of 5 to 10 mg are quite suitable for a single dc arc analysis. Large quantities of solids, powders, or liquids may be required for trace or ultratrace analysis, especially if separation and/or concentration techniques are necessary. The maximum amount of sample taken is limited usually by difficulties in sampling and manipulation. Some materials are available only in limited quantities because of their rareness or value; lunar rock and soil samples are an ideal example. On the other hand, the appropriateness of using a 1-g sample to represent the composition of a 300-ton melt of steel or a trainload of coal must be evaluated by suitable statistical sampling procedures. Ordinarily a sample of 0.1 to 10 g is employed for determination of low ppm or ppb level elements. The requirements for sample size reflect the accuracy of analysis. For survey qualitative analysis, enough sample to insure excitation of elements at the suspected levels is required. For quantitative results, an optimum sample size may be calculated from the operating range of the concentration calibration curves developed for the particular sample. The calibration curves may differ and the optimum sample size may change with the matrix or host material, but common matrix procedures are often used to reduce these effects.

Some practical sample sizes used in standard ASTM methods (ASTM, 1971) include for dc arc powder techniques 1–2 g of metal samples (e.g., Ag, Au, Al, W, Pb, Zn, Mo, Ta, U) or for carrier and other arc techniques 0.2–0.6 g (e.g., Cu, Ni, Ce, Al_2O_3, Sn, In). Samples of slag may be up to 5 g although only 10–100 mg of the sample is used for one exposure. Similar amounts (1–2 g) are used in arc methods of solution residues and most spark solution methods. As little as 10 mg of gold or as much as 20 g of lubricating oil represent spark solution size limitations. The solution spark residue technique (e.g., the graphite spark), on the other hand, requires 0.04- to 100-mg samples. Spark and arc techniques with briquetted pelleted samples may employ from 0.5 to 2 g. For spark excitation from metal casting or pieces, 40–500 g may be prepared, although much of the sample mass is not used in the analysis. Solid metal pins or rods require proportionally less sample. Sample size requirements for plasma excitation techniques with solutions are comparable to those used in direct solution spark or combustion flame methods.

3.2.3 Sample Matrix

The environment in which the sample exists can affect the selection of an emission spectrometric method, although the matrix is often modified to suit the other conditions of Table 6.

If no pretreatment is conducted, the matrix may determine the dispersion (i.e., the size of the spectrograph) necessary to complete the analysis, as illustrated in Table 10. During excitation elements may be separated from the matrix as in the carrier-distillation or gases-in-metal techniques. The matrix effect may be reduced by dilution of the sample with a material that will be common to all samples and sufficient in quantity to minimize the in-

TABLE 8

DETECTION LIMITS (μg/ml)

Element	Wave-length[a] (Å)	Induction plasma emission[b]	Flame emission	Atomic absorption	Atomic fluorescence[c]	Stabilized arc or plasma jet[d]	Microwave plasma	DC arc[e]	DC arc cathode layer[f]	Copper spark[g]
Elements which form stable monoxides, $D_0(MO) > 7.0$ eV										
Ce	4186.6	0.007	10	95	—	—	20	500	10	0.3
Hf	3399.8	0.01	75	15	—	—	—	200	—	0.25
La	4086.7	0.003	0.1 (LaO)	2	—	0.07	—	—	5	0.02
Nb	4058.9	0.01	0.6	5	—	—	—		10	0.10
Ta	3012.5	0.07	18	6	—	—	—	—	—	0.3
Th	4019.1	0.003	150	—	—	—	4	—	10	0.5
Ti	3349.4	0.003	0.2	0.04	—	0.7	1	0.5	1	0.01
U	4090.1	0.03	10	12	5	0.5	—	—	—	2
W	4008.8	0.002	0.5	3	—	0.8	0.2	2	—	0.4
Y	3710.3	0.0002	0.06	0.3	—	0.008	—	0.3	2	0.005
Zr	3438.2	0.005	3	5	—	—	15	0.1	1	0.01
Elements with excitation potentials >4.0 eV										
As	2288.1	0.1	6	0.2	0.1	—	4	1.0	3	5
B	2496.7	0.03	0.3 (BO)	3	—	0.05	0.03	4	0.1	0.5

Cd	2288.0	0.03	2	0.001	10^{-6}	0.003	0.5	1	0.3	1
Co	3453.5	0.003	0.05	0.005	0.01	3	—	0.06	1	0.05
P	2136.2	0.1	3 (PO)	—	—	1.1	—	0.3	5	4
Pb	4057.8	0.008	0.2	0.004	0.5	0.03	1	0.02	0.3	0.1
Sb	2598.1	0.2	1.5	0.2	0.05	—	0.6	0.15	0.5	2
Zn	2138.6	0.009	50	0.0005	4×10^{-5}	0.01	0.1	2.0	5	0.5
				Elements with excitation potentials <4.0 eV						
Al	3961.5	0.002	0.01	0.02	5	0.1	0.03	21	0.1	0.05
Ba	4554.0	0.0001	0.001	0.05	—	0.3	1	1	1	0.02
Ca	3933.7	—	0.0001	0.001	0.002	0.003	—	3	1	0.05
Cr	3578.7	0.001	0.005	0.002	0.01	0.003	—	0.15	0.2	0.05
Fe	3719.9	0.005	0.05	0.002	0.02	0.005	0.5	0.8	0.1	0.5
Li	6707.8	—	3×10^{-6}	0.005	—	0.0008	—	40	0.2	0.001
Ni	3524.5	0.006	0.03	0.005	0.003	0.003	0.3	0.06	0.5	0.05
Sr	4077.7	0.00002	0.00001	0.01	0.03	0.07	—	200	1	0.002
V	4379.2	0.006	0.01	0.02	—	0.2	0.1	0.3	0.5	0.02

[a] Wavelengths used are not always those used for results shown.
[b] Dickinson and Fassel (1969).
[c] Winefordner *et al.*, (1970).
[d] Valente and Shrenk (1970).
[e] Addink (1957; 1971).
[f] Avni (1969), Avni and Boukobza (1969).
[g] Faris (1962).

TABLE 9

ANALYTICAL LINES BACKGROUND EQUIVALENTS AND DETECTION LIMITS USING HIGH-VOLTAGE SPARK FOR ALUMINUM[a]

Element	Wavelengths of suitable lines (Å)	Concentration range (%)	Background equivalent (%) Spectrometer 1.5-m	2-m	Detection limit[b] (ppm)	Shape of analytical curves
Si	2516.12	0.001–14.0	—	—	5	nonlinear
	2881.58	0.001–14.0	0.05	0.03	5	nonlinear
	3905.53	0.50–14.0	1.2	—	—	linear
Fe	2382.04	0.001–4.0	0.02	—	4	nonlinear
	2395.62	0.001–4.0	—	0.02	—	nonlinear
	3020.64	0.01–1.0	—	—	—	nonlinear
Cu	2247.00	0.01–5.0	—	—	—	nonlinear
	3273.96	0.001–1.5	0.01	0.01	3	nonlinear
	5105.54	0.05–30.0	0.75	0.54	—	linear to 14%
Mn	2593.73	0.001–8.0	0.01	—	2	nonlinear
	3460.33	0.05–8.0	—	0.10	—	linear
Mg	2795.53	0.001–1.5	—	—	—	nonlinear
	2852.13	0.001–1.5	0.003	—	0.6	nonlinear
	5167.34	0.05–11.0	—	—	—	linear to 8%
	5172.70	0.05–11.0	—	—	—	linear to 8%
	5183.62	0.05–11.0	0.08	0.04	—	linear to 8%
Cr	2766.54	0.10–4.0	—	—	—	nonlinear
	4254.35	0.001–4.0	0.05	0.02	5	nonlinear
Ni	2316.04	0.10–10.0	0.05	—	—	nonlinear
	3414.76	0.001–3.0	0.05	—	3	nonlinear
	3515.05	0.001–3.0	0.12	—	10	nonlinear
Zn	2138.56	0.001–0.5	—	—	—	nonlinear
	3345.02	0.001–10.0	0.10	—	10	linear
	4810.53	0.01–8.0	0.13	0.06	—	linear
Ti	3372.80	0.001–0.5	0.02	—	3	linear
	3685.20	0.01–1.0	—	—	3	linear
V	3183.41	0.001–0.05	—	—	—	linear
	4379.24	0.001–0.05	0.08	—	10	linear
Pb	4057.82	0.002–0.7	0.08	—	10	linear
Sn	3175.02	0.001–7.5	0.15	—	10	linear

Table 9 continued

Element	Wavelengths of suitable lines (A)	Concentration range (%)	Background equivalent (%) Spectrometer 1.5-m	2-m	Detection limit[b] (ppm)	Shape of analytical curves
B	2497.73	0.001–0.05	0.01	—	2	linear
Be	2348.61	0.001–0.05	—	—	—	linear
	3130.42	0.001–1.2	—	—	1	linear
Na	5889.95	0.001–0.05	0.002	—	0.4	nonlinear
Ca	3933.67	0.001–0.2	0.002	—	0.4	nonlinear
Bi	3067.72	0.001–0.7	0.07	—	10	linear
Ga	2874.24	0.001–0.05	—	—	10	linear
	2943.64	0.001–0.05	—	—	10	linear
Zr	3391.98	0.001–1.0	0.020	—	6	linear
	3438.23	0.001–1.0	0.041	—	10	linear
Cd	2288.02	0.01–2.0	0.03	—	100	nonlinear
	5085.82	0.01–2.0	0.10	—	—	linear
Co	3453.50	0.001–2.0	0.01	—	10	nonlinear
	3465.80	0.001–2.0	0.17	—	10	nonlinear
Ba	4554.04	0.001–0.05	0.007	—	10	linear
	4934.09	0.001–0.05	0.016	—	10	linear
Ag	3280.68	0.001–5.0	0.005	—	5	nonlinear
Al	2567.99	internal standard	—	—	—	—

[a] Adapted from American Society for Testing and Materials (1971).
[b] Detection limit equals standard deviation at 0.001% concentration level.

fluence of the original matrix. Fusion techniques, dilution with spectrographic-grade graphite, and buffers are common in powder dc arc methods. In addition, samples may be dissolved and subjected to chemical treatment including separation techniques to remove the matrix influence.

For all emission analyses the sample and the standard matrices must be carefully matched to ensure quality results.

TABLE 10

SPECTROGRAPH REQUIREMENTS FOR VARIOUS GROUPS OF ELEMENTS AND ALLOYS[a]

Matrices requiring medium-size spectrograph[b]	Matrices requiring large-size spectrograph[c]
Alkali metals	Transition elements
Copper, silver, gold	Rare earths
Boron family	Thorium and uranium
Alkaline earths	Cobalt alloys
Copper alloys	Nickel alloys
Aluminum alloys	Steel
Magnesium alloys	
Zinc	
Tin alloys	

[a] From Mossotti (1970).

[b] Reciprocal dispersion: 20 A/mm; focal length: 1 m.

[c] Reciprocal dispersion: 5 A/mm; focal length: 3 m.

3.2.4 Sample Form

Spectrometric techniques can accept samples in a diversity of forms; some will require little or no sample preparation prior to excitation, while others will need extensive modification to meet the varying conditions and requirements of Table 6. Materials processed in conventional emission spectrometry fall into three main main cateogories: solids, liquids, and powders. Gases in gases and gases in metals are also treated, but under somewhat special conditions. Because these classifications are not rigid, given sufficient time and reasonable demand, the experienced spectrochemist will interconvert the sample into an appropriate form, as dictated by the elements to be determined, their concentration range and distribution, the matrix of the sample, and the quality of the required results. Figure 3 illustrates the types of interconversions used among the three classes.

3.2.5 Solid Samples

Some of the important variables of solid metal samples include size, shape, composition, metallurgical condition, particle size, melting and casting temperatures, rolling direction, degree and type of chill, interme-

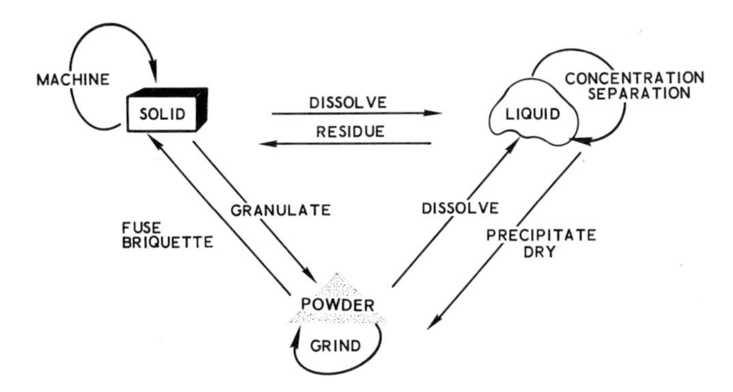

Fig. 3. Interconversion steps among sample forms. Solids can be prepared by surface grinding or machining for direct excitation, dissolved for solution excitation, or converted into a powder for arc excitation. Powders need to be ground to correct size before excitation, or fused, briquetted, or pelleted. Solutions can be dried to a solid residue or precipitated, heated, and ground to form powders.

tallic compounds, heat treatment, contamination, homogeneity, surface roughness, cracks, and surface chemical condition (ASTM, 1971).

Other than gases and halogens, major and minor constituents of elements in solids can be treated almost routinely by arc and spark methods. The spark technique provides a rapid survey or qualitative analysis for samples such as pins, rods, wires, plates, chips, or other material forms which can be held in or on an appropriate stand. Quantitative analysis requires complete control of the sample preparation, and a series of casting molds have been developed (ASTM, 1971) for solid sampling. Molds are designed to produce castings that are homogeneous, free from voids or porosity in the region to be sparked, and representative of the melt. Although casting molds are used extensively in foundries for metal analyses, chips, wires, drillings, or other pieces can be remelted, cast, or pelleted for analysis (ASTM, 1971). After some relatively minor surface finishing, solid disks obtained from these molds are analyzed by spark or spark-ignited arcs in the "point-to-plane" method while being held on a stand. The disk is the plane and the point is a counterelectrode, usually ¼-in.-diameter graphite or silver rod (especially in vacuum spectrometers). Disklike samples for the point-to-plane method may also be obtained as portions cut from ingots, slabs, bars, rods or other mill products. Special caution must be used with these samples owing to differences in metallurgical and mechanical variables developed during processing.

If a solid sample is not homogeneous or cannot be readily recast or otherwise reshaped into a suitable form for analysis, samples may be removed and dissolved. The sensitivity of the spark point-to-plane method is not as good as other methods, such as the dc arc, and routine analysis of high-purity materials is not often conducted with this method. If the sample is not homogeneous, the point-to-plane method may require that the sample be moved during sparking to average the location of the spark sample during analysis.

A number of the limitations of the point-to-plane spark technique may be reduced by a technique conceived and developed by Barnes (1967, 1969, 1974). In this modified point-to-plane method, a thin liquid layer is applied across the sample area during sparking. The liquid-layer solid-sample spark method significantly increases the amount of material removed from the sample and simultaneously increases the intensity of the emission of analysis lines.

The problem of standardization constitutes the most unattractive feature of emission spectrometric analysis of solid samples. For the occasional quantitative analysis of a solid material, methods other than direct arcing or sparking are selected because of the time necessary to develop procedures, operating conditions, and calibration curves. The occasional sample is often better analyzed by more "universal" methods employing solutions or dc arc powder techniques.

a. Pin and Rod Electrodes. The cast pin or rod, prepared for analysis in many cases by being turned on a lathe or having the end ground to a suitable shape by an abrasive wheel, are being replaced in metallurgical analysis by the cast disks in the point-to-plane technique (Hurwitz, 1971). If they are not properly cast or if the casting mold is not correctly designed, cast pins often show severe surface irregularities and contain voids or centerline shrinkage cavities. With pins and rods, both sample and counterelectrode can be identical for spark and arc excitation. Conditions of excitation and type of material must withstand the temperature generated by the discharge without melting or burning. In order to obtain replicate analysis, resurfacing is necessary.

3.2.6 Powders

Many samples, such as minerals, cements, and chemicals are obtained in powder form. Other powder samples are produced by sample pretreatment procedures, such as precipitation or evaporation of solutions, grinding of solids, or ashing of organic materials. Samples must be ground to small particle sizes (100–150 μm) before analysis to insure even volatilization and representative sample aliquots. Powder samples can be treated in a number of ways. They may be dissolved and treated by a solution method or compressed into a pellet for direct excitation in a spark or pulsed arc. Powders have been blown or poured into arcs, sparks, and plasma discharges or carried on a tape through a spark gap (Boumans, 1972a). The most widely used powder method is excitation in an arc discharge after the powder has been packed into a cup electrode (usually graphite). Although there exist a number of good arc techniques using sample forms other than powders, the majority of arc methods are based upon powder samples.

Powdered samples may be analyzed directly or may first be blended with other materials such as buffers, internal standards, carriers, and most often spectrographic-grade graphite powder. All materials added to the sample powder must be pure enough not to contribute contaminants. When several materials are added to each sample, they are usually mixed beforehand in larger quantities.

a. Arc Analysis of Powders. Once the sample has been ground, blended, mixed, and otherwise prepared for analysis, it must be loaded into an electrode for arc excitation. Figure 1 provides a schematic representation of an idealized arc. The most common procedure is to pack the sample (0.5–100 mg) into a crater drilled into the end of a graphite electrode. Some typical shapes of graphite electrodes are given by the American Society for Testing and Materials (ASTM, 1971). Some of the advantages graphite offers as an electrode material are given in Table 11.

i. Sample Volatilization. Upon initiation of the arc current, the electrode heats thermally, and elements from the sample are vaporized into the arc gap. The order and rates of volatilization of elements in an arc have been studied and reported in detail (Boumans, 1966, 1972a). Figure 4 gives a somewhat idealized illustration of volatilization curves for different ele-

TABLE 11

Advantages of Graphite as an Electrode Material

1. Available in high purity form at reasonable cost
2. Easily machined
3. Fair electrical and thermal conductivity
4. No absorption of moisture
5. Nontoxic hot or cold
6. Provides chemically reducing atmosphere
7. Emits few spectral lines or bands
8. Produces high excitation potential arc (11 eV)
9. Porosified by heating
10. Chemically inert at room temperature, and not wetted at arc temperatures
11. Sublimes (3600°C) rather than melts

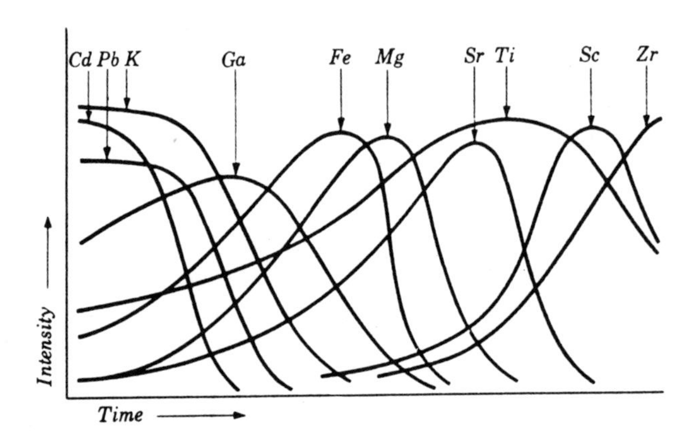

Fig. 4. Somewhat idealized family of volatilization curves illustrating selective volatilization (fractional distillation) from a mixture in an arc. Curves are obtained during arcing by moving photographic plate or with multichannel direct-reading spectrometer. [From L. H. Ahrens and S. R. Taylor, "Spectrochemical Analysis," second edition, 1961. Addison-Wesley, Reading, Massachusetts.]

ments. Table 12 summarizes the order of volatilization of elements in an arc. If the lower electrode is the anode and contains the sample, the vaporization of the sample is called "anode excitation." Samples are volatilized more rapidly from the anode than from the cathode, thus permitting larger

TABLE 12

ORDER OF VOLATILIZATION OF ELEMENTS IN THE ARC[a]

Tendency to volatilize in arc				
		As oxides		
As elements	As sulfides	Volatile	Medium volatile	Nonvolatile
Hg > As > Cd > Zn > Sb ≥ Bi > Tl > Mn > Ag > Sn, Cu > In, Ga, Ge > Au > Fe, Co, Ni > Pt ≫ Zr, Mo, Re, Ta, W	As, Hg > Sn, Ge ≥ Cd > Sb, Pb ≥ Bi > Zn, Tl > In > Cu > Fe, Co, Ni, Mn, Ag ≫ Mo, Re	As, Hg > Cd > Pb, Bi, Tl > In, Ag, Zn > Cu, Ga > Sn > Li, Na, K, Rb, Cs >	Mn > Cr, Mo?, W?, Si, Fe Co, Ni > Mg > Al, Ca, Ba, Sr, V >	Ti > Be, B?, Ta, Nb > Sc La, Y, and many rare earths > Zr, Hf

[a] From Boumans 1972a, by courtesy of Marcell Dekker, Inc., as adapted from L. H. Ahrens and S. R. Taylor "Spectrochemical Analysis," second edition, 1961. Addison-Wesley, Reading, Massachusetts.

samples to be used for analysis. Cathode excitation, however, provides greater sensitivity than anode excitation. The transport mechanisms for anode and cathode excitation have been described in detail by Boumans (1966, 1972a).

The separation of elements by volatility in anode excitation is referred to as selective volatilization or fractional distillation. Selective volatilization has been exaggerated in the determination of trace elements in line-rich refractory materials, such as uranium, by the "carrier-distillation" method. One of a variety of carrier mixtures is used to promote the rapid removal of trace elements prior to the vaporization of the line-rich matrix. As a result, background and line interferences from the matrix material are absent during the exposure for trace elements, and a significant line-to-background improvement is obtained. Many of the studies have been referenced by Boumans (1972a), and the electrode arrangement in Fig. 5 bears the names of the method originators, Scribner and Mullen (1946).

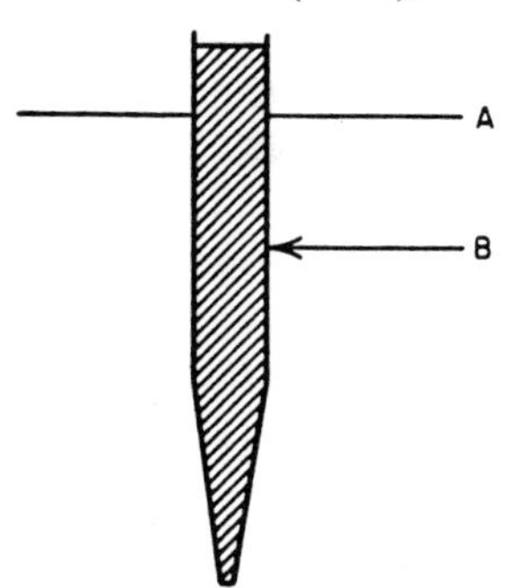

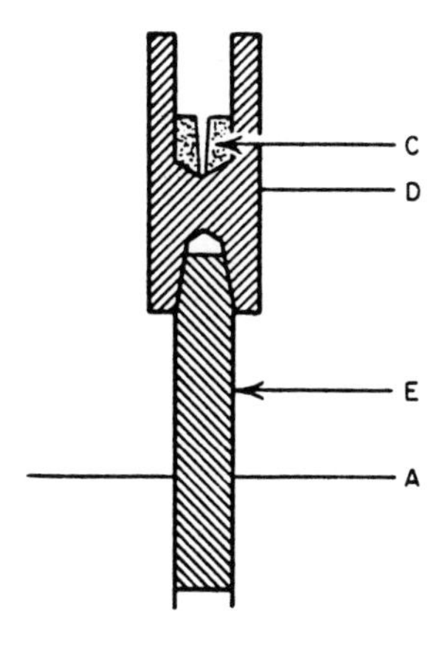

Fig. 5. Electrode configuration for carrier-distillation method. The sample (C) is tamped into the anode (D) and mounted on a graphite pedestal (E) for thermal isolation. A small vent hole is pressed into the center of the sample charge to improve the smoothness of the burn. The electrode holders are represented by A, and B is the cathode. [From Scribner and Mullin, 1946.]

ii. Spectroscopic Buffer. Except for a few methods, including carrier distillations, the procedures developed for the analysis of powder by arc techniques minimize fractional volatilization. The excitation conditions in the arc plasma change drastically as the composition of material volatilized from the electrode crater changes. A spectroscopic buffer compound is added (5–50% of final weight) in practical spectrochemical analysis to make excitation conditions virtually independent of the properties of substances to be analyzed (Boumans, 1972a). In the ideal case the buffer element maintains a constant concentration in the discharge throughout arcing to provide a constant arc temperature and electron concentration. All elements of the sample, therefore, are uniformly excited regardless of fractional volatilization. Buffers promote smooth, regular volatilization and stabilize arc temperatures. The use of buffers requires, in turn, a number of practical compromises. To obtain a maximum quantity of sample entering the discharge while minimizing the disturbing effects of the sample on the arc, an empirical selection of electrode shape and grade, buffer compound, weight ratio of buffer to graphite, particle size of sample, packing of sample, and composition of gaseous atmosphere must be conducted (Boumans, 1972a; Rosza, 1972).

The appropriate choice of buffer, for example, may compromise the limit of detection of different elements. On the other hand, buffers reduce the need for many different internal standard compounds. Typical buffers include CuO, LiF, and Li_2CO_3.

In addition to a buffer, a diluent, usually graphite or carbon powder, consisting of approximately 50% of the final mixture reduces the specimen to a uniform graphite-buffer matrix, minimizes vaporization variations, and provides a smooth consumption of sample.

Although the original powder sample may be diluted by as much as 1:200 by these procedures, the sensitivity of the arc and the reduction of differences among samples more than compensate in the analysis.

iii. Internal Standard. The improvement of the accuracy and precision of arc methods for powders, as well as for other emission techniques, can be obtained by the application of the principle of internal standards. Internal standards are employed to eliminate or minimize the influence of fluctuations in excitation processes on spectrochemical results. Added to the sample is an element which will be influenced, in the ideal situation, by the fluctuations in the discharge to the same extent as the elements under analysis. The ratio of line intensities of analysis and internal standard elements should not depend upon excitation conditions. The compromises in the selection of internal standards have been treated in some detail (Barnett *et al.*, 1968), and most routine methods take advantage of the internal standard approach.

iv. Atmosphere Control. A further development in the analysis of powders in arc discharges is the control of the arc discharge atmosphere. When a free burning arc is operated in air, the anode spot wanders, the graphite electrode is evaporated, and the carbon reacts with nitrogen from the air to produce extensive interfering emission from cyanogen molecules. An effective method by which arc wander is reduced and cyanogen band emission eliminated is to operate the arc in a controlled atmosphere of argon–oxygen. Figure 6 illustrates one of many atmosphere control devices, commonly called the Stallwood jet (Boumans, 1972a). In this device an axial gas stream emerges from an opening concentric with the lower electrode.

3.2.7 Solutions

The emission spectrometric analysis of solutions provides a number of inherent advantages over that of solids or powder, which may account for the widespread use of solutions for analysis. Solutions are homogeneous samples which provide easy manipulations (i.e., concentration, dilution, separation) for standard preparation and are readily supplemented by an internal standard and/or buffers. A diversity of solvents can be used to dilute the sample. When samples are received as solutions, analysis can be rapid, but dissolution or other manipulations slow the overall analysis.

A variety of solution methods in emission spectrometry have developed, as summarized in Table 13. Solutions may be treated directly, nebulized into a discharge, or deposited directly or indirectly as a residue, which is

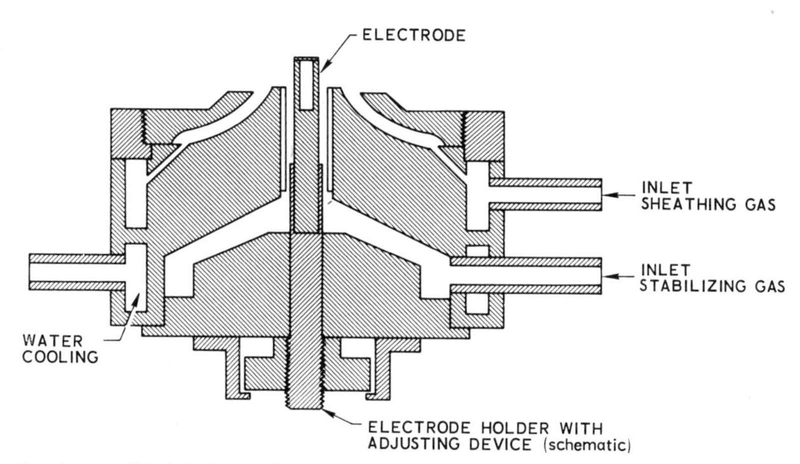

Fig. 6. Modified Stallwood-jet device with two independently controlled gas flows. The inner flow stabilizes the arc and controls the atmosphere of the discharge. The outer flow sheaths the discharge from the ambient air and eliminates the need for an enclosure commonly found with atmosphere jets having only the inner flow. [From Boumans, 1972a, by courtesy Marcel Dekker, Inc.

TABLE 13

SOLUTION TECHNIQUES IN EMISSION SPECTROMETRY[a]

Impregnate or residue	Nebulized	Direct
Copper/graphite spark (S, A)	Spark-in-spray (S)	Rotating disk (S, A)
Rotating platform (S)	Fulgurator (S)	Capillary electrode (S)
Gordon arc (A)	Aerosol spark (S)	Vacuum cup (S)
Hollow cathode (D)	Tube electrode (S)	Dropping electrode (S)
Fusion dip (S)	Spark-in-flame (S)	Porous cup (S)
Hot porous electrode (S, A)	Combustion flames (F)	Plastic/lucite cup (S)
	Plasma jets (A)	Wick electrode (S)
	Plasma discharges	Filter paper (S)

[a] S: spark; A: arc; D: discharge; F: flame excitation source.

later sampled by the discharge. The solution residue techniques have provided some of the most sensitive and reliable emission analysis. Dispersing the solution into fine droplets and carrying the droplets into a discharge (i.e., nebulization) is one of the most convenient methods of solution analysis. The direct approach to solution analysis encorporates solution sampling with discharge excitation. The rotating disk, porous cup, and vacuum cup techniques are popular. Because of the diversity of solution methods, reviews (Gusarskii, 1968; Szakacs, 1964; Young, 1962; Scribner and Margoshes, 1965) should be consulted for applications and key references.

a. RESIDUE TECHNIQUES. *i. Copper, Graphite Spark.* A small volume of solution (50 μl) is evaporated onto the flat ends of ¼-in. diameter copper rods and dried; then the rods are sparked against each other. Absolute detection limits are in the range of 0.2 to 2000 ng for 64 elements with accuracy of about $\pm 5\%$ (Fred and Nachtreib, 1947; Faris, 1962). Replacement of the copper electrodes with silver and platinum yields similar results.

The substitution of graphite for metal electrodes also affords a very suitable residue technique, providing the penetration of solution into the relatively porous graphite is limited by waterproofing the electrode top (Scheibe and Rivas, 1936). Machining a very shallow depression in the top of the electrode prevents solution from flowing over the edge.

ii. Rotating Platform Electrode. The rotating platform electrode was developed (Rozsa and Zeeb, 1953) in order to provide an increased area onto which to evaporate solution. The solution residue is deposited in a circular groove in the lower electrode, rotated, and sparked in a point-to-plane configuration during one revolution. Thus fresh sample is always presented to the discharge, and excitation conditions are more uniform than with stationary electrodes. Compared to other electrode methods, the "rotrode" provides somewhat lower limits of detection with good reproducibility for both aqueous and nonaqueous solvents (Baer and Hodge, 1960).

iii. Hollow Cathodes. The hollow cathode discharge is applied in emission spectrometry for the analysis of small samples, solution residues, and gases, and for the sensitive determination of nonmetals including the halogens, arsenic, selenium, phosphorus, and sulfur (McNally *et al.*, 1947; Harrison and Prakash, 1970). For convenience, the cathode is usually demountable. Figure 7 illustrates schematically one commercial demountable hollow cathode arrangement. Samples (20 mg) are readily introduced into the graphite or metal cathodes as solutions, and a residue remains on the inner walls of the cathode upon drying. Material in the hollow cathode is sputtered with the cold cathode discharges into the cathode space, where it is excited. The sensitivity is good because the sample is retained in the cathode field, little material is lost, and stable long-lived signals are obtained. Precisions of ± 3 to 5% have been obtained for analysis of trace elements in solution residues (Harrison and Prakash, 1970).

iv. Gordon Arc. Gordon and Chapman (1970) have developed an automated arc method, based upon a 10-μl-solution residue, to provide rapid,

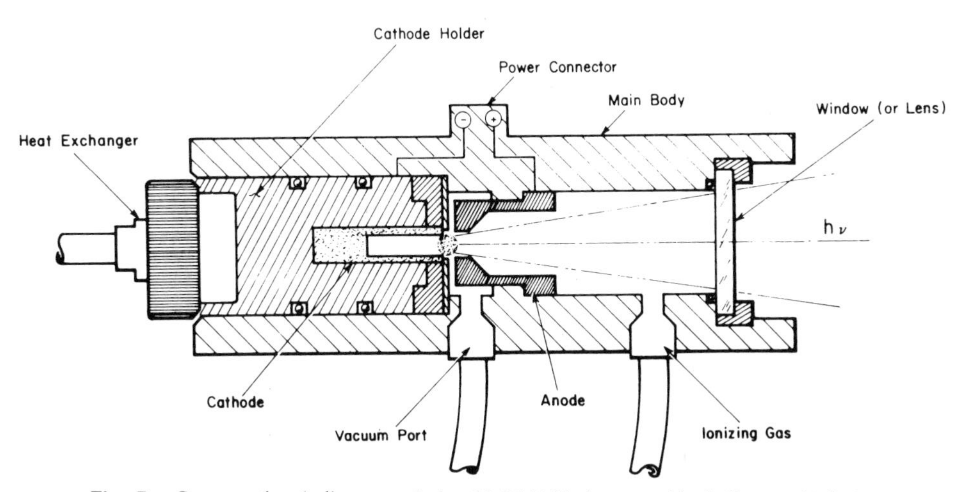

Fig. 7. Cross-sectional diagram of the GLOMAX demountable hollow cathode lamp. Rapid interchange of either metal or graphite cathodes and variable anode-to-cathode spacings are provided. Very high currents may be used with water cooling. [From Technical Data Bulletin, updated February 1973, courtesy of Barnes Engineering Company, Stamford, Connecticut.]

accurate, precise, simultaneous quantitative analysis of random-sample compositions for 20 elements without an internal standard. The approach illustrates the state of the art in applications of arc emission spectrometry.

A block diagram of the dc arc system is given in Fig. 8 and a detailed illustration of the arc chamber and electrode arrangement is given in Fig. 9. The procedure is based upon excitation of solution residues in a common buffer with a highly stabilized arc discharge column. The radiation from the arc chamber (Fig. 9) is dispersed with a multichannel spectrometer, and signals from 20 elements and the silver buffer element are monitored. The electrical signal from the silver channel is compared to a predetermined, programmed arc light intensity profile. Any difference in this comparison causes the dc arc discharge current to alter (between 11–36 A) in order to follow the programmed light intensity and to insure high reproducibility from sample to sample. Stability of the arc column is obtained with the argon atmosphere and a special tantalum-tipped cathode electrode, which eliminates entirely all cathode spot wander. The specially developed carbon anode electrode provides quantitative vaporization of de-

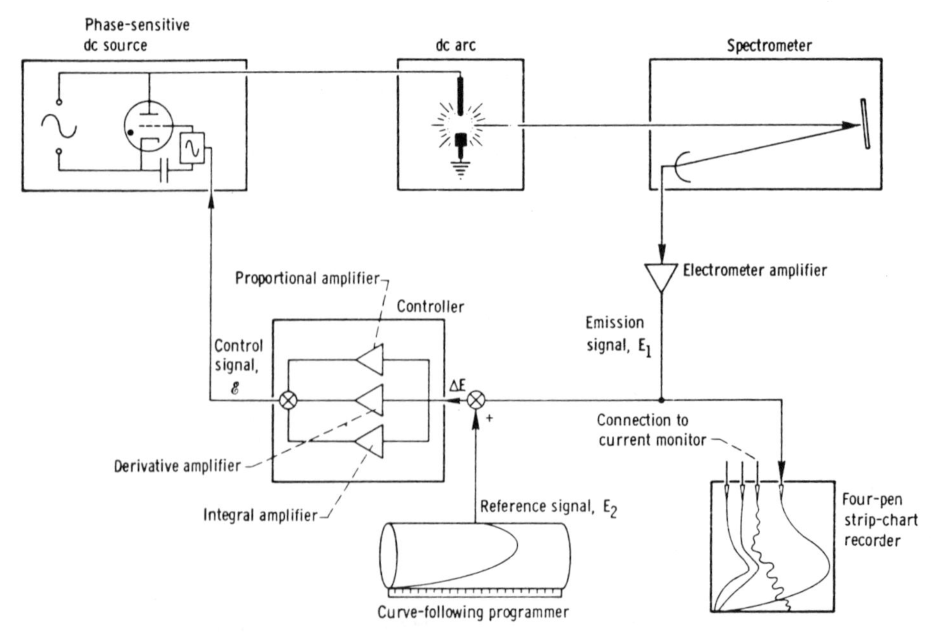

Fig. 8. Block diagram of Gordon arc control loop. Emission from the arc is collected by a spectrometer and compared to a preprogrammed reference signal. The difference adjusts the phase-sensitive dc arc source through a controller. Simultaneous recording of four spectrometer channels or arc current provides visual readout. [From Gordon, 1968.]

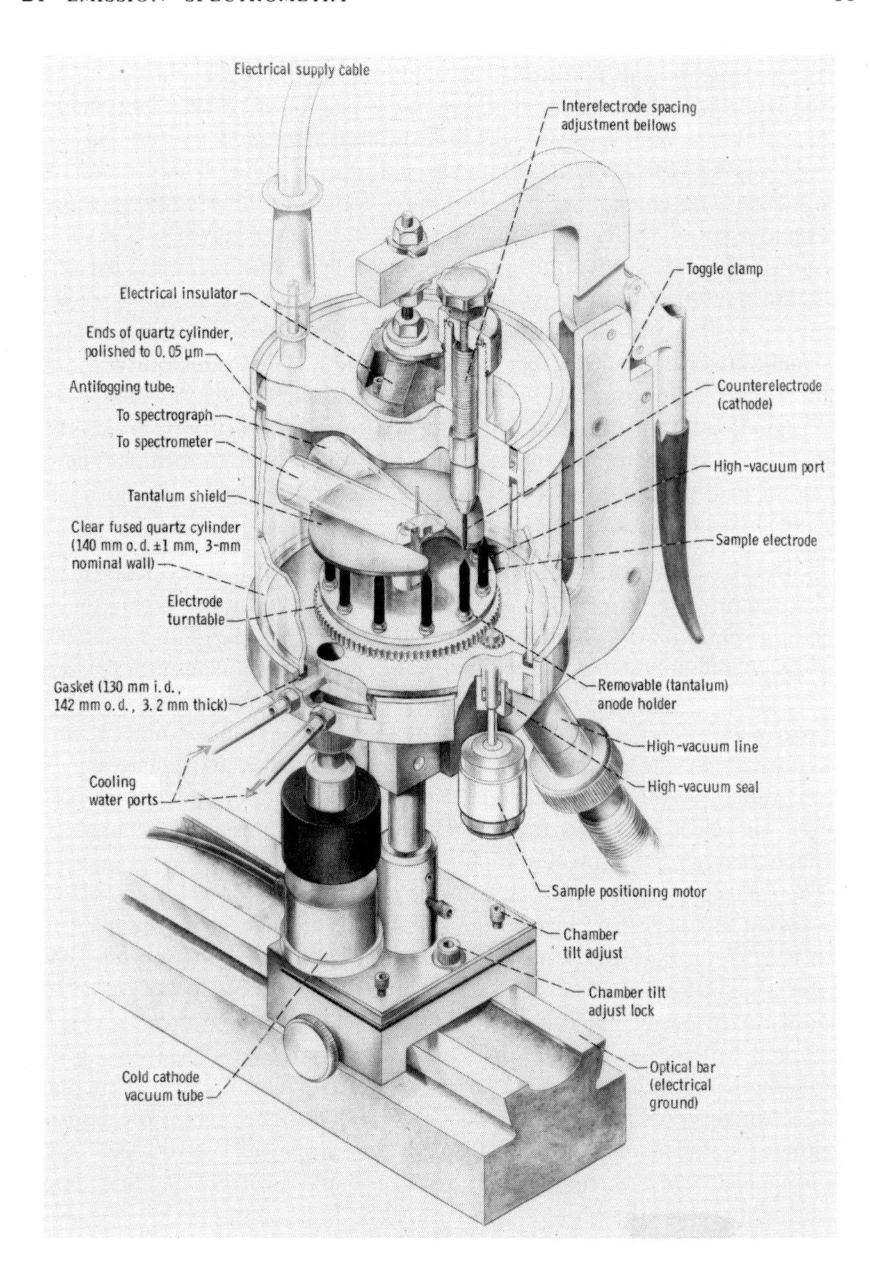

Fig. 9. Controlled-atmosphere arc chamber. Eleven prepared electrodes are sequentially positioned axially with tantalum bead cathode, and radiation is viewed simultaneously by spectrograph and spectrometer. Automated evaluation, flushing, and backfilling with argon takes place as new electrodes move into position. [From Gordon and Chapman, 1970.]

posited elements and additional arc column stability. Standards are prepared from pure materials, and element calibration curves are obtained independently of sample types with an automated readout system. A computer program provides simultaneous calibration data, background, and line interference corrections for 20 elements. Calibration precision averages 7% relative standard deviation.

Samples in the form of alloys or miscellaneous specimens are dissolved in acids. A 400:1 dilution with the common matrix AgCl allows routine analysis of random samples without serious matrix interferences. No internal standard is needed because of the method reproducibility and arc control, and calibration curves of absolute amounts of each element span at least three concentration decades. After sample dilution, about 4 min/sample are required for the analysis through data recording. The arc chamber in Fig. 9 contains 11 samples, which are automatically positioned for each analysis.

b. DIRECT SOLUTION ELECTRODES. The rotating disk, porous cup, and vacuum cup electrodes are popular direct solution techniques in emission spectrometry. Although each electrode has a particular advantage and disadvantage, they perform remarkably alike in the excitation process.

The porous cup electrode (Feldman, 1949) is a hollow graphite tube with a porous floor through which solution seeps. Since only solution which passes through the electrode floor reaches the spark, particulates in solution remain inside the electrode. The porous cup electrode is simple to use, if the electrode jaws are protected from the possibility of solution seepage through the electrode walls.

The rotating graphite disk, in contrast, is not limited to solutions without suspended particulates, because the sample is transported by the electrode as it rotates through the solution reservoir (Pagliasotti and Porsche, 1951). Material is transported mainly on the surface, although Nickel (1969) demonstrated notable soaking effects into the electrode. A fresh surface is presented to the spark as the disk rotates into position. The rotating disk electrode is widely used for analysis of wear and contamination elements in lubricating oils (ASTM, 1971). Disks of copper have been used for determination of halogens and sulfur (Beruzin and Popovich, 1970). As in the rotating platform electrode system, a motor and electrode-holding device must be fitted into the spark stand.

The vacuum cup electrode uses a Teflon reservoir to hold the solution, which enters the spark or ac arc after passing through a central hole in the electrode (Zink, 1959). For viscous solutions, the central hole is enlarged. Particulates in the solution may block the hole.

The lucite cup electrode is similar to the vacuum cup except that the solution filters through the graphite electrode (Zink, 1960). The reservoir-cup-center-post electrode also requires the solution to seep through the graphite center post (Flickinger *et al.*, 1958). The solution is stored in a reservoir fitted to the electrode, which must be drilled to permit solution to pass to the inner section.

Baer and Hodge (1960) compared five different solution methods for precision and accuracy of analysis, but few other comprehensive studies have been conducted, so that selection of a solution electrode is not always definitive.

In a comparison of the porous cup, vacuum cup, and rotating disk electrodes excited in a high-voltage spark discharge, Barnes used time-resolved techniques to observe solvent decomposition and solute propagation phenomena. The solvent apparently decomposes early during the spark discharge at the electrode surface, and the solute is sampled and propagated into the spark gap at a velocity identical to a solid sample of similar composition. This suggests that the sampling and excitation mechanisms for these solution electrodes is similar to that described for solid samples by Walters (1968a).

c. SPRAY METHODS. Although a limited number of direct spray methods are used in conjunction with the conventional arc and spark electrode arrangement, developments using the solution spray from a nebulizer have employed steady-state arc and high-frequency plasma discharges for excitation. These techniques include a number of plasma jet designs and microwave plasma discharges, which combine the simplicity of solution operation with the high energies and temperatures of arc and plasma discharges.

i. Spark-in-Spray. A simple solution excitation technique, the spark-in-spray directs the solution sprayed from a commercial nebulizer between two horizontal or vertical electrodes across which a high-voltage spark is discharged (Malmstadt and Scholz, 1955). The solution does not contact the electrodes when properly used, and a spray collector is positioned opposite to the nebulizer to minimize splashing in the spark chamber. In a typical application, precisions of 3.2 and 4.4% were obtained for titanium in steel and hafnium in zircon (Kobayashi and Kanno, 1971).

ii. Plasma Jets. Combining the convenience and advantages of solution methods with the sensitivity of arc excitation, a number of plasma jet arrangements (Fig. 10) have been developed since 1959 (Boumans, 1972a) and produced commercially (Mitteldorf and Landon, 1963). The plasma jet results from the flow of gas into a closed arc chamber, containing anode and annular cathode electrodes. The arc generated in the flowing gas forms a stream or jet of hot, partially ionized, radiating gas flowing from the arc chamber through the cathode orifice. Solutions nebulized pneumatically or ultrasonically are injected into the arc, atomized, and excited in the dis-

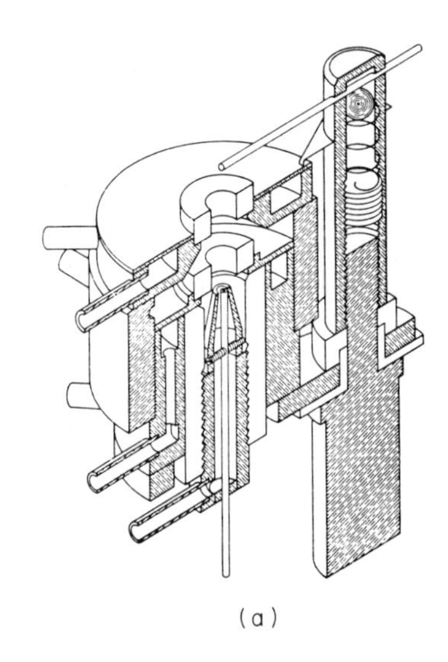

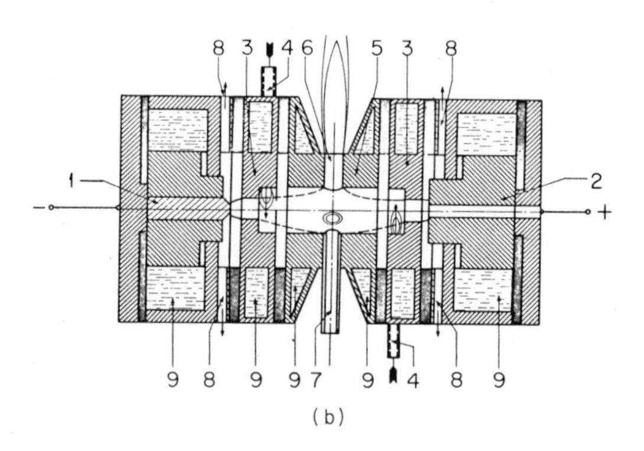

Fig. 10. (a) Schematic drawing of plasma jet developed by Margoshes and Scribner (1959), modified by Owen (1961), and Mitteldorf and Landon (1963). [From Owen, 1961.] (b) Sectional view of Kranz plasma jet. 1: Cathode (carbon, tungsten, or copper); 2: anode (carbon, tungsten, or copper); 3: stabilizing disks (copper or brass); 4: tangential gas inlets; 5: central part (copper); 6: orifice for jet; 7: inlet for carrier gas and aerosol. [From Kranz, 1964].

charge. Atomic and molecular emissions are recorded from the flamelike jet for emission-spectrometric analysis. Fine powders can also be introduced in much the same way.

The original plasma jet designs lacked stability of the discharge within the chamber, and later changes employ either wall stabilization (Fig. 10b) or an external electrode configuration for improved operating stability (Fig. 10a). In the external electrode configuration, a gas-stabilized arc is produced with current conduction and other properties similar to the conventional dc arc (Boumans, 1972a). The plasma jet can be used with aqueous solutions, organic solvents of low flashpoint, and petroleum products. Nonmetallic elements such as N, O, Cl, S, H, and C have been determined, for example, by diatomic molecular band emission (Heemstra, 1970).

iii. High-Frequency Plasma Discharges. Plasmas generated at high frequencies provide excitation energies in excess of combustion flame discharges and comparable to those in arcs without the spectral interferences observed from combustion products or arc electrodes and atmosphere (e.g., cyanogen bands). Although initial studies with high-frequency discharges were reported in the 1940s and 1950s, not until the mid-1960s had much interest been generated for analysis applications (Boumans, 1972a). From results obtained during the past few years, high-frequency discharges appear to be very competitive in accuracy, precision, and sensitivity compared to other emission spectrochemical methods (Dickinson and Fassel, 1969).

The energy of a high-frequency oscillator operating between 0.5 MHz and 30 GHz can be transferred to a plasma discharge by means of coaxial cavities, waveguides, induction coils, or capacitor loops with little or no contact by electrodes. "Single-electrode" or "practically electrodeless" discharges have been developed and operated for spectrochemical analysis in frequencies from 27 MHz to 2.45 GHz. At microwave frequencies power from a magnetron is supplied to a torchlike coaxial cavity (Mavrodineanu and Hughes, 1963). The plasma torch is made into a resonance cavity and fitted with a tuning plunger to provide adjustment of impedance matching (Fig. 11).

At radio frequencies a similar single-electrode plasma torch is used for spectrochemical analysis in which the central electrode is part of an oscillator LC tank circuit (Mavrodineau and Hughes, 1963) (Fig. 12). Although in both the rf and microwave arrangements the single metal electrode is in the gas stream and plasma chamber, little, if any, reaction occurs with the electrode tip, since the production of the plasma results only from the intense electrical field at the tip and not the direct sampling of the electrode. As a result, the high electrical field strengths needed for electron generation and acceleration can be obtained at relatively low power (e.g., 500 W).

Microwave plasma discharges have been applied to determinations of various elements including (1) mercury in water; (2) molecular impurities in argon; (3) boron, tungsten, and molybdenum in binary ferrous alloys; and

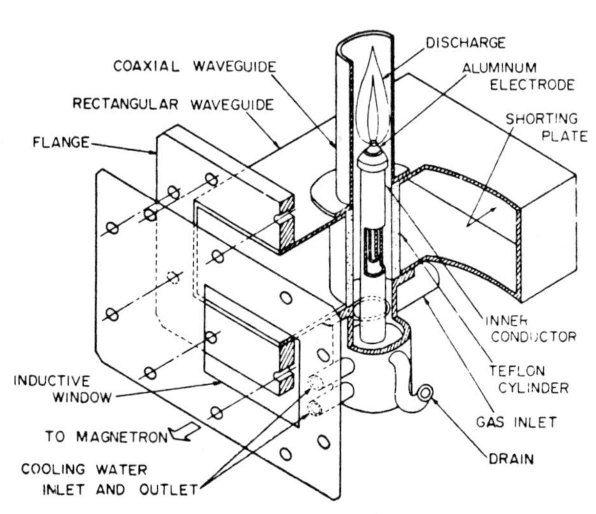

Fig. 11. Waveguide configuration of microwave discharge source. [From S. Murayama, 1968, *J. Appl. Phys.* **39,** 5479.]

(4) microgram quantities of sulfur (Barnes, 1972). Limits of detection are compared to other methods in Table 8, and their precisions, expressed as relative standard deviations, appear to be ± 3–7%.

Discharges without electrodes are produced with rf and microwave fields through inductive and capacitive coupling or with suitable waveguide configurations external to the discharge confinement tube. Radio-frequency fields in the 0.25-kHz to 0.15-GHz range can be coupled with an inductive coil (solenoid) or through a capacitive electrode at atmospheric pressure. The former appears more useful for spectrochemical analysis and has been

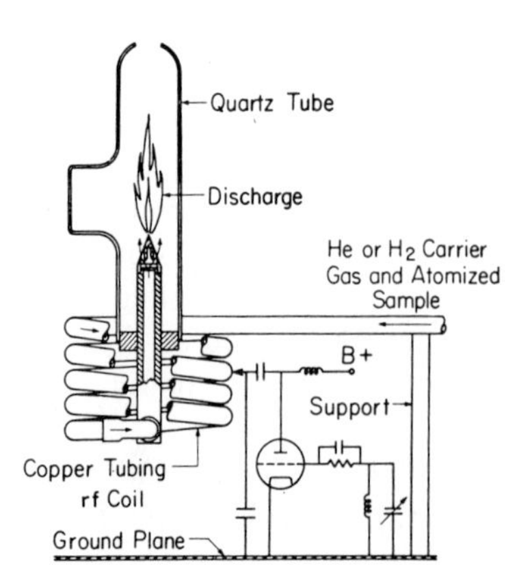

Fig. 12. Schematic diagram of an rf discharge plasma, including oscillator circuit with induction coil and sample introduction. Sample droplets from a pneumatic nebulizer are carried through the coil and the four small holes in the molybdenum discharge tip into the discharge. [From Mavrodineanu and Hughes, 1963.]

the configuration most commonly used during the development of rf electrodeless discharges. Figure 13 illustrates an experimental arrangement used for analysis of solutions with the inductively coupled plasma. The sample is nebulized ultrasonically and carried through a heated chamber to remove the solvent from the droplets. The dried particles are carried into the main discharge region by an argon stream.

During the early development of the rf induction plasma, the sample traveled around the main discharge region as schematically illustrated in Fig. 14a, and only moderate success for analysis was obtained. A comparison of the limits of detection among various investigators using this configuration

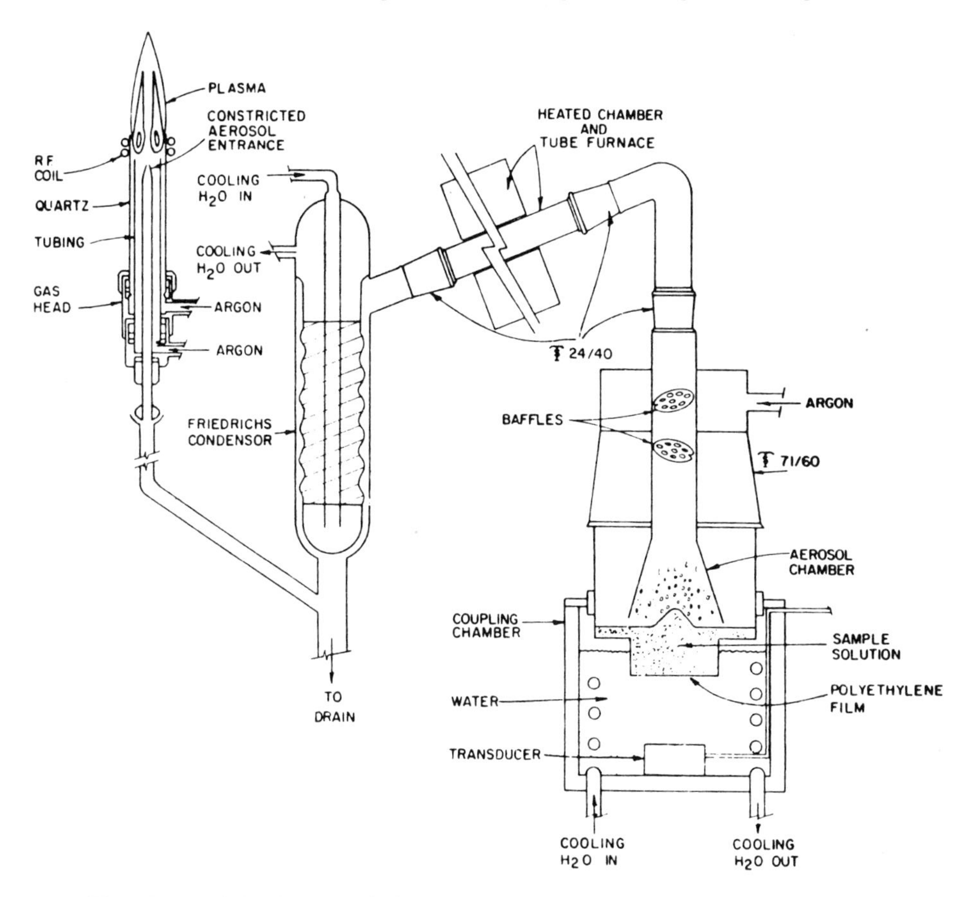

Fig. 13. An induction-coupled discharge arrangement showing one type of ultrasonic nebulization and desolvation system. Desolvated sample particles are carried into the central region of the plasma discharge. Additional argon streams cool the quartz tubing and support the discharge. [From Dickinson and Fassel, 1969.]

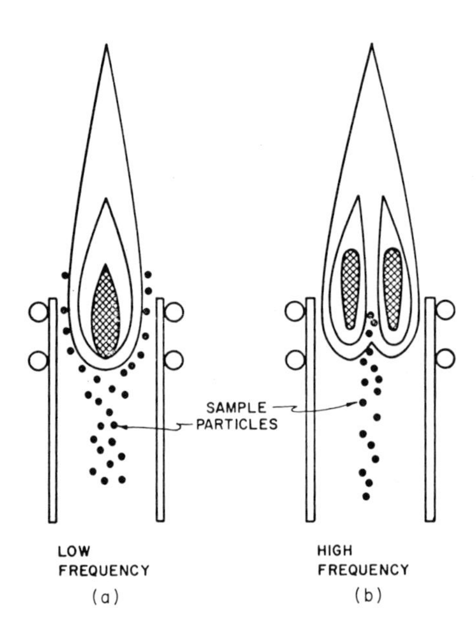

Fig. 14. Highly pictorial representation of ellipsoidal (low-frequency) and annular (high-frequency) plasma shapes illustrating proposed sample paths. Annular shapes are obtained with a critically directed high-velocity central flow stream. At lower frequencies (4 MHz) the plasma resists forming annular shape. [From Dickinson and Fassel, 1969.]

indicated an inefficient use of the energy available in the plasma discharge. Dickinson and Fassel (1969) improved results significantly by developing the plasma discharge configuration illustrated in Fig. 14b. The analyte enters the center of the discharge region instead of traveling along the periphery of the high-temperature discharge. The flow rates and location of sample transport injection probe are critical in establishing this plasma configuration. Figure 15 indicates that if the sample passes along the edge of the discharge instead of along the center, it will experience temperatures thousands of degrees lower. Introduction of the sample into the high-temperature region also increases the sample density in the plasma by approximately an order of magnitude. Limits of detection, compared in Table 8, increased by more than an order of magnitude over the old configuration. In addition, values for some elements appear to be significantly better than any other emission spectrometry method because of the high-temperature, inert argon discharge. Although relatively few analyses have been reported for the induction-coupled plasma in this configuration, the results indicate an absence of low-energy interelement effects which plague combustion flame methods.

The range of use of the inductive plasma is not limited to solution samples because powders (Dagnall *et al.*, 1971) or nebulized metals from molten sources are readily injected into the discharge (Fassel and Dickinson, 1968).

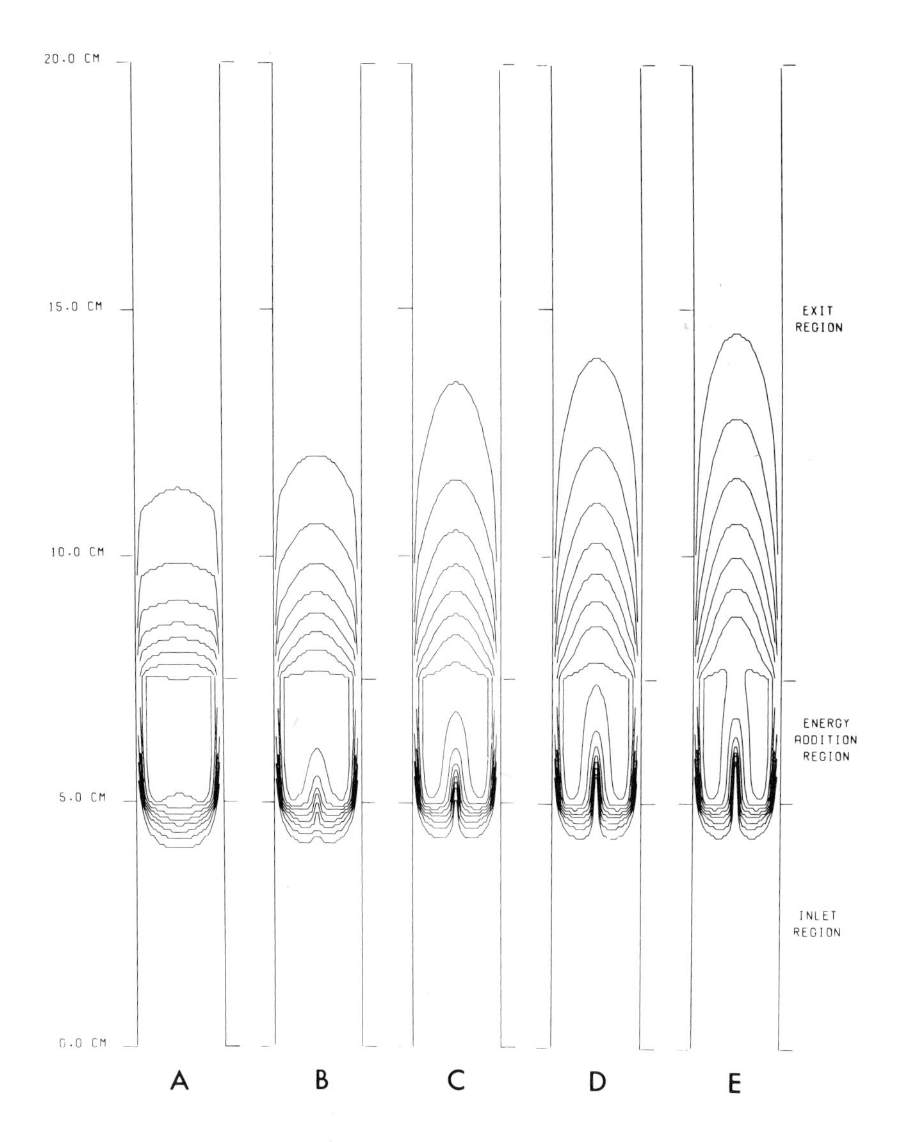

Fig. 15. Computer-generated temperature profiles in 1000°K steps for a 27-MHz, 1.7-kW induction plasma in an arrangement described by Scott *et al.* (1974). Contours represent constant temperatures ranging from 3000 to 10000°K. The central gas flow rate is increased from 0 to 2.0 liter min^{-1} (A to E) in 0.5 liter min^{-1} steps. Samples traveling with the central stream appear to penetrate the discharge and experience temperatures approaching 10000°K. Samples flowing around the discharge region undergo temperatures thousands of degrees lower. Decomposition of a 10-μm aluminum oxide particle is complete 4 cm into the energy addition region at 1.5 liter min^{-1}. [R. M. Barnes and R. G. Schleicher, *Spectrochim. Acta,* 29**B**, 1974 (in press).]

3.2.8 Gases

The emission-spectrometric analysis of gases is much less routine than the analysis of metals or metalloids. The inherent difficulty in the emission analysis of gases is the high excitation potential required and the location of most emission lines to the ground state in the vacuum ultraviolet region.

A variety of excitation sources have been used to provide the necessary energy for emission excitation. Gases contained in hollow cathode lamps are readily excited (Boumans, 1972a). High-frequency discharges in flowing and contained gases also give rapid and straightforward analysis. A microwave plasma discharge was employed in the determination of carbon, nitrogen, hydrogen, and oxygen in hydrocarbons, nitrogen compounds, and pure argon at the ppm level with precision of better than $\pm 10\%$ (Taylor *et al.*, 1970). A simple rf (8 MHz) emission detector for fixed gases (air, CO, CO_2, SO_2, NH_3, NO, NO_2, N_2, and CH_4) in a helium carrier, was developed and may be suitable as a gas-chromatographic detector (Boos and Winefordner, 1972).

Fassel and co-workers developed an arc extraction–excitation method for gases in metals (Fassel and Gordon, 1958). The metal sample is supported in a special graphite electrode and arced as the anode in argon. The sample melts in the electrode, dissolves the graphite electrode wall, and forms a molten globule on the electrode support. The dissolved graphite reacts with the oxides nitrides, and hydrides to release oxygen, nitrogen, and hydrogen. The evolved gases are determined as impurities in the supporting argon atmosphere by means of their visible-range wavelengths. For some materials, notably zirconium, titanium, thorium, and vanadium, another arrangement is used in which a pressed platinum wafer (500 mg) is placed under the metal sample (100 mg) before the determination of oxygen. This method has been extended and combined with gas-chromatographic detection for simultaneous analysis of oxygen and nitrogen in metals (Winge and Fassel, 1969).

3.2.9 Concentration Distribution

Significant developments in materials science have made the determination of the distribution of elements almost as important as the identification of the element itself.

A few decades ago, the emission spectroscopy laboratory faced with the determination of microimpurities in a material such as an alloy had few choices. If a selective chemical or electrochemical dissolution of the matrix to leave the microinclusion was possible, then the microimpurity could be analyzed by conventional arc or spark methods (ASTM, 1971).

Not only have a number of emission methods now become available for sampling regions of less than 50 μm, but other powerful techniques such as electron and ion microscopy have been developed. Two emission methods for *in situ* microanalysis include (a) the laser microprobe (Brech and Cross, 1962) and (b) the microspark (Vogel and Kneip, 1962; Chaplenko *et al.*, 1966).

Concentration distributions in depth are often as important as those at the surface of a sample. Although materials may be cut and polished to expose depth variations, some concentration distributions change within a few hundred or thousand angstroms. There have been developed a number of emission spectrometric devices such as the glow discharge lamp and the dc arc aerosol generator, which permit the determination of concentration distributions in depth.

a. MICROSPARK. Originally designed by Vogel (1967) and Kneip (1962) and developed into a commercial instrument by Chaplenko *et al.*, (1966), the microspark uses a bare tungsten wire counterelectrode electropolished to a point with a radius of approximately 2 μm and positioned 50 to 20 μm from the sample spot by means of a modified metallurgical microscope. The electrode holder is mounted on top of the objective to permit rapid indexing (positioning) in order to obtain spark crater diameters of 5 to 10 μm. The microspark is limited to electrically conducting samples and inclusions but provides smaller crater diameters than the laser microprobe.

b. LASER MICROPROBE. Because of the coherence of laser radiation, a laser beam can be focused to an image with radiance greater than the source. The focused laser beam can vaporize both conducting and nonconducting material from a spot as small as a 10 to 25-μm diameter in a single pulse, and the plasma produced may be used for spectrochemical analysis of the material directly or after further excitation in an independent discharge. Figure 16 gives a generalized diagram of one commercial laser microprobe, in which the laser radiation is focused through a special microscope objective onto a sample located on the microscope stage. Normal crater diameters vary between 10 and 50 μm, although 5-μm spots have been obtained with biological samples. Sample masses of less than 1 μg can be obtained, although the operational mode of the laser determines the amount of material sampled, the shape of the vapor plume, and the vapor residence times (Felske *et al.*, 1970). Maximum sensitivities of 10^{-12} to 10^{-14} g have been reported for a number of elements (Rosan, 1965; Yamane and Matsushita, 1971).

Laser microsampling with a normal-pulse ruby laser in the original laser microprobe required further excitation of the vapor plume by a spark discharge located above the sample site for sufficient intensity in analysis. The development of Q-switched ruby and Nd-glass lasers has provided sufficiently intense plumes to make spark excitation unnecessary for photo-

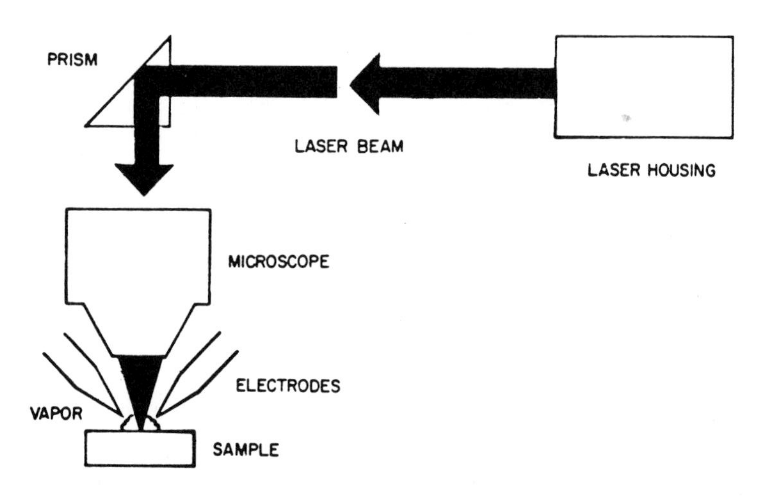

Fig. 16. Schematic drawing of a laser microprobe. Electrodes mounted near the sample on a microscopic stage provide spark excitation of the vapor produced by the focused laser beam. [Courtesy of Jarrell-Ash Division, Fisher Scientific, Waltham, Massachusetts.]

electric detection. Comparisons with and without cross excitation have shown, however, that line intensities can be increased and the character of the line spectrum completely controlled by the spark discharge process (Runge *et al.*, 1964; Rasberry *et al.*, 1967; Piepmeier and Malmstadt, 1969).

Laser vaporization offers advantages for analysis of small inclusions and other microsamples. Nonconducting materials can be examined without difficulty, and living specimens have been sampled without injury to the animals. Current books (Moenke and Moenke-Blackenburg, 1971, Ready, 1971) and reviews (Barnes, 1972) provide descriptions of recent developments and details.

For inclusions of 10 μm or larger, the laser microprobe provides results at a fraction of the cost of an electron microprobe analyzer.

c. Glow Discharge Lamps. In a glow discharge operated under abnormal cathode fall conditions, the removal of atoms by random ion bombardment of the sample cathode (i.e., cathode sputtering) provides location concentration analysis in depth rather than at microlocations along the surface (Grimm, 1967, 1968; Boumans, 1972b; Dogan *et al.*, 1971, 1972; El Alfy *et al.*, 1973).

Glow discharge lamp configurations (Fig. 17) differ in the anode spacing, amount of cathode shielding, and electrical operating conditions (Johnson and Gram, 1972). The solid sample is held against the lamp body, and a small area (0.28–0.5 cm^2) is exposed as the cathode to ion bombardment from the low-pressure, noble gas glow discharge. The mass sputtered in unit time per unit current strength is a linear relationship with the operating voltage (Boumans, 1972b). The sputtered material is excited and emits within a few millimeters of the sample (Dogan *et al.*, 1971).

The Grimm glow discharge configuration (Grimm, 1967, 1968) has been applied to bulk analysis of metals, gases in metals, and nonconducting materials mixed and pressed into conducting binders. Linear calibration

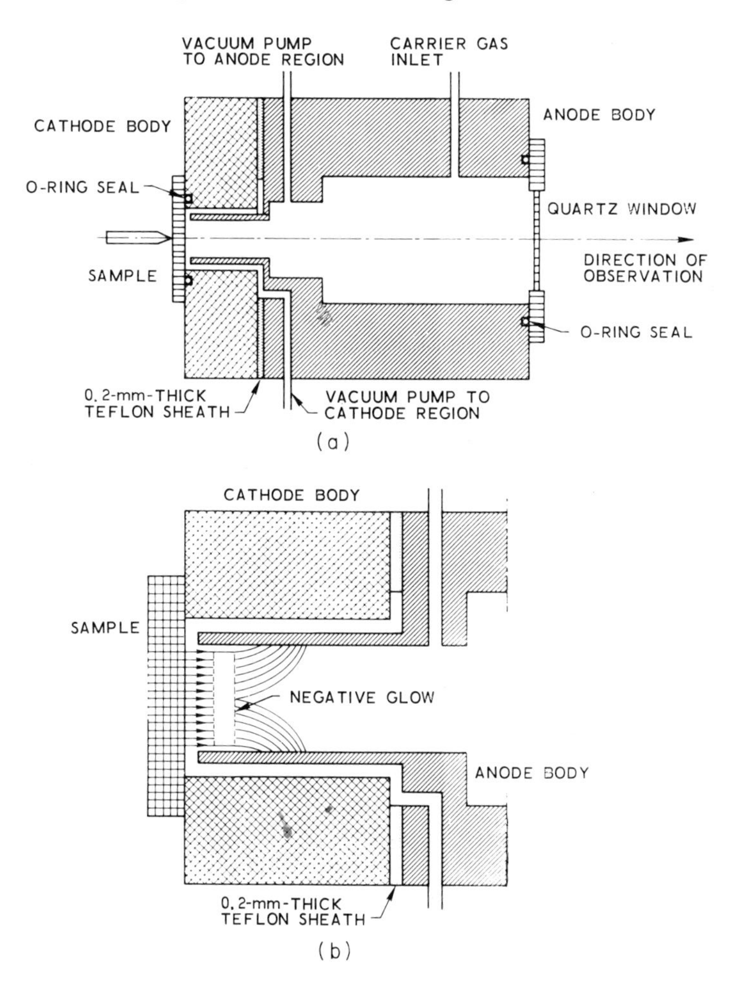

Fig. 17. (a) Diagram of Grimm glow discharge lamp illustrating (b) the obstructed discharge observed through the quartz window. [From Boumans, 1972b.]

curves, good reproducibility, little or no self-absorption and pressure broadening, low background, and wide concentration coverage appear as major advantages of the Grimm source.

d. ARC AEROSOL GENERATOR. A confined dc arc discharge device operated at atmospheric pressure in argon–nitrogen is illustrated schematically in Fig. 18. The sample forms the cathode of a dc arc discharge, and as the cathode spot of the discharge moves rapidly over the electrode surface, submicron particles (droplets) are removed uniformly from the surface and carried out of the discharge by a continuous flow of argon (Jones *et al.*, 1971). The surface is sampled by a series of microcraters. Although the device is specifically designed to separate sampling and excitation steps in the spectrochemical analysis, the average composition of the sample surface is obtained when the aerosol particles are introduced and excited in a discharge source (Jones *et al.*, 1971; Winge *et al.*, 1971).

The arc device does not have the control available in the glow discharge device for the layer-by-layer analysis of materials, but it does permit localized sampling.

3.3 ANALYSIS TIME

One of the major advantages of emission spectrometry is its speed. The methods are fast compared to classical chemical procedures primarily because separation steps are not needed. The radiation emitted by different

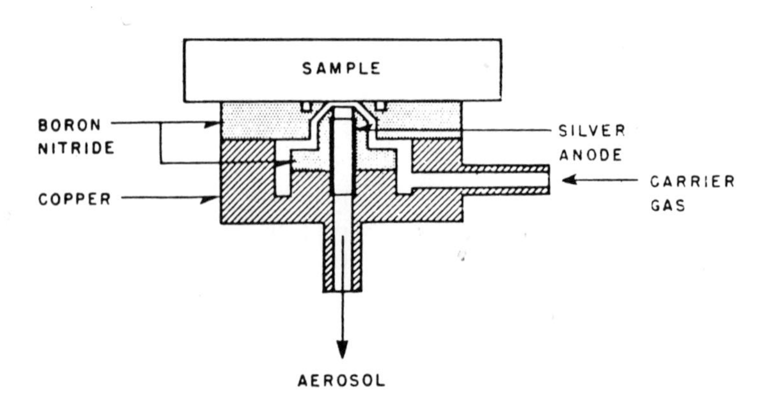

Fig. 18. Diagram of dc arc aerosol sample generator. The discharge between sample cathode and silver anode produces sample aerosol carried by the discharge gas to an external excitation source. [From Jones *et al.*, 1971.]

elements is separated by the dispersive device in the spectrometer. The fastest procedures are developed for those samples which can be analyzed as they are received. The steps in an emission-spectrometric analysis are given in Table 3. Although the actual analysis time is short, considerable development time is required. Standard instrumental operating conditions and calibration curves must be established prior to the routine analysis described (ASTM, 1971; Rozsa, 1972). Once the routine is established, however, calibration takes only a few minutes. An analysis for carbon, sulfur, phosphorus, manganese, and 12 other elements to be found in a 300-ton basic oxygen steel furnace takes less than two minutes after the sample arrives in the laboratory. In steel, aluminum, magnesium, zinc, and brass industries, more than 75% of all analyses are done spectrometrically (Rozsa, 1972).

For the unique or individual sample a number of rapid semiquantitative and quantitative methods are available. The Gordon arc method (Gordon and Chapman, 1970) requires only four minutes once the sample is dissolved. Only a few laboratories are prepared to use the Gordon method, however. Other approaches reviewed by Wang *et al.* (1972) may take a number of hours or a day.

If a completely new quantitative analysis must be established for the routine determination of a new material or product, depending upon experience, as much as six months or more may be required to develop the operating conditions and procedures. Research is actively under way in numerous laboratories to describe and characterize the fundamental processes in excitation sources in order to provide meaningful guidelines for the rapid development of new routine procedures.

3.4 Accuracy and Precision of Results

The precision and accuracy requirements for an elemental composition determination drastically affect the approach and time involved in an emission-spectrometric analysis. A qualitative analysis requires a minimum effort, whereas semiquantitative and quantitative analyses require increasingly greater care and control of the conditions of the analysis. If a representative sample has been obtained, errors of bias in the method of analysis, bias in the individual laboratory, and random errors further contribute to inaccuracies (ASTM, 1971). Spectrochemical equipment is largely responsible for the random errors that influence precision, and both method and individual laboratory bias influence the accuracy. In addition, relative precision and accuracy depend upon concentration levels. A 1% precision at the 0.1% concentration level may well increase to more than 100% at the 1-ppm level. A comparison of precisions obtained with a

TABLE 14

Precision in direct-reading spectrometry[a]	Relative standard deviation (%)
Electronic signal	±0.1–0.2
White light or mercury lamp	±0.2–0.3
Nonsegregated sample, measured for concentration levels above about 0.5%	±0.5–1.5
Sensitive elements in a nonsegregated sample, measured for concentration levels in the range 0.1–0.01%	±2–5
Nonsegregated sample, for concentrations at limit of photographic detectability	±25–100

[a] From R.F. Jarrell (1960). In "The Encyclopedia of Spectroscopy" by G. Clark, © 1960. Reprinted by permission of Van Nostrand Reinhold Company.

multichannel direct reading spectrometer under different conditions is given in Table 14. Although electronic stability provides precisions (expressed as relative standard deviations) of +0.1 to 0.2%, the precision obtained from a good sample concentration signal degrades as the concentration decreases. To obtain high precision and accuracy, the level of control and standardization increases with the quality of result and the decrease in concentration.

In the assessment of an emission spectrometric method, evaluation of individual errors contributed by each of the factoıs outlined in Table 15 will assist in improving the result. Many evaluations occur during the development of a method, and standard methods include a value of merit for the analysis when performed under the controlled conditions specified. Table 16 illustrates the accuracy and precision studies obtained for chromium from a low-alloy steel method for 16 elements using a vacuum spectrometer (ASTM, 1971).

Since emission spectrometric methods are direct rather than comparative and relative rather than absolute, calibration with concentration standards is required. Any alteration in the conditions which affect the consistency of results from one exposure to another or from one day to another necessitates recalibration. Thus, for quantitative analysis, the quality of standards is critical. Standards and samples must be matched carefully in physical and chemical composition, and for those techniques in which standards are poor or unavailable, the range of application is seriously limited.

TABLE 15

FACTORS AFFECTING ACCURACY AND PRECISION[a]

A. Sampling	E. Optical, spectroscopic effects Resolution, dispersion Scattered light Alignment Aberrations Stability
B. Sample Composition Matrix Physical structure Metallurgical history Shape (thickness)	F. Weather
C. Operating conditions Cleanliness Cross contamination Electrodes Standard materials	G. Detector effects Line Selection Emulsion calibration Microphotometer error Linearity Background correction Stability
D. Source Electrical parameters Discharge characteristics Discharge location, stability Interactions with sample Gap geometry Line-to-background ratio	H. Calculations, data evaluation

[a] Data from Arrak and Mitteldorf (1960).

Differences occurring during excitation may be minimized through the proper use of an internal standard. If, however, the sample itself is not representative of the overall material, even the best method provides little useful information. For this reason, particular care must be exercised in obtaining samples, whether they be from a steel foundry or in an environmental pollution study. Suitable sampling procedures and sample preparation techniques must be developed for the analysis in order to minimize variables. Thereafter, procedures must be closely followed unless the impact of a modification is studied.

The accuracy required for an emission-spectrometric analysis determines the approach to be applied. Because of the extremely broad applications of emission spectrometry, three categories of accuracy are commonly distinguished: qualitative, semiquantitative, and quantitative analyses. Table 17 outlines the accuracy limits used in a number of definitions. Clearly, the limits are not critically fixed, and a method which is "quantitative" for a number of elements may be semiquantitative for others.

TABLE 16

PRECISION AND ACCURACY OF CHROMIUM IN STEEL[a]

		Precision		
	Average concentration (%)	Standard deviation (%)	Relative (%)	Number of analyses
	0.85	0.008	0.94	9
	0.35	0.0063	1.8	15
	0.05	0.0045	8.9	15
		Interlaboratory precision		
Laboratory:				
1	0.76	0.0048	0.62	
2		0.0026	0.34	
3		0.0046	0.61	
		Accuracy		
	True value (%)	Average spectrometer value (%)	Deviation (%)	Number of laboratories
	0.85	0.85	0.00	3
	0.37	0.35	−0.02	5
	0.05	0.05	0.00	5

[a] Adapted from American Society for Testing and Materials (1971): ASTM E 415.

To obtain quantitative emission-spectrometric analyses with accuracies better than ±10%, every step of sampling, sample preparation, excitation, data recording, retrieval, and analysis must insure elimination of errors. Sampling must be representative, the sample prepared correctly, the standards obtained and used properly, the excitation controlled, and the data recording, processing, and retrieval standardized. In contrast to semiquantitative analysis, either a direct-reading spectrometer or careful densitometry of spectrographic plates is necessary. A high degree of technique and competence is required for the development of good quantitative methods, and a good laboratory technician can provide quality routine and special analyses.

Qualitative analysis is based upon the identification of unique spectral emission lines, often the five most persistent ("raise ultimes"), for about 70 elements in routine methods. Groups of lines also assist the experienced

TABLE 17

ACCURACY RANGES FOR ANALYSIS (%)[a]

	Mitteldorf[b]	Rozsa	Harvey	Ahrens
Qualitative	> ± 150	> ± 300	> ± 200	> ± 200
Semiquantitative	15 ~ 150	20 ~ 300	30 ~ 200	10 ~ 200
Quantitative	< ± 15	< ± 20	< ± 30	< ± 10

[a] Adapted from Wang *et al.* (1972).
[b] From Nohe and Mitteldorf (1965).

spectroscopist in the rapid identification of common elements by visual inspection of the spectral plate or film. Although not essential, most qualitative analyses are performed by dc arc methods because of the sensitivity available. The techniques used for element identification apply equally well to other modes of excitation, however.

Determining the elements present in a sample is useful, but most qualitative analysis also incorporate some type of quantitative estimate. In the simplest form, designation of "major," "trace," and "absent" provides limited information. However, no standards or concentration estimates are necessary. To improve estimates, quantities can be divided into a series of percentage-range groups: 100–10% (VS), 10–1% (S), 1–0.1% (M), 0.1–0.01% (W), 0.01–0.001% (VW), 0.001–0.0001% (T), and less than 0.0001% (FT). The codes represent very strong, strong, moderate, weak, very weak, trace, and faint trace. This approach requires more sample preparation than for qualitative analysis. As often as not, a semiquantitative analysis procedure is followed, but only qualitative estimates are reported. As the need arises, the spectrographic plate may be later reevaluated to upgrade the accuracy of results if a semiquantitative procedure has been used. Details of qualitative analysis line identification and quantitative estimates are presented by Wang *et al.* (1972).

Semiquantitative methods represent a compromise between accuracy and precision, and the time and technique required to obtain acceptable results. Often samples are not satisfactorily characterized by qualitative estimates but do not require the accuracy of a quantitative method. Thus, a semiquantitative approach may be selected. Semiquantitative analyses have tended to be more or less universal methods. This may change as understanding of spectroscopic sources improves. Universal methods need no longer be semiquantitative with development of techniques like the Gordon arc method (Gordon and Chapman, 1970).

The semiquantitative, universal methods commonly used are modifications of spectrographic dc arc techniques. Methods in which no standards are employed require reference to empirical tables developed previously through painstaking effort. Methods not using empirical tables depend upon some type of standard, prepared either in the laboratory or purchased commercially. The spectra of the standards are compared visually for intensity coincidences with the unknown samples. One set of commercial standards contain 49 elements in 10 matrices (Mitteldorf, 1962). These methods are often improved when conventional microphotometric recording is used; this in turn increases the time and effort involved in the analysis.

The semiquantitative methods of Harvey, Rudolph, and Wang have been reviewed (Wang *et al.,* 1972), and the methods of Jaycox (1955, 1958), Addink (1971), and Long (1971) should be considered in the selection of a semiquantitative analysis technique.

A number of "universal" methods have been developed for miscellaneous materials, such methods provide accuracies and precision values within the $\pm 10\%$ limit (Thompson and Bankston, 1969; Gabler and Peterson, 1970; Gordon and Chapman, 1970).

Acknowledgments

The preparation of this chapter was supported in part by the National Science Foundation (Grant GP 25909) and the Petroleum Research Fund, administered by the American Chemical Society (Grant 2055-G3).

References

Addink, N. W. H. (1957). *Spectrochem. Acta* **11,** 168.

Addink, N. W. H. (1971). "DC Arc Analysis." Macmillian, New York.

Ahrens, L. H., and Taylor, S. R. (1961). "Spectrochemical Analysis," 2nd ed. Addison-Wesley, Reading, Massachusetts.

Alkemade, C. T. J. (1969). *In* "Flame Emission and Atomic Absorption Spectrometry" (J. A. Dean and T. C. Rains, eds.), Vol. 1, pp. 101–150. Dekker, New York.

Alkemade, C. T. J., and Zeegers, P. J. T. (1971). *In* "Spectrochemical Methods of Analysis" (J. D. Winefordner, ed.), pp. 4–125. Wiley (Interscience), New York.

American Society for Testing and Materials (ASTM) (1971). "Methods for Emission Spectrochemical Analysis: General Practices, Nomenclature, Tentative Methods, Suggested Methods," 6th ed. ASTM, Philadelphia, Pennsylvania.

Arrak, A., and Mitteldorf, A. J. (1960). *In* "Encyclopedia of Spectroscopy" (G. L. Clark, ed.), pp. 99–106. Van Nostrand Reinhold, Princeton, New Jersey.

Avni, R. (1969). *Spectrochim. Acta* **24B,** 133.

Avni, R., and Boukobza, A. (1969). *Spectrochim. Acta* **24B,** 515.

Baer, W. K., and Hodge, E. S. (1960). *Appl. Spectrosc.* **14,** 141.
Barnes, R. M. (1967). NASA Tech. Memo. NASA TM-X-1429.
Barnes, R. M. (1969). NASA Tech. Memo. NASA TM-X-1753.
Barnes, R. M. (1972). *Anal. Chem.* **44,** 122R.
Barnes, R. M., and Jarrell, R. F. (1971). *In* "Analytical Emission Spectroscopy" (E. L. Grove, ed.), Vol. 1, Pt. 1, pp. 209–322. Dekker, New York.
Barnes, R. M., and Malmstadt, H. V. (1974). *Anal. Chem.* **46,** 66.
Barnett, W. B., Fassel, V. A., and Kniseley, R. N. (1969). *Spectrochem. Acta* **23B,** 643.
Berneron, R., and Romand, J. (1964). *Mem. Sci. Rev. Met.* **61,** 209.
Beruzin, I. A., and Popovich, N. E. (1970). *Ind. Lab.* **36,** 1718.
Boos, L. E., and Winefordner, J. D. (1972). *Anal. Chem.* **42,** 876.
Boumans, P. W. J. M. (1966). "Theory of Spectrochemical Excitation." Plenum Press, New York.
Boumans, P. W. J. M. (1972a). *In* "Analytical Emission Spectroscopy" (E. L. Grove, ed.), Vol. 1, Pt. II, pp. 1–254. Dekker, New York.
Boumans, P. W. J. M. (1972b). *Anal. Chem.* **44,** 1219.
Boumans, P. W. J. M., and Maessen, F. J. M. J. (1969). *Spectrochim. Acta* **24B,** 611.
Brech, F., and Cross, L. (1962). *Appl. Spectrosc.* **16,** 59.
Candler, C. (1964). "Atomic Spectra." Van Nostrand Reinhold, Princeton, New Jersey.
Chaplenko, G., Landon, D. O., and Mitteldorf, A. J. (1966). *Spex. Speaker* **11(3),** 1.
Dagnall, R. M., Smith, D. J., and West, T. S. (1971). *Anal. Chem.* **54,** 397.
Devlin, J. J. (1971). *In* "Analytical Emission Spectroscopy" (E. L. Grove, ed.), Vol. 1, Pt. 1, pp. 73–130. Dekker, New York.
Dickinson, G. W., and Fassel, V. A. (1969). *Anal. Chem.* **41,** 1021.
Dogan, M., Laqua, K., and Massmann, H. (1971). *Spectrochim. Acta* **26B,** 631.
Dogan, M., Laqua, K., and Massmann, H. (1972). *Spectrochim. Acta* **27B,** 65.
El Alfy, S., Laqua, K., and Massmann, H. (1973). *Fresenius' Z. Anal. Chem.* **172,** 1.
Faris, J. P. (1962). *Proc. Conf. Anal. Chem. Nucl. Reactor Tech., 6th* TID-76655, Gatlingburg, Tennessee.
Fassel, V. A., and Dickinson, G. W. (1968). *Anal. Chem.* **40,** 247.
Fassel, V. A., and Gordon, W. A. (1958). *Anal. Chem.* **30,** 179.
Faust, W. H. (1971). *In* "Analytical Emission Spectroscopy" (E. L. Grove, ed.), Vol. 1, Pt. 1, pp. 131–208. Dekker, New York.
Feldman, C. (1949). *Anal. Chem.* **21,** 1041.
Felske, A., Hagenah, W.-D., and Laqua, K. (1970). *Naturwissenschaften* **57,** 428.
Flickinger, L. C., Polley, E. W., and Galletta, F. A. (1958). *Anal. Chem.* **30,** 502.
Fred, M., and Nachtreib, N. H. (1947). *J. Opt. Soc. Amer.* **37,** 279.
Gabler, R. C., and Peterson, M. J. (1970). *Appl. Spectrosc.* **24,** 354.
Gaydon, A. G. (1968). "Dissociation Energies and Spectra of Diatomic Molecules," 3rd ed. Chapman and Hall, London.
Gordon, W. A., (1968). NASA Tech. Note NASA TN D-4769.
Gordon, W. A., and Chapman, G. B. (1970). *Spectrochim. Acta* **25B,** 123.
Grimm, W. (1967). *Naturwissenschaften* **54,** 586.
Grimm, W. (1968). *Spectrochim. Acta* **23B,** 443.
Gusarskii, V. V. (1968). *Ind. Lab.* **34,** 1767.
Harrison, G. R. (1969). "Massachusetts Institute of Technology Wavelength Tables," 2nd ed. MIT Press, Cambridge, Massachusetts.
Harrison, W. W., and Prakash, N. J. (1970). *Anal. Chim. Acta* **49,** 151.
Heemstra, R. J. (1970). *Appl. Spectrosc.* **24,** 568.
Hurwitz, J. K. (1971). *Can. Spectrosc.* **16,** 1.

Jarrell, R. F. (1960). *In* "Encyclopedia of Spectroscopy" (G. L. Clark, ed.), pp. 138–148. Van Nostrand Reinhold, Princeton, New Jersey.

Jaycox, E. K. (1955). *Anal. Chem.* **27,** 347.

Jaycox, E. K. (1958). *Appl. Spectrosc.* **12,** 87.

Johnson, J. D., and Gram, H. R. (1972). Pittsburgh Conf. Analytical Chemistry and Applied Spectroscopy, Cleveland. Paper 160.

Jones, J. L., Dahlquist, R. L., and Hoyt, R. E. (1971). *Appl. Spectrosc.* **25,** 628.

Kayser, H., and Ritschl, R. (1939). "Tabelle der Hauptlinien der Linienspektren aller Elemente," 2nd ed. Springer, Berlin.

Kelly, R. L. (1968). Atomic Emission Lines below 2000 Angstroms. Hydrogen Through Argon. Naval Res. Lab. Rep. 6648. U.S. Gov. Printing Office, Washington, D.C.

Kobayashi, S., and Kanno, T. (1971). *Bunskei Kagaku* **20,** 582.

Kranz, E. (1964). *In* "Emissionsspektroskopie," p. 160. Akademie-Verlag, Berlin.

Kuba, J., Kucera, L., Plzak, F., Dvorak, M., and Mraz, J. (1965). "Coincidence Tables for Atomic Spectroscopy." Elsevier, Amsterdam.

Long, T. S. (1971). *Appl. Spectrosc.* **25,** 37.

McNally, J. R., Harrison, G. R., and Rowe, E. (1947). *J. Opt. Soc. Amer.* **37,** 93.

Malmand, F. (1970). *Method Phys. Anal.* **6,** 349.

Malmstadt, H. V., and Scholz, R. G. (1955). *Anal. Chem.* **27,** 881.

Margoshes, M., and Rasberry, S. D. (1969). *Spectrochim. Acta.* **24B,** 497.

Margoshes, M., and Scribner, B. F. (1959). *Spectrochim. Acta* **15,** 138.

Mavrodineanu, R., and Hughes, R. C. (1963). *Spectrochim. Acta* **19,** 1309.

Meggers, W. F., Corliss, C. H., and Scribner, B. F. (1961). Tables of Spectral-Line Intensities, Pts. I, II. Nat. Bur. Std. Monogr. 32, U.S. Gov. Printing Office, Washington, D.C.

Mitteldorf, A. J. (1962). *Spex Speaker* **7,** 1.

Mitteldorf, A. J., and Landon, D. O. (1963). *Spex Speaker* **8(1),** 1.

Moenke, H., and Moenke-Blackenburg, L. (1971). "Laser Micro-Emission Spectroscopy." Hilger, London.

Moore, C. E. (1950). An Ultraviolet Multiplet Table, Sect. 1. Nat. Bur. Std. Circ. 488. U. S. Gov. Printing Office, Washington, D.C.

Moore, C. E. (1952). An Ultraviolet Multiplet Table, Sect. 2. Nat. Bur. Std. Circ. 488. U.S. Gov. Printing Office, Washington, D.C.

Moore, C. E. (1962). An Ultraviolet Multiplet Table, Sect. 3, 4, 5. Nat. Bur. Std. Circ. 488. U.S. Gov. Printing Office, Washington, D.C.

Moore, C. E. (1971). Atomic Energy Levels, Vols. I, II, III, NSRDS-NBS 35. U.S. Gov. Printing Office, Washington, D.C.

Moore, C. E. (1972). A Multiplet Table of Astrophysical Interest. NSRDS-NBS 40. U.S. Gov. Printing Office, Washington, D.C.

Mossotti, V. G. (1970). *In* "Modern Analytical Techniques for Metals and Alloys" (R. F. Bunshah, ed.). Wiley, New York.

Murayama, S. (1968). *J. Appl. Phys.* **39,** 5478.

Nickel, H. (1969). *Z. Anal. Chem.* **245,** 250.

Nohe, J. D., and Mitteldorf, A. J. (1965). *Spex Speaker* **10(2),** 1.

Owen, L. (1961). *Appl. Spectrosc.* **15,** 150.

Pagliasotti, T. P., and Porsche, F. W. (1951). *Anal. Chem.* **23,** 198.

Pearse, R. W. B., and Gaydon, A. G. (1963). "The Identification of Molecular Spectra," 3rd ed. Wiley, New York.

Piepmeier, E. H., and Malmstadt, H. V. (1969). *Anal. Chem.* **41,** 700.

Rasberry, S. D., Scribner, B. F., and Margoshes, M. (1967). *Appl. Opt.* **6,** 81.

Ready, J. F. (1971). "Effects of High-Power Laser Radiation." Academic Press, New York.
Rosan, R. C. (1965). *Appl. Spectrosc.* **19,** 97.
Rozsa, J. T. (1972). *In* "Analytical Emission Spectrometry" (E. L. Grove, ed.), Vol. 1, Pt. II, pp. 451–531. Dekker, New York.
Rozsa, J. T., and Zeeb, L. (1953). *Petrol. Process.* **8,** 1708.
Runge, E. F., Minck, R. W., and Bryan, F. R. (1964). *Spectrochim. Acta.* **20,** 733.
Sacks, R. D., and Walters, J. P. (1970). *Anal. Chem.* **42,** 61.
Scheibe, G., and Rivas, A. (1936). *Angew. Chem.* **49,** 443.
Scribner, B. F., and Margoshes, M. (1965). *In* "Treatise on Analytical Chemistry" (I. M. Kolthoff and P. J. Elving, eds.), Vol. 6, Pt. I, pp. 3347–3461. Wiley (Interscience), New York.
Scribner, B. F., and Mullin, H. R. (1946). *J. Res. Nat. Bur. Std.* **37,** 369.
Scott, R. H., Fassel, V. A., Kniseley, R. N., and Nixon, D. E. (1974). *Anal. Chem.* **46,** 75.
Szakacs, O. (1964). *Magy. Kem. Foly.* **70,** 516.
Taylor, H. E., Gibson, J. H., and Skogerboe, R. K. (1970). *Anal. Chem.* **42,** 876.
Thompson, G., and Bankston, D.C. (1969). *Spectrochim. Acta* **32,** 339.
Valente, S. E., and Schrenk, W. G. (1970). *Appl. Spectrosc.* **24,** 197.
Vogel, R. S. (1967). *Method Phys. Anal.* **3,** 131.
Vogel, R. S., and Kneip, T. J. (1962). Process Develop. Quart. Rep. MCW 1479, AEC R and D Rep.
Walters, J. P. (1968a). *Anal. Chem.* **40,** 1540.
Walters, J. P. (1968b). *Anal. Chem.* **40,** 1672.
Walters, J. P. (1969). *Appl. Spectrosc.* **23,** 317.
Walters, J. P. (1972a). *Appl. Spectrosc.* **26,** 17.
Walters, J. P. (1972b). *Appl. Spectrosc.* **26,** 323.
Walters, J. P., and Bruhns, T. V. (1969). *Anal. Chem.* **41,** 1990.
Walters, J. P., and Malmstadt, H. V. (1965). *Anal. Chem.* **37,** 1477.
Walters, J. P., and Malmstadt, H. V. (1966). *Appl. Spectrosc.* **20,** 80.
Wang, M. S., Cave, W. T., and Coakley, W. S. (1972). *In* "Analytical Emission Spectroscopy" (E. L. Grove, ed.), Vol. 1, Pt. II, pp. 395–450. Dekker, New York.
Winefordner, J. D., Svoboda, V., and Cline, L. J. (1970). *Crit. Rev. Anal. Chem.* **1,** 233.
Winge, R, K., and Fassel, V. A. (1969). *Anal. Chem.* **41,** 1606.
Winge, R. K., Fassel, V. A., and Kniseley, R. N. (1971). *Appl. Spectrosc.* **25,** 636.
Yamane, T., and Matsushita, S. (1971). *Bunseki Kagaku* **20,** 1202.
Young, L. G. (1962). *Analyst* **87,** 6.
Zaidel', A. N., and Shreider, E. Ya. (1970). "Vacuum Ultraviolet Spectroscopy." Ann Arbor-Humphrey, Ann Arbor, Michigan.
Zaidel', A. N., Prokof'ev, V. K., Raiskii, S. M., Slavnyi, V. A., and Shreider, E. Ya. (1970). "Tables of Spectral Lines." Plenum Press, New York.
Zink, T. H. (1959). *Appl. Spectrosc.* **13,** 4.
Zink, T. H. (1960). *In* "Encyclopedia of Spectroscopy" (G. L. Clark, ed.), pp. 297–307. Van Nostrand Reinhold, Princeton, New Jersey.

CHAPTER 22

Flame Photometry

Juan Ramírez-Muñoz

Beckman Instruments, Inc.
Irvine, California

Introduction

Flame methods have provided one of the most interesting and attractive fields in analytical chemistry, since they fulfill one of the requirements most sought after by analytical chemists when developing a new analytical methodology: Signals obtained for each analyte under test are the measurements of a physical variable, characteristic of the analyte. This involves, at the same time, a qualification and a quantification of the analyte. Signals are the result of measuring emissions or absorptions of luminous radiations; these emissions or absorptions occur at characteristic wavelengths (i.e., characteristic to the analyte sought), and the intensity of the radiation emitted or the percentage of the radiation absorbed is closely related to the concentration of the analyte in the sample examined.

Thus, flame methods present to the analytical chemist a twofold capability: *the possibility of obtaining both qualitative and quantitative information.*

Since the very beginning of the application of flame methods to analysis, another characteristic was observed—an analytical characteristic that strongly attracted the attention of analytical people: *sensitivity.* Many of the existing methods of analysis had a limited applicability because of the lack of sensitivity, which made them unusable for determining or identifying analytes in the lowest regions of the concentration scale. However, flame methods possess great sensitivity, and minimum amounts or concentrations of some desired analytes may be identified and/or determined, even in the presence of some macrocomponents that might be heavily interfering in other analytical procedures or methods.

Analytical chemists also desire two other characteristics in new analytical methods: *accuracy* and *precision.* In general, instrumental methods cannot reach the same levels of accuracy and precision normally found in classical (wet) methods of analysis. However, careful work with flame methods allows the operator to reach very acceptable levels of accuracy and precision, which, considering the low levels of concentration determinable and/or identifiable by the use of the flame, recommends the method as one with high-sensitivity providing reasonable accuracy and precision for most of the tested applications.

At the very beginning, flame methods were restricted to *emission methods* (measurement of emitted light by entities containing analytes) and to photographic recording or signals. This was the type of application widely known as *flame spectrography*. The use of photographic plates is a time-consuming technique, subject to operator error during the processing of the plates, and requiring afterward an operator's effort to perform the readings of the recorded spectrograms. Such readings consist of line identification and the measurement of blackenings (densities) of each selected line, which is correlated to emission intensity, and consequently, to concentration. This is the reason why the method rapidly gravitated toward *direct reading* of the lines, first by means of photocells, and later by means of phototubes and photomultiplier tubes in conjunction with proper amplification systems and readout devices. The term *flame photometry* was introduced to describe these direct measurement methods.

Many authors have made efforts to find some evidence of the use of flames in early publications where flames might have been mentioned as a means of obtaining analytical information. They have tried to credit early scientists for pioneer work with flames and for having described or applied the analytical capabilities of flames in their own experimentation. However, in reality the interest of analysts in flame methods was a result of the publications—presently considered as classical flame-spectrographic work—by Lundegårdh (1929, 1934), Mitchell, (1964), and Mavrodineanu and Boiteux (1965). The paper by Barnes *et al.* (1945) made a strong impact on the analytical field, since it was the first publication that clearly demonstrated the capabilities of direct-reading procedures in flame methods. In the mid-1940s the first commercial flame photometers appeared, and reports on experimental work with both commercial and home-made instruments began to appear in the scientific literature. The first book published exclusively on the topic of flame photometry was that by Burriel-Martí and Ramírez-Muñoz (1955), printed in Spain (in Spanish). The same authors published the first book in English on flame photometry (Burriel-Martí and Ramírez-Muñoz, 1957). This book was later translated into Russian. The English-language text is still considered an appropriate collection of information for beginners and users of flame-photometric methods in the various areas of applied analysis. Some time later, another book devoted exclusively to these methods was published by Dean (1960), also in English. Other volumes have appeared in the United States, Germany, Hungary, and Russia. Ramírez-Muñoz (1969, 1971) has also contributed special chapters on flame photometry and flame-photometric fundamentals to multi-author treatises. These publications plus the increasing number of papers appearing in the literature form a good information background of the methodology and applications of flame photometry. The collection *Atomic Absorption and Flame Emission Spectroscopy Abstracts* printed in Great Britain by the Science and Technology Agency, London, periodically presents abstracts of many of the papers published on different topics in this field. The collection is edited by Masek *et al.* and distributed by the Science & Technology Agency.

In a field like flame photometry, with more than 25 years of constant development and acceptance as an established method of analysis, a researcher will find it very difficult to trace each and every paper dealing with the subject. Some authors include analytical data in their papers to support their findings without mentioning that these data have been obtained by means of flame photometry. Others merely mention that the results were obtained with a flame photometer without including details of the techniques, instrumentation, and particularities of sample preparation and measurements.

In spite of the introduction of new instrumentation, including many modern items to make the operator's work easier, the original features of flame photometry remain unchanged. A flame is still used, test solutions are still introduced into the flame in the form of a spray (mist or nebula), emissions are selected, and these emissions measured. Finally, the signals are displayed or recorded. Ancillary devices for semiautomated sample feeding, autocompensation of some variabilities of the instrument, and numerical display or printing are just aids to the fundamental operation and achieve only a partial liberation of the attention and/or hands of the operator during laboratory work and postoperational interpretation of results.

Analytical Fields of Flame Photometry

Flame photometry has been extended to several other fields. All of them use flames as a basis for their operations, but phenomena produced in the flames are handled in different ways, as evidenced by the initial subdivision of the field into *emission* and *absorption* flame-photometry.

In 1955, the same year in which the first book on flame photometry was published, it was announced by Alkemade and Milatz (1955), in Holland, and by Walsh (1955), in Australia, that the absorption of luminous radiations by free atoms liberated in flames might be used as another analytical flame method. This was the beginning of the field known as *atomic-absorption flame photometry,* which since 1960 has attracted considerable attention and has gained in acceptance. The novelty of the application of flames to another methodology, the attractive collection of results published by the pioneers of the method, and the prompt availability of commercial instrumentation caused flame photometrists to turn rapidly to atomic absorption. It was thought that the new field, rather than just overlapping the established method in applicability, might very soon make emission flame photometry obsolete. Many papers commented without restraint upon the absence of interference when, in reality, they were referring to the absence in the new method of highly disturbing spectral interferences. Also mentioned were its capabilities of determining magnesium, zinc, cadmium, manganese, iron, cobalt, and other elements with high sensitivity. For example, the determination of zinc and magnesium, so interesting in the study of living matter, was not an easy task in emission flame photometry. It was very soon understood that both methods might and should coexist. Some element can be very easily determined by emission, and interference, if any, can be easily corrected. Emission instrumentation is much simpler and available at lower cost. Some commercial companies have offered and still produce double-purpose instruments, giving the analyst the opportunity of working with the same unit in both emission and absorption at his will and convenience.

The term *spectrophotometry* (flame spectrophotometry) became popular in light of the fact that photometric measurements are made *after spectral selection.* However, there have been frequent attempts to relate flame photometers to filter photometers and flame spectrophotometers to instruments with monochromators. Such a distinction is misleading since both filters and monochromators are included as part of the whole instrumental system for performing spectral selection (at wider or narrower spectral bandwidths). The term *spectrometry* (flame spectrometry) is acceptable since direct measurements are made after spectral selection. However, *flame photometry* is a widely accepted term and it is used in many publications to denote emission flame-spectrometric methods.

The fact that atoms or molecules can be active entities in flame-photometric processes provides the basis for a new subdivision of flame photometry:

Emission

(a) Atomic-emission flame photometry (or spectrometry).
(b) Molecular-emission flame photometry (or spectrometry).

Absorption

(a) Atomic-absorption flame photometry (or spectrometry).
(b) Molecular-absorption flame photometry (or spectrometry).

Atomic- and molecular-emission methods and atomic-absorption methods seem to have been the most popular and widely accepted flame methods up to now. As of the present moment few works have been communicated in which molecular-absorption processes are involved.

Another field of flame photometry is that known as *chemiluminescence flame photometry* (or spectrometry). It is an emission method subdivision in which certain processes can be used for analytical purposes owing to the contribution of some ionic species. These species can contribute excitation energy beyond that normally considered feasible within the limits of the thermal energy supplied by the flame. In these processes, the active species can help to dissociate monoxides formed by the analyte (with the liberation of the free atoms of the analyte). Their dissociation might be very difficult or impossible under the normal thermal conditions of the flame used. Gilbert (1963) has studied the behavior of many analytes that present chemiluminescence phenomena.

A recent development in flame analysis is *atomic-fluorescence flame photometry* (or spectrometry). It is in reality an emission method, but it needs an auxiliary radiation source to illuminate the population of free atoms formed in the flame. The atoms are excited by means of the luminous energy from this source and reemit luminous radiation (fluorescence emission).

In addition, the flame has been replaced by other means of atomization (furnaces, heated carbon filaments, vapor flow cells, etc.). This has led to the development of methods that utilize existing instrumentation with some modifications in the atomization system. These methods constitute the group of so-called "nonflame methods" (of atomic-absorption or atomic-fluorescence spectrometry).

Contents of This Chapter

This chapter is devoted in particular to emission methods (atomic- and molecular-emission flame photometry or spectrometry). Atomic-absorption and atomic-fluorescence methods are described in Volume III, Chapter 30.

The flame and the processes involved in the flame are included in this chapter, as well as some comments on instrumentation and interference. Details concerning flames and processes in the flame, instrumentation, and interference are not repeated in Chapter 30 except when necessary for clarity. The same applies to other aspects of methodology that are common to the different branches of flame spectrometry.

1 Fundamentals

The whole process of flame photometric identification or determination has as its basis the excitation produced by the flame on species constituted by the analyte or containing the analyte. The flame is used for several purposes in flame photometry, but the main one is to provide sufficient thermal energy to produce an excitation. The species receive the thermal energy of the flame, are brought from their normal energy level to a higher energy level, and afterward return that energy increment in the form of luminous radiation of a given wavelength. The same process is repeated if they remain in the hottest region of the flame. In reality, other species of the same type come to this hot region, are excited, emit energy, and so on.

It seems like a continuous process by is actually a succession of energy changes in which new entities enter into action.

The wavelength of the emitted radiation is characteristic of the analyte under examination and constitutes a qualitative variable that can be used to locate and identify an analyte on the wavelength scale. The intensity of the radiation depends mainly on the concentration of the analyte-containing entities in the flame in a given instant. This concentration depends on the concentration of the analyte in the test solution introduced into the flame.

The active entities entering into the above mentioned process can be *neutral atoms, ionized atoms,* or *molecules* (molecules or molecular-like associations of one or several atoms with atoms of the analyte).

The radiation emitted by these entities form a spectrum that can be instrumentally dispersed, conveniently selected (spectral selection), and measured (photometric measurement). This spectrum can be very simple or rather complicated. Actually all depends on the thermal excitation energy supplied by the flame; this determines the nature and concentration of the gaseous components of the flame. The spectrum can be atomic, ionic, molecular, a continuum, or some combination thereof. Atomic and molecular spectra have been widely used in flame-photometric work. However, flame spectroscopists prefer to use atomic spectra (line spectra) whenever possible. Atomic spectra are produced as atoms of the analyte (free atoms) return to the ground state after having been excited to higher energy levels.

The energy returned as luminous radiation satisfies the equation $\Delta E = h\nu$, where ν is the frequency of the emitted radiation. The process follows transitions governed by selection rules.

Not all the atoms of the analyte present in the flame are involved in these phenomena. Only a fraction of them are excited. The fraction of free atoms excited depends on the flame temperature. If the flame is considered to be in a state of thermal equilibrium, the number of excited atoms can be represented by

$$n_j = n_0(g_j/g_0)\exp(-E_j/kT) \tag{1}$$

where

n_j is the number of atoms in a given excited state;
n_0 is the number of atoms in the ground state;
g_j and g_0 are the statistical weights of the atoms in the excited state and in the ground state;
E_j is the energy of excitation;
k is the Boltzmann's constant; and
T is the absolute temperature.

It must be emphasized that this situation (thermal equilibrium) is an ideal case and can be disturbed by any factor influencing the degree of excitation. The number of neutral atoms is of great importance, since the intensity of radiation depends on that number. Intensity also depends on temperature and on the probability of the transition from an excited state to a state (or states) of lower energy.

1.1 Temperature Factor

If the temperature factor is now considered, spectra obtained for a given analyte by using different gas mixtures will be different. For instance, spectra obtained with low-temperature flames will show hardly any lines corresponding to high-energy-level transitions. Examples of low-temperature flames are air–natural gas, air–propane, air–coal gas, air–butane, and oxygen–propane flames. Spectra with these flames will be simple, showing only a few analyte lines.

More complicated spectra are obtained with mixtures of gases that produce flames of higher temperatures; examples of these are the oxygen–cyanogen, air–acetylene, oxygen–acetylene, and nitrous oxide–acetylene flames.

1.2 Line and Band Spectra

Line spectra are preferred in flame work whenever possible. In work with line spectra, it is desirable to use the resonance line (or lines) for qualitative and quantitative tests. Resonance lines are those lines which correspond to transitions to the ground state. Higher intensities appear in lines of long wavelength. Lines of short wavelength (lines corresponding to high-energy-level transitions) will be weak or nonexistent in spectra obtained by using low-temperature flames. It is not advisable to work with resonance lines, however, if there are spectral interferences by neighboring lines and/or bands. In such cases other lines should be used, even if they are lines of lower intensity.

Low-intensity lines, however, may be advantageous for the analyst, since he can use them when dealing with samples containing analytes at high concentration.

Some elements (alkali metals, for instance) present doublet spectra. This type of spectrum can be predicted from the electron spin associated with its orbital motion. When alkali-earth metals are under examination, singlets and triplets may be expected. Doublets, triplets, and multiplets can be identified as such if sufficient resolution is available in the instrumentation used. On the other hand, low-resolution instruments (filter instruments, for instance) will perceive doublets, triplets, or multiplets as a single-emission peak.

1.3 Ionization Effects

Atoms which can be easily ionized show greatest ionization effects when they are present in the test solutions at low concentrations. The ionization rate is about the same at low or at high concentrations, but the recombination with electrons generally proceeds slowly at low concentrations. Ionization effects are noted in practice as a deviation from linear response in the analytical curves. Lines from single-ionized atoms of some analytes will only appear when high-temperature flames are used. The degree of ionization of a given analyte can be controlled to some extent by adjusting the temperature of the flame. It also is possible to do something in this direction by adding considerable amounts of an easily ionizable element to the test solution. The new component, being present in the flame at the same time as the element under analysis, will increase the abundance of electrons (electron concentration) and produce some shift of equilibrium. Elements that are added to produce such an effect are called counterionization agents.

1.4 Self-Reversal and Self-Absorption Effects

In the case of self-reversal, radiation produced in the hottest zones of the flame travels through the cooler outer zones of the flame, and is partially absorbed by atoms in their ground state, which are present in these cooler zones. As a result the radiation, as measured by the instrument, appears lower in intensity. When solutions with high concentrations of the analyte are used, the population of absorbing atoms in the cooler regions of the flame with be higher. In this case, the effect will appear more intense on the high-concentration side of the concentration scale, and as a result, the analytical curves will be nonlinear. Self-absorption is a very similar phenomenon, but it occurs even when the flame has the same temperature at all measured points.

1.5 Molecular Emission

Molecular entities present in the flame and containing the analyte may come from undissociated compounds (due, in some cases, to insufficient flame temperature) or from partial association of the analyte atoms with other entities present in the flame. The formation of any of these molecular forms tends to decrease the concentration of free atoms, and, consequently, the intensity of the free-atom emission.

Molecular forms, if excitable and excited, can also be used for analytical purposes. Band spectra are emitted. These molecular-type entities undergo excitation processes different from those of the free atoms. Vibrational and rotational effects result in the band structure of the spectra. Bands spread out from the frequency that corresponds to the electronic transition. Frequently, bands observed in flame spectra are detected as envelopes of a degraded band. Even in the case of a degraded band, measurements of analytical value can be made at the band head.

In some cases the flame photometrist may search for the conditions necessary to produce band emissions, as they may constitute the best or the only way to determine certain elements (nonmetals, for instance).

1.6 Continuum

A continuous spectrum can also be obtained as a result of unquantized electronic transitions. A continuum pattern may also be expected in cases in which thermal radiation is produced from incompletely vaporized particles or from recombination processes. (One particular case can be the recombination equilibria established between electrons and positive ions.) Flame photometrists try to avoid, if possible, the production of continua, since they result in an increase of background radiation.

2 The Flame

Flame structure, functions of the flame in the atomization process, and some details about different types of flames are included in this section.

2.1 Flame Structure

The flame consists of an external mantle and an inner cone. When observing a flame in the laboratory it is difficult, in many cases, to distinguish the inner cone clearly; a boundary zone appears as a diffuse zone at the base of the flame. The maximum temperature is reached in the flame slightly above the inner cone. Therefore, the best way to find this hot zone is to localize it by means of a series of experimental readings to determine the highest signal (i.e., the most intense emission). The localization of this zone varies from one flame to another as a function of the gas flow for a given combination of gases. The search of this hot zone (selection zone, selected zone, observation zone, or viewing zone) not only involves the observation of signal size, but also the value of the background and noise at each setting chosen for photometric reading. In order to have a visual representation of the response of the flame it is advisable to construct so-called "flame profiles." A flame profile is a representation obtained by plotting the vertical position of the burner or burner elevation (vertical axis) versus the signal size (horizontal axis). This representation gives a good idea of the distribution of the active emitting entities along the vertical axis of the flame.

If a complete representation of the entire distribution is desired, the horizontal distribution should also be studied. A symmetric distribution should be expected. *Flame patterns* are obtained in this way. Such a study of the horizontal distribution at each burner elevation chosen is time-consuming, as it involves a point-by-point examination of the entire front section of the flame.

Flame profiles and flame patterns are normally different for each analyte and for a given flame; they also are different for a given analyte if the flame changes (i.e., in composition, gas flows, gas pressures, burner geometry, etc.).

The flame is produced by the combustion of one gas (fuel or combustible gas) or mixture of gases (fuel mixture) with the help of another gas (oxidizer) which supports the combustion. At the base of the flame the gases are heated by conduction or radiation. The combustion itself takes place in the inner cone, which is the primary reaction zone. Above this zone the gases of the flame consist of molecules such as CO, CO_2, N_2, H_2 and H_2O, and also species such as atoms (H, O) or molecular-type forms (OH); in this area thermal equilibrium is reached. The gases begin to lose temperature as they go to the mantle, where gases from the area surrounding the flame are entrained. If the flame is produced in the open air, the oxygen present may help to oxidize products of incomplete combustion from the inner cone.

The support gas is often used as spraying gas (gas which helps to produce the spray from the test solution). If this gas is not an oxidizing gas (nitrogen, argon), it does not support the combustion at all. Combustion is then supported by the surrounding air. This type of flame is called an *entrained-air flame* (as, for instance, a hydrogen–argon entrained air-flame). If the surrounding gas is an inert gas instead of air, a clear separation can be observed between the primary combustion zone and the upper parts of the flame. Flames of this nature have been referred to as *separated flames*. A flow of inert gas is constantly supplied by proper ducts all around the top of the burner.

Two kinds of flames are normally used in flame photometry: (a) turbulent flames and (b) premix flames (premixed-gas flames).

For the first type of flames, burners are used in which both combustion gases meet at the tip or top of the burner (tip in the small cylindricoconical burners; top in the flat upper section burners). For the second type, burners are used in which gases are premixed before reaching the upper open end of the burner. Laminar-type flames are obtained in this case.

It is quite difficult to establish a sharp distinction between premixed and diffusion flames. Even when both gases come separately to the combustion area, they diffuse there just before any combustion occurs, and the turbulence itself, at this point, results in gas mixing. In laminar (premix) burners a laminar-type flame should be expected; however, some turbulence is produced at the open end of the burner.

2.2 Functions of the Flame

The function of the flame as a thermal source supplying the necessary excitation energy has already been mentioned in this chapter. Before the flame can act as an excitation source, it accomplishes a series of thermal processes to convert the test solution into active entities and maintain them thus as long as possible.

The most frequent type of analysis encountered in flame photometry is the examination of aqueous solutions of saline compounds. The test solution is divided into small droplets (by means of sprayers or nebulizers), and these are introduced into the flame. The droplets of the spray lose solvent (water in the case of aqueous solutions) by evaporation, and the remaining solid particles (salt particles) are vaporized. The compounds are partially dissociated at this stage to release neutral atoms of the analyte. Some fraction of the metallic atoms liberated during this process may react with components of the burned flame gases, and even with components—some of them already in the form of radicals—introduced into the flame by the aerosol. Molecular or atomic hydrogen and oxygen, as well as neutral or charged hydroxyl radicals, may exist in the mixture of burned gases. Metal atoms tend to form molecular entities of the type AO or AOH, as is frequent in the case of calcium, aluminum, lanthanides, and other analytes. The tendency to form hydroxides, for example, may be expected if the energy of formation of the gaseous hydroxide is comparable with the available energy of the flame. Lanthanum and elements of the lanthanide group will hardly show line emissions in low-energy flames. Contrariwise, line emissions of alkali metals are very intense owing to the low energy necessary to decompose the alkali metal hydroxides. As a matter of fact, at this stage atoms or molecular compounds can be excited, and they behave as described earlier in this chapter.

Aerosol formation (spraying or nebulization) and *atomization* are two different processes. Sprayers are no longer called "atomizers" because spraying and atomization are not synonymous, but two steps of a logical sequence. Spraying involves only the division of the test solution into small droplets, and the sprayer merely acts as a device for sample-size reduction. The sprayer can be helped by a spray chamber in which further sample-size reduction can be achieved by drop-size selection. Spoilers, dispersal balls, dispersal plates, baffles, traps, reflux chambers, or condensing chambers are other devices which help in this physical process. Once the fine spray has been formed and selected, the flame will convert the fine spray into a gaseous state; decompose, to some extent, the molecular species into atoms; and excite the free atoms or simple molecular-type entities, thus allowing them to emit characteristic radiation.

The final transformation of the original spray into a fine mist before it enters the flame involves a dilution of the analyte in the aerosol and also the loss of some part of the analyte through the drain when large drops are separated and eliminated. In order to avoid this decrease of concentration, heated chamber burners are used; in these burners the mist is evaporated, the vapor of the solvent is partially condensed and separated, and then dry (or almost dry) particles containing the analyte are brought into the flame, accompanied only by a small fraction of solvent vapor. In this way, not too much analyte is lost by the drain, and the gaseous mixture reaching the flame shows a higher relative concentration of the analyte in comparison with cold-chamber instruments.

The picture is different when sprayer burners are considered. A heterogenous drop-size mist is brought into the flame. No preselection is made. The process of vaporization of the solvent is left to the flame. Only a fraction of the spray will be really evaporated, and only a fraction of the dry particles will be vaporized and decomposed. Thus, a considerable part of the spray passes on through the flame and the efficiency is poor.

A numerical (or graphical) correlation can be established between the physical variable measured in the flame photometer and the concentration of the analyte in the test sample solution. This correlation is the result of the coexistence of several efficiency coefficients that represent the efficiency of (a) the sprayer, (b) the spray chamber, (c) the burner, and (d) the flame. In the flame itself several other coefficients, which might be thought of as several equilibria, coexist: (1) dissociation equilibrium, (2) ionization equilibrium, and (3) excitation equilibrium. The probability of transition toward the lower states and the energy of the light quantum might be also considered at this point. High-efficiency coefficients are needed in flame photometry to achieve good sensitivity in terms of large signal and low noise (adequate signal-to-noise ratio). Good sensitivity will allow the flame photometrist to detect analytes at very low concentrations in test solutions and to perform quantitative determinations with sufficient accuracy and precision at low concentration levels.

2.3 Distribution of Active Entities in the Flame

The preparation and study of flame profiles and flame patterns give an idea of the distribution of the active entities in the flame as seen from the measuring section of the instrument. If these representations are compared with the temperature distribution pattern of the flame, a close correlation can be found. Actually, more active species—emitting species—will be found in zones of the flame where the temperature is higher.

In the upper parts of the flame, the temperature decreases, and consequently the number of excited species also decreases. There will be a higher tendency toward a reassociation of analyte atoms with combustion products. The concentration of emitting species rapidly decays. A maximum (corresponding to a maximum signal) will be found in the center of the flame a little above the inner cone. The decay will be more rapid for elements which easily form oxides and/or hydroxides, such as iron, calcium, and strontium, among others.

The addition of solvents—organic solvents capable of mixing with the aqueous test solution—can change both the flame temperature and the general picture of the physical and chemical equilibria. This double effect procedures a drastic change in the molecular and atomic distribution in the flame, and changes in the flame profiles and flame patterns can be also observed.

The use of solvents as additives can increase the efficiency of the spraying processes, atomization and molecularization; such additives can act chemically insofar as their molecules intervene in overexcitation processes.

2.4 Types of Flames

A wide choice of flame temperature is available because of the diversity of gas mixtures which can be used to produce flames of acceptable analytical characteristics. Some instruments still operate with low-temperature flames for easily excited analytes. Others use high-temperature flames, which have, in addition to high temperature, other convenient features such as low burning velocity and reducing characteristics. A few details will be given here for some of the most popular types of flames used in the flame laboratory.

2.4.1 Low-Temperature Flames

Low-temperature flames produced with air and city gas, coal gas, propane, butane, etc., are used in simple flame photometers. They easily excite such analytes as alkali metals. These instruments also meet the requirements of laboratories in which other types of fuels (acetylene, for example) are not allowed. Some of the so called "Na–K flame photometers" operate with compressed air and propane stored in small tanks. Small-size tanks convert these instruments into portable units, conveniently used in the laboratory and in field tests.

Low-temperature flames cannot be recommended for other elements, especially when the analytes must be determined in the presence of complicated concomitants as, for example, a matrix containing phosphates, aluminum, or other concomitants which can produce molecular species difficult to dissociate in the flame.

2.4.2 Hydrogen Flames

Air–hydrogen flame and oxygen–hydrogen flame constitute a first step toward the use of higher temperature flames. They are very convenient because both produce very low flame background. Mixtures of air, oxygen, and hydrogen can be used—even oxygen and nitrogen mixtures containing up to 60% oxygen. This last mixture has been used with success in the determination of cesium, where it is imperative to avoid the formation of CsOH and also to reverse the ionization of the cesium.

When working with refractory elements, it is possible to use fuel-rich flames (enriched in hydrogen) or to work with oxygen–hydrogen flames fed with solutions containing ethyl alcohol (up to 99%).

Hydrogen is very suitable for use in burners in which argon is used as the spraying gas. This results in the argon–hydrogen entrained-air flames, which are very suitable for determination of magnesium, calcium, strontium, barium, cobalt, nickel, copper, manganese, chromium, gallium, thallium, indium, silver, and cadmium.

2.4.3 Acetylene Flames

The support gases, air, oxygen, and nitrous oxide, have been used in combination with acetylene to achieve higher temperatures. The reducing action of acetylene may also be introduced into the flame process; this action can be used to full advantage if the acetylene/support gas ratio is increased. This fuel-rich flame is bright, luminous, and containing unburned carbon particles. Fuel-rich oxygen–acetylene flames with aqueous solutions or with solutions containing ethyl alcohol have been used very successfully to determine cobalt, iron, barium, antimony, tin, boron, vanadium, rare earths, scandium, yttrium, niobium, rhenium, titanium, molybdenum, and tungsten.

An oxygen–acetylene flame was first used only with sprayer burners, but it has since been also used with premix burners designed especially to handle this high-burning-velocity flame.

The nitrous oxide–acetylene flame was introduced to work in atomic absorption to achieve the atomization of some analytes requiring high-temperature and reducing conditions in the flame. It has been widely accepted because of the relatively low burning velocity. It has also been used in emission work with notable results in terms of sensitivity.

The high temperature of acetylene flames may induce the appearance of ionization phenomena. However, the use of counterionization agents may help to reduce ionization and restore expected sensitivity.

2.4.4 *Other Flames*

Some uses have been reported of other flames such as the oxygen–cyanogen flame, fluorine–hydrogen flame, and perchloryl fluoride– hydrogen flame. These have not had such wide application among flame photometrists as those already mentioned in Sections 1–3.

2.4.5 *The Effect of Flame Background*

The radiation seen by the instrument is the result of the emission of the analytes, of the flame, and of some concomitant species. Many flames produce a continuum and bands. Even flames that are characteristic because of their low background, e.g., hydrogen flames, normally show OH and O_2 bands. Acetylene flames have a higher background which varies according to the support gas used. OH bands are normally seen in acetylene flame spectra. The presence of these bands and of bands due to C_2, CH, CN, CO, and NH interferes in some cases with the lines or bands chosen for analytical work. In addition to this interference, the background also flickers, thereby contributing noise in addition to the phototube shot noise; these latter two factors must be considered overriding when determining the detectability of analytes under a given set of working conditions. Too much noise, i.e., a low signal-to-noise ratio, reduces detectability. In such cases it is necessary to choose other spectral intervals or to use higher dispersion.

3 Instrumentation

There is available a variety of commercial instruments, from simple instruments prepared for only one or two elements to the most sophisticated research instruments. Some researchers still build and use homemade instruments, especially for work with special flames or for multichannel operation. Basically the philosophy behind flame photometers has not changed too much, but modern instrumentation has adopted digital readouts, sample feeders, and automatized sample preparation. Schematically, the basic components of a flame photometer can be represented as shown in Fig. 1.

The instrument is prepared by combining several instrumental subsystems whose functions will be briefly described below.

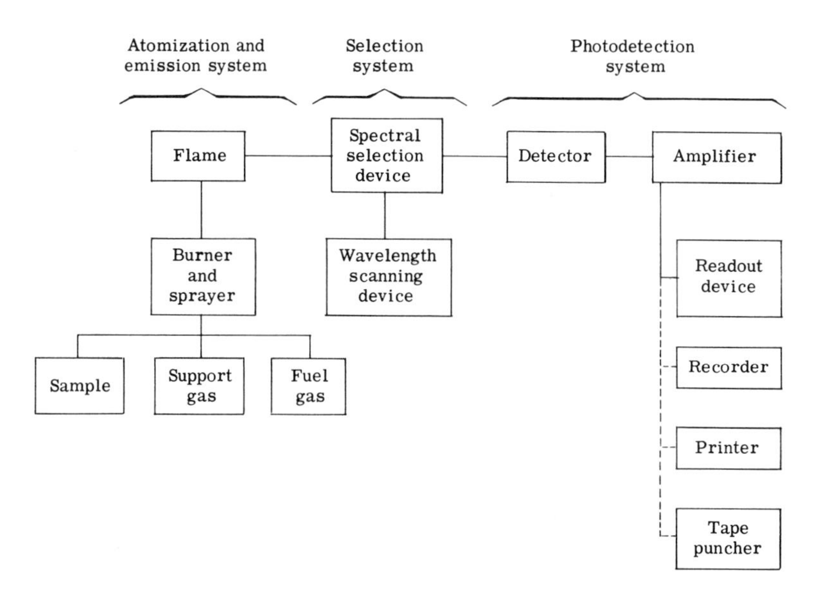

Fig. 1. Block diagram of flame photometer and readout accessories.

3.1 Atomization and Emission System

The atomization and emission system is fundamental to the flame photometer. In this system the test solution is converted into active entities and these entities are excited. Moreover, the system has a device for converting the test solution into a spray (sprayer or nebulizer) and another device (burner) to produce the flame into which the spray will be introduced. The flame takes care of the last step of converting the spray, mist, or nebula into the final entities which will undergo excitation.

Sprayer and nebulizer seem to be the most accepted terms, having completely superseded the old term "atomizer," since the real atomization or molecularization is never performed by the sprayer itself. Even if the sprayer is combined with the burner, i.e., a sprayer–burner, no atomization is produced until the flame is in operation. If the flame is extinguished, the sprayer may still produce an aerosol of droplets in a gas or gaseous mixture. Also the term "aspirator" does not convey the complete image, since some sprayers act by gravity or they are force-fed by a motor-driven syringe for constant sample-consumption rate.

The combination of sprayer and burner has been often referred to as a total-consumption sprayer–burner, or, simply, a total-consumption burner.

Actually the test solution is totally consumed, even if not completely atomized. If the sprayer is mounted separately from the burner, only part of the solution sprayed reaches the burner. For this reason such an arrangement is called a partial-consumption system. The part of the solution that does not reach the flame is lost through the drain after collision with spoilers, or the walls of the spray chamber. It should be remembered that this type of device, i.e., the partial-comsumption system, was first introduced in atomic-absorption work, where it was desired to feed premix flames with a well-selected, fine mist. Later the system was also used in emission work when premix flames were tested and used for emission purposes. Heated spray chambers have also been used in emission in spite of the fact that they had originated and been used for the first time in atomic absorption.

Two or three gases are supplied to the burner, according to the type of flame chosen. All gases have to be regulated by means of pressure regulators, flow regulators, or both. Gas regulation is critical, as slight variations may produce changes in the shape and characteristics of the flame with the result that changes may be produced in the final analytical readings. Gas regulation settings should be verified at the beginning of and during each working session to eliminate this source of instrumental errors. The pressure and/or flow of the support gas, if used as spraying gas, can change the sample-feeding rate as well.

The flow rate of the test solution is another important factor in the performance of the instrument. Too much solution can produce cooling effects in the flame. Too little solution will greatly decrease the signal and sensitivity.

If separated flames are to be used, a shielding gas and the necessary ducts should be incorporated. This new gas also requires regulation and constant flow during operation.

Ultrasonic spraying systems have also been proposed, described, and used in emission work; on the other hand, torches have been substituted for flames. Excitation phenomena have been achieved in this case by means of radio-frequency and induction-coupled plasmas.

The atomization and emission systems also require other devices such as burner elevation mechanisms, back reflection mirrors, diaphragms, focusing optics, and sheaths.

The most accepted flames shape for emission work is cylindrical or pseudocylindrical (from round burner tops). Flat flames obtained with longitudinal-slot premix burners have also been used, especially in work with nitrous oxide–acetylene flames. The flame in this case should be viewed from one of the narrow edges.

3.2 SELECTION SYSTEM

Filters were used in most of the early models of flame photometers. They are still used in some simple models in which the instruments determine one, two, or perhaps three analytes. Instruments for more versatile work include prism or grating monochromators. Any of these methods may be used to select a narrow wavelength band of the emitted spectrum which includes the line or band to be measured by the photodetection system.

The selection system is comprised of the spectral selector (filter, filters, or monochromator) and all the necessary elements to accomplish the spectral selection of the desired wavelength band: entrance slits, exit slit or slits, wavelength selector and indicator, etc. In instruments designed to scan the spectrum, the scanning mechanism should be considered part of the selection system. Band-pass filters are used in grating monochromators to avoid overlapping orders.

The selection system should have a reasonable effeciency because the instrument must accept low-luminosity sources. When radiant power is too low, flame photometrists must work under adverse noise conditions. On other hand, if they desire good line/background ratios, they need as much dispersion as possible. Dispersion in prism instruments is limited in the high-wavelength range.

Monochromators have been fully accepted, since the need for good selectivity and resolution recommends their use in spite of the fact that they are elevated in price, and transmit a smaller fraction of the flame emission. They also necessitate an electronic amplification system to deal with the weak signals reaching the photodetection system.

It is desirable to use the narrowest slits possible because of the different behavior of lines and background with slit-width variations, and because of the advantage gained in resolution. A decrease in slit width, however, produces noise and thus amplifier gain must be increased accordingly.

3.3 PHOTODETECTION SYSTEM

The photodetection system is formed by the photodector and the electronic elements, which provide a usable signal correlated with the concentration of the analyte in the test solution. Signals can be read directly from a meter or a digital readout, or they can be permanently recorded by means of a recorder (analog or bar-diagram type), a printer, or a tape puncher. A teletype is most convenient for the latter purpose.

Photovoltaic and photoemissive cells have been supplanted in most instruments by photomultipliers. There exists a wide variety of photomultipliers, and they offer not only good sensitivity in different regions of the spectrum, but also the capability for varying their response by changing their dynode voltage.

The weak currents the photodetector are amplified using either dc or ac amplifiers. For ac operation it is necessary to interrupt the light coming from the emission source with a chopper and to tune the amplifier to the same frequency. The narrow-band amplifier is normally followed by a linear rectifier. If the amplifier is not the narrow-band type, then the amplifier should be followed by a phase-sensitive rectifier or a lock-in amplifier. Integration techniques are becoming very popular in single- or multichannel instruments.

Multichannel operation is popular because several elements may be determined from a single test solution. It requires a multiselection system and multiphotodetectors or a single selector with a set of different prelocated photodetectors. Simultaneous or successive readings may be performed.

3.4 Auxiliary Equipment

Most of the auxiliary equipment proposed, described, and/or used are feeding devices (multisample feeders) and data-processing equipment; both these tend to minimize personal attention in routine analyses. Automatic instrumentation for sample preparation has also been introduced, especially for biological and clinical analyses.

Flame sensors, flame ignition devices, and other types of devices can be included under the general heading of auxiliary equipment.

4 Interference

Interference is any kind of disturbance which tends to decrease the performance and/or reliability of analytical flame-photometric work. Many factors have been pointed out as causes of interference, and special attention has been devoted to the interference problem since the inception of flame photometry as an analytical tool.

Before any formal attempts were made to classify logically the different kind of interference found in flame work, many publications reported disturbances observed during the course of work with analytical systems exposed to unsuspected sources of interference. Such disturbances have been reported in tables outlining the effects of concomitants on different analytes, and comparative tables summarizing the data reported by different authors have been complied. Owing to small variations in the nature of the flame or in operating conditions, very surprising and disjunctive results have been reported by authors working with the same or different types of instrumentation and the same analytical system. This situation was conducive to the formulation of different theories and tentative explanations. Some of these theories have actually helped to avoid, diminish, or correct interference and have provided the base for many well-established procedures now used in flame photometry.

4.1 Classification of Interference

There are many classifications of interference in flame photometry. The classification used here is very simple but covers most of the cases frequently encountered in emission work.

4.1.1 Spectral Interference

Spectral interference is produced by other radiation at the same wavelength as the analytical line or band sought in qualitative work or chosen for quantitative measurement. Continua, bands, and/or lines overlap, and they may come from the flame background, the solvent used to maintain the analyte in solution, analytes present in the sample, or just concomitants, also present in the sample but not usually determined. The interference may vary from one instrument to another, as the effect of the interference is a function of the spectral bandwidth of the instrument used.

In some cases simple dilution helps to some degree in decreasing spectral interference effects. In other cases the effects are so marked that the analyst must effect preliminary chemical separations.

Several remedies can be applied: (a) stop filters, such as the didymium filter, used to determine analytes in the presence of high concentrations of sodium; (b) background correction (if the interference is due to background, a correction can be made from a careful measurement of the background at one or both sides of the line, as in general practice); (c) correction factors, used as empirical factors once the analytical system has been previously studied by means of suitable synthetic solutions; and (d) radiation suppressors, which may depress the effect of some concomitants.

In extreme cases preliminary chemical separation is mandatory, or it is necessary to look for another line or band of the analyte far away from the interference. Addition methods can also be used.

4.1.2 Physical Interference

Test solution components that might alter the physical properties of the solution are the main causes of physical interference. Viscosity and surface tension differences between test solutions and comparison standard solutions are highly disturbing.

The most common remedies are (a) if possible, correcting standards to match the physical properties of the test solutions by equalization of the chemical composition; (b) simple dilution; (c) addition methods; and (d) internal standard methods, under the supposition that physical differences will affect the element added and used as an internal standard in the same way and to the same extent.

4.1.3 Chemical Interference

Chemical interference involves changes in the chemical nature of species existing in the flame and containing the analyte, whether combined or not. Two kinds of chemical interference are normally found: (a) condensed-phase and (b) vapor-phase.

Condensed-phase interference corresponds to inhibiting the vaporization of the solid-phase entities that contain the analyte. Should the vaporization phenomena occur at a slower rate than expected, the signals measured will be smaller, owing to the decreased number of emitting entities available, if the active entities are produced after vaporization of the original solid-phase entities.

Vapor-phase interference results from disturbances in the vapor phase equilibria, for example, in the dissociation of vaporized molecules or in the excitation of gaseous entities. In this case it is possible to find either signal depression or enhancement, as equilibria can be shifted in either way. Overexcitation, ionization, and other similar phenomena should come under this classification.

A variety of remedies have been recommended and used to overcome chemical interference: (a) application of correction factors, resulting from the testing of synthetic solutions, to the system under analysis; (b) equalization, by equalizing the composition of standards to that of test solutions; (c) simulation, by trying to arrive at a standard composition as close as possible to that of the samples, when equalization is not possible; (d) internal standards; (e) addition methods; (f) the use of releasers, which liberate the analyte by combination of the releaser with the disturbing component; (g), the use of protectors, where the protectors protect the analyte from its association with disturbing concomitants; and (h) the use of hot flames, which help to dissociate many compounds formed between the analyte and certain concomitants (for instance, those of an anionic nature: phosphates, sulfates, silicates).

The use of hot flames has been the best solution for many chemical interferences of the condensed-phase type. Their use in combination with added releasers provides a good prospect for a successful analysis.

Difficult cases in interference will require preliminary separations. This should be kept as a final possibility if other ways fail. If preliminary separation is chosen, extractions should be tried first because of their high efficiency, selectivity, and rapidity.

5 Analytical Characteristics of Flame Photometry

The first analytical characteristic that should be pointed out is the applicability of flame photometry to an extensive number of elements (metals and nonmetals). The introduction of hot flames and the possibility of using certain bands have helped to include many elements which were ignored in the first attempts to extend flame photometry to elements other than alkali metals and alkaline earth metals.

In Fig. 2 a periodic table is presented in which elements detectable under and over 1 ppm are shown. This table not only gives an idea of the extent of the applicability of flame photometry, but also a comparative picture of the high degree of detectability which may be expected in the application of emission flame methods.

As a consequence of the applicability of flame photometry, the number of cases in which the flame photometer can be used to obtain analytical information has increased to an extent not imagined at the time the methods

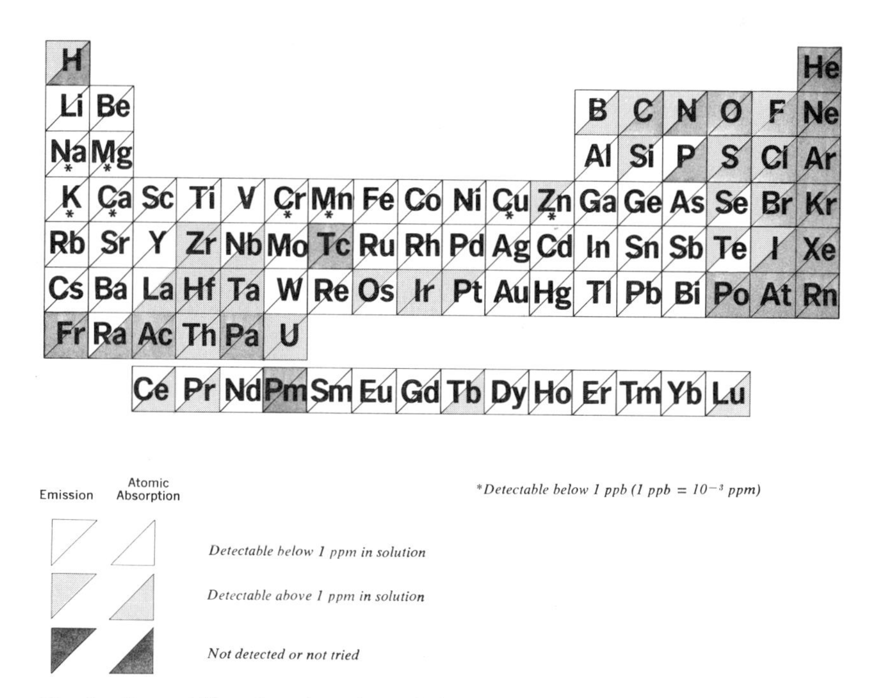

Fig. 2. Detectability of analytes by emission and atomic-absorption flame photometry expressed in terms of fluctuational concentration limits. [Courtesy Beckman Instruments, Inc.]

were proposed as a new analytical tool. At present the flame photometer is used in one of three ways: (1) as an instrument specifically selected as the best choice for solving a given isolated analytical problem; (2) as an alternative method (because it can get an analytical answer with less sample preparation, in spite of the fact that concomitants may effect accuracy and/or precision); or (3) as a routine-type analytical instrument constantly used for the determination of one or several analytes within a fixed and established analytical schedule.

Flame photometry can be used for qualitative and quantitative analyses. After qualitative identification of analytes, it is possible to estimate the level of concentration of some (or all) of them, which permits the analyst to stop the investigation at this point if he considers the semiquantitative estimation of concentration sufficient. Or flame photometry can serve as a basis for choosing the best intervals of concentration for quantitative determination.

5.1 Qualitative Analysis

Qualitative analysis by flame photometry is, as in other instrumental methods, a means of determining the composition of samples by providing an identifiable signal for the analytes present. Identification is not merely detection. The signal should be detected over the background, and also should be identified as corresponding to a given analyte.

Qualitative analysis can be performed element by element. This is the only method for use with simple flame photometers employing filters. The test is limited to a search for a given signal at a specific wavelength interval. With a more sophisticated instrument the analyst can follow the same procedure by looking for signals at certain positions on the wavelength scale and carrying out the search element by element after setting the corresponding characteristic wavelengths. The method is time consuming and is advised only in those cases for which a report about the presence or absence of a few analytes must be given.

Instruments with monochromator and wavelength-scanning capabilities provide a better approach. The sample is sprayed into the flame and the whole spectrum is scanned and recorded, preferably with an analog recorder. Then the analyst can try to identifiy lines and bands by using some of the available guides, such as that included in the book by Burriel-Martí and Ramírez-Muñoz (1957), or with the help of spectral tables.

The scanning procedure is not so easy as it seems at first glance. It is mandatory to have enough sample to feed into the flame during the entire scanning process. It is advisable to repeat the scanning at different instrument sensitivities because at low instrument sensitivity only the high-intensity lines and/or bands will be identified, and many other emissions will hardly be detected over the background. When working at high-sensitivity instrument settings, the intense lines and/or bands will often appear as solid blocks; however, some weak lines or band heads will show.

The procedure requires skill and a good knowledge of the possible types of interference to be expected in accordance with the nature of the sample (concomitants from the sample matrix). High resolution is recommended.

5.2 Semiquantitative Analysis

Semiquantitative analysis is similar to qualitative analysis. Scanning is preferred. Levels of concentration can be estimated by visual comparison of the recorded line and/or band intensities with standard spectra (spectrograms) previously obtained for the same analytes under the same operating conditions and at various concentration levels.

The procedure can be recommended for routine types of analyses. The analytes to be encountered are known in advance and intervals of concentration are also known. Therefore, with a few sets of standard spectrograms, the visual comparison can be easily accomplished in a few minutes. Errors from ± 20 to $\pm 50\%$ (perhaps higher) should be expected. However, these errors may be permissible in semiquantitative estimations at the part-per-million level.

5.3 Quantitative Analysis

Quantitative analysis is oriented toward the accurate analytical determination of each analyte under the ideal operating conditions for that particular analyte in a particular sample. Thus, the analyst does not accept any compromises, as he would normally when examining several analytes at the same time in qualitative and semiquantitative analyses. In qualitative and semiquantitative analyses he must often sacrifice conditions convenient for one or several analytes in order to operate in an average mode applicable to the whole set of analytes. This is the reason why in some qualitative analytical schedules it is recommended that different flames be used for different scannings. Even so, flame richness, slit width, and burner elevation must be kept invariable for a given set of analytes to save time and effort.

Most of the analytical determinations done by flame photometry are direct determinations. There are others, however, that are indirect. A few details about them will be given in the following sections.

5.3.1 Direct Analysis

All procedures based on the measurement of radiation emitted from entities containing the analyte should be considered direct analysis procedures or direct determinations. This procedural category includes not only determinations by measurement of line and band emissions (bands due to molecular forms normally originate in the flame), but also determinations by measurement of emissions from molecular forms whose formation is purposely induced. The analyte included in these specially formed molecules would never offer any usable emission if uncombined. The determination of chlorine by association with copper or indium might be mentioned in this connection.

5.3.2 Indirect Analysis

Indirect analysis procedures (indirect determinations) are reserved for analytes which do not produce utilizable emissions for direct measurement. These analytes may be elements or even radicals.

Two precipitation procedures have been used in conjunction with flame analysis. In the first procedure the analyte reacts with an excess of a precipitating agent followed by the flame-photometric determination of the unreacted precipitating agent in the supernatent liquid or filtrate. An example of this is the analysis of chlorides, bromides, etc., by precipitation with Ag^+ to form insoluble silver salts; the remaining silver ion is determined by flame photometry. From this determination the concentration of the reacted Ag^+ and hence the anion of interest may be calculated. The second precipitation method involves the precipitation and quantitative recovery of the precipitate followed by resolution and flame-photometric analysis of a constituent of the precipitate. An example of this method is the analysis of phosphorus by precipitation as barium phosphate. The precipitate is recovered, and redissolved again providing Ba^{++} which is subsequently analyzed by flame photometry. From this analysis the original phosphorus concentration may be determined.

5.3.3 The Special Case of Nonmetallic Elements

It is desirable to mention the analysis of nonmetallic elements by flame photometry, and many indirect methods have been studied and proposed, based mainly precipitation procedures.

In order to achieve indirect determinations some advantage can be obtained by using the chemical interference that depresses the signals of certain analytes. For example, the interference of phosphates on calcium has been used for the determination of phosphorus.

Bands of CuCl were first used to determine chlorine, but these bands overlap the CuH and CuOH bands. Subsequently, indium was substituted for Cu. The InCl spectrum does not overlap emissions of In, InH, and InO, and thus, is suitable for the determination of chlorine.

Phosphorus and sulfur can now be determined in fuel-rich air–hydrogen flames by measuring emissions corresponding to HPO, PO, and S_2 bands. Phosphorus can also be determined by use of non-fuel-rich flames in solutions containing 10% ethyl alcohol.

5.4 Limitations in Flame Photometry

All sources of interference should be considered limitations to work in flame photometry. As a matter of fact, an excess of concomitants (very saline solutions, for example) may make the determination of some analytes impossible. Through dilution of the original samples, the analytes may remain at too low a concentration in the final solutions to permit a reliable determination. The same applies to samples of high viscosity.

Other limitations should be mentioned: (a) existing nondeterminable elements (see Fig. 2) (nonexcitable or too unstable to be handled); (b) elements existing in the samples at concentrations far below the limits needed for a reliable determination, even before dilution of the sample; and (c) limited sample size insufficient to start or maintain a flow to the sprayer for the time required for a reading. Dilution of the sample to increase the available volume results in (b) above.

5.5 Sensitivity

Sensitivity is perhaps one of the most attractive analytical characteristics of flame photometry. Much has been published on sensitivity in flame methods, but unfortunately some of the concepts disseminated are misleading.

The concept of sensitivity in flame methods is essentially the same as the concept of sensitivity in general analytical chemistry. Sensitivity is simply a relationship between effect and cause, between response and stimulus. In flame methods the response is a reading and the stimulus a concentration. Thus, sensitivity is expressed by the reading–concentration ratio. Sensitivity is not a concentration as is often stated in scientific papers. Sensitivity has the inverse dimensions of a concentration.

For practical purposes some idea of the sensitivity achieved in flame-photometric analytical operation can be expressed by the concentration needed to produce a given signal (a small signal). This concentration coincides with the concept of concentration limit.

Thought of as a relationship or ratio, sensitivity might be expressed by the slope of the calibration curve or analytical curve. If the slope is small, a greater concentration of analyte will be needed to produce a given signal.

The size of the signal can be varied in flame photometry by various instrumental means (instrumental sensitivity, gain, dynode voltage, slit width). Then, it will be necessary to express the sensitivity for each set of operating conditions. A variation of any one parameter may result in a variation of sensitivity. For instance, sensitivity for calcium determinations can be varied simply by changing the nature of the flame or the fuel richness of the flame used.

If a scale of 0 to 100 were used, it might be possible to speak about the *percentual sensitivity,* and the concentration needed to produce a signal of 1 division on the scale (1%) will become the *percentual concentration limit.* If the sensitivity of the instrumental system is now increased, the sensitivity of the analytical determination will increase as well, and the percentual concentration limit will become smaller.*

Many workers are interested in the smallest possible concentration that can be identified for a given analyte. In this case another analytical characteristic is looked for, i.e., *detectability.* The lowest useable concentration is a function of the characteristics of the element in a given set of operating conditions and the response of the analytical chemical system under these conditions. In other words, on the one hand there is the excitability of the analyte and the intensity of the emitted radiation; on the other hand there is the intensity of the background, noise level, and degree of interference.

Under a given set of conditions, with a given analyte, it is possible to calculate detectability as the ratio of the signal identifiable over the noise level and the concentration necessary to produce such a signal. Once again, some idea of the detectability can be given by specifying the concentration needed to produce that signal, which also provides us with the concentration limit. But since in this case the limit is given as a function of the noise fluctuations, it may be called *fluctuational concentration limit.* The expression, detection limit, sometimes found in the literature, is not complete unless it is defined as the *concentration detection limit, dilution detection limit,* or *dilution ratio detection limit.*

Regarding the matter of detectability, many discrepancies are found among authors since each one calculates his results in accordance with the operating conditions found to be optimum for his own particular instrumental setup. For this reason the fluctuational concentration limits shown in Fig. 2, should be considered as informative. Perhaps these detectability limits will never be achieved with certain flame photometers.

In flame photometry the calculation of detectability or the corresponding concentration limit is based on the signal-to-noise balance achieved by the

* The adjective "percentual" is used here to define a concentration limit or a sensitivity calculated as a function of a percentage of the measuring scale.

selection of the instrumental gain and slit width. These parameters should be adjusted to obtain the best signal-to-noise ratio. This means that the flame, sprayer, and other parameters have already been adjusted to produce the lowest noise contribution possible.

Some data for fluctuational concentration limits have been collected in Table 1. For complete collections of recent data, the reader should consult the publication by Ramírez-Muñoz (1971) where data are tabulated for a number of elements under different experimental conditions.

TABLE 1

FLUCTUATIONAL CONCENTRATION LIMITS (FCL)[a]

Analyte	Analytical wavelength (nm)	Flame	Burner	FCL (ppm)	pA[b]
Li	670.8	AHF	TFB	0.0010	9.0
	670.8	OHF	TFB	0.00020	9.7
Na	589.0	AHF	TFB	0.00030	9.5
	589.0	OHF	TFB	0.00020	9.7
K	766.5	AHF	TFB	0.0010	9.0
	766.5	OHF	TFB	0.00030	9.5
Ca	422.7	AHF	TFB	0.040	7.4
	422.7	OHF	TFB	0.010	8.0
	622.0[c]	AHF	TFB	0.020	7.7
	622.0[c]	OHF	TFB	0.0040	8.4
Cu	324.8	AHF	TFB	0.16	6.8
	324.8	OHF	TFB	0.10	7.0
	327.4	AHF	TFB	0.40	6.4
	327.4	OHF	TFB	0.10	7.0
Mo	390.3	NOAF	LFB	0.10	7.0
Ti	399.9	NOAF	LFB	0.20	6.7
V	437.9	NOAF	LFB	0.010	8.0

[a] In this table AHF is the air–hydrogen flame; OHF, the oxygen–hydrogen flame; NOAF, the nitrous oxide–acetylene flame; TFB, the turbulent flow burner; and LFB, the laminar flow burner.

[b] The pA corresponds to the exponent of concentration of analyte: $pA = -\log C$ where C is expressed in grams per milliliter. Values of pA give an idea of *relative sensitivity* (fluctuational sensitivity).

[c] Data given for Ca at 622 nm correspond to the band CaOH.

5.6 Accuracy

For quantitative determinations accuracy is required in flame photometry as in any other field of analytical chemistry. Accuracy in flame photometry depends on the quality of comparison standards prepared for each analytical chemical system. Standards should be prepared in such a way that they match as perfectly as possible the chemical composition and physical properties of the final solutions prepared from sample materials.

Variability in sample composition (matrix variations from sample to sample) is one of the common causes of low accuracy. In this case accuracy is lost because of chemical interferences or spectral interferences.

Careful work and good sets of standards help to keep errors low enough to achieve an accuracy of 1% or better. Normally accuracy of 1 to 2% is acceptable. Larger errors may be accepted for determinations of analytes at trace level. Duplicate or triplicate readings for standards and test solutions also result in better accuracy, especially in cases where high variability of readings is expected.

5.7 Precision

Precision is affected in flame work by fluctuations of the flame, variations in the sprayer, and contributions from the electronics of the instrumental system. In order to achieve the best precision possible, care should be taken to maintain as invariable as practicable all experimental parameters. Gas regulation deserves special attention, as any variation in the regulation section may lead to changes in the flame (variations of temperature, flame pattern of emitting-species disturbution, variation of interferences).

Precision can be represented by calculated values of relative standard deviation (RSD) from a series of repeated readings (obtained by repeating the readings of a standard or a test solution). Relative standard deviation (also known as coefficient of variation) may normally vary from 1 to 3%.

Noisy systems may contribute to an increase in RSD values of up to 5% or more. If this occurs, attention should be given to possible causes of variability in the instrument; repeated readings should be obtained for each solution (duplicate or triplicate); and averages should be calculated to calculate final results.

5.8 Linearity

Linearity is very desirable in flame photometry. However sometimes a lack of linearity or deviations from linearity are observed. In previous section some causes of deviations from linearity have been mentioned (ionization, self-reversal). A typical case is found in potassium determinations, where S-shaped calibrations curves are encountered because of superimposition of ionization and self-reversal phenomena.

Better linearity can be achieved by diluting the solutions, as is normally done in sodium determination. Excessive dilution may, however, require an excessive increase of instrumental sensitivity to obtain signals of sufficient size; this produces, at the same time, a notable increase of noise.

Work with a less intense line may help to obtain analytical curves with better linearity, but less sensitivity (less slope).

Good linear response is desired for those instruments in which scales are prepared to read concentrations directly. In such instruments the zero is carefully set, and the reading from a given standard is adjusted to correspond to the known concentration of the standard. In simple instruments used to determine sodium and potassium in clinical samples, this is done by taking advantage of the linearity of potassium at low potassium concentration and of the proximity to linearity of sodium.*

6 Analytical Methodology

Some information on the general analytical methodology recommended for work in flame photometry will now be included.

6.1 Standards and Blanks

In flame photometry standard solutions have the same meaning that they have in so many instrumental methods in which sample solutions are compared with a set of carefully prepared standards.

Standards must be as similar as possible in chemical composition and physical properties to samples. Even slight changes in composition may result in changes of the chemical interference pattern or changes in background. Another important point is that standards should be read *in each working session* under the same operating conditions chosen to measure the test solution. Readings performed one day can not be used to establish calibration curves for succeeding days, or even for other sets of samples read at a different working session.

Blanks are solutions which contain all the components of the test solution except the analyte. Acids, salts, organic matter, etc., should be added in the same amounts to blanks.

* This is obtained when the sodium-containing solution has been diluted and samples are interpolated between two close standards. The portion of the calibration curve between these standards may be considered as essentially linear.

These guidelines for standards and blanks can be less rigidly followed if original samples are to be highly diluted; in such case many concomitants will have little or no effect.

Reagent grade chemicals should be used in all cases, except in special instances where specpure chemicals might be desirable.

Dilutions from stock solutions should be made with care. Dilutions may follow linear or logarithmic functions to end in standards containing, for instance, 1, 2, 3, 4, 5, etc., ppm of analyte or 0.1, 0.2, 0.5, 1, 2, 5, etc.

At this point it is necessary to note that additions introduced into the final standard solutions may produre some loss of sensitivity (lowering the calibration curve slope), as compared to curves obtained with pure uncompensated standards, but the response may appear closer to that of the original samples containing all concomitants (see Fig. 3).

If many additions must be included in the final standards, it is a very convenient practice to prepare a general addition solution containing double the concencentration of each component of the concentration desired in the final working standards. From the stock solution of each analyte, some intermediate solutions are prepared containing the analyte at double the concentration desired for the standards. The final working standards are prepared by mixing equal parts of the addition solution and the corresponding intermediate standards. See Table 2.

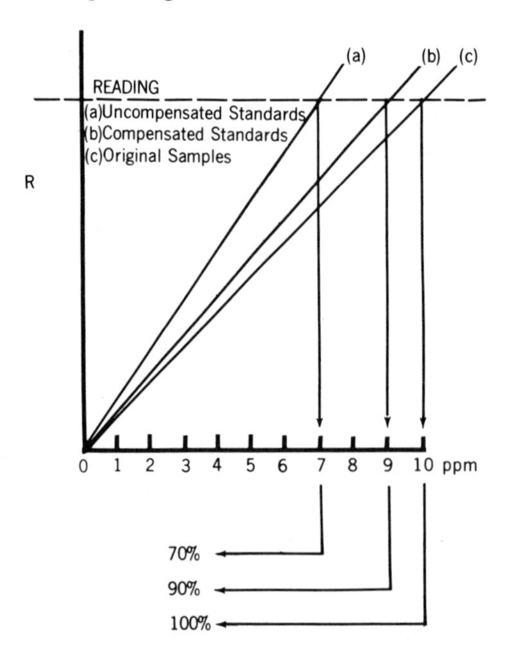

Fig. 3. Variation of slope of linear calibration curves from uncompensated standards to original samples. The slope of standards prepared by adding some of the concomitants present in samples approaches the slope of the curve found for the analyte disturbed by all the concomitants present. [Modified from *Flame Notes* **4**, No. 1, 13 (1969). Courtesy Beckman Instruments, Inc.]

6.2 Calibration

Calibration in flame photometry is based on a very simple equation:

$$R = \varphi(C) \tag{2}$$

where R is the reading and C the concentration of the analyte in the test solution. Alternatively,

$$R = \varphi_1(C_1) \tag{3}$$

where C_1 is the concentration of the analyte in the original sample.

Emission intensity readings can be in arbitrary units as on a scale or meter of the instrument. If the background is subtracted, the operator can work with *net emission* (ΔE).

The relation between C and C_1 is a pure concentration conversion factor depending on the dilution achieved from the original solution up to the final test solution.

Concentration can be expressed as desired. It has been very popular to use parts per million (ppm), and even parts per billion and trillion (ppb and ppt) for cases of high detectability. Other authors prefer to use milligrams per liter, milligrams per milliliter, micrograms per milliliter, percent, and even milligrams/100 ml, and milliequivalents per milliliter or liter, as frequently is seen in clinical analysis.

For very diluted solutions the expressions milligrams per liter (mg/liter), micrograms per milliliter (μg/ml), and parts per million (ppm) can be taken as equivalent. For other concentration units conversion factors may be applied. See the tables published by Burriel-Martí and Ramírez-Muñoz (1957).

A logical universal expression might be grams per milliliter (g/ml). This expression includes two units of the same order of magnitude. Then, 10^{-6} g/ml might be taken as equivalent to 1 ppm or 1 μg/ml.

Calibration curves are prepared with a series of standards, and these curves are used as working curves (analytical curves) to interpolate readings corresponding to samples (or averages of reading if duplicate or multiple readings are obtained; see Fig. 4).

If the calibration curve is linear, it might be possible to work only with one standard and the blank. If there is curvature, a series of standards is mandatory.

Interpolation can also be used between two close standards even in the case of calibration curves with curvature, if the interval between these close standards is considered to be linear.

The addition method is another possibility which has frequently been used in flame photometry for a small number of samples (see Fig. 4). The

TABLE 2

PREPARATION OF COMPENSATED WORKING STANDARDS

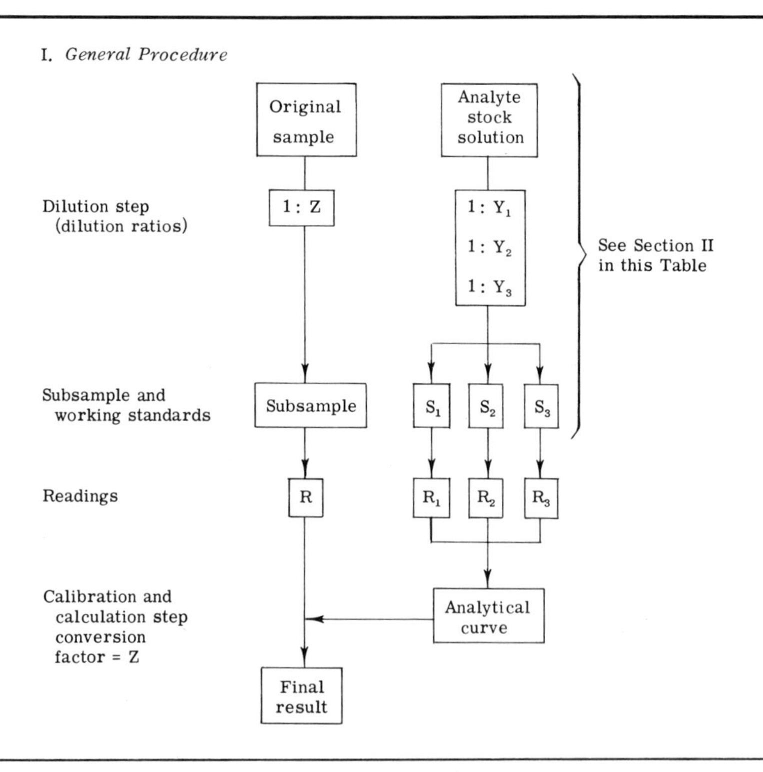

addition method is not recommended if the number of samples is large, as many subsamples must be prepared for each run.

The dilution method can be also used, as described by Ramírez-Muñoz (1968).

6.3 EVALUATION OF RESULTS

Interpolation has been discussed in the preceding section in terms of graphical interpolation. Interpolation can also be accomplished by numerical calculation. This applies to (a) readings obtained at fixed (analytical) wavelengths, or (b) readings obtained by scanning the spectrum along a short wavelength interval (within limits situated at both sides of the analytical wavelength).

Table 2 continued

II. *Addition of Concomitants to Standards*

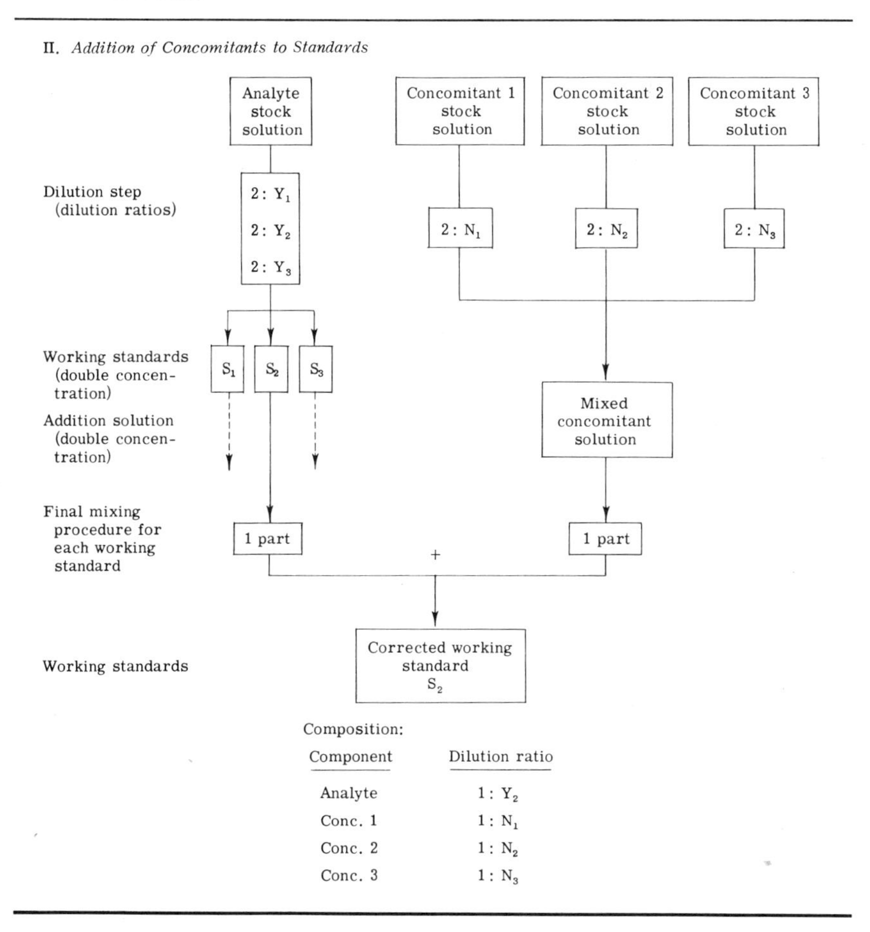

Composition:

Component	Dilution ratio
Analyte	$1 : Y_2$
Conc. 1	$1 : N_1$
Conc. 2	$1 : N_2$
Conc. 3	$1 : N_3$

The easiest type of calculation is that corresponding to linear-response analytical systems, as concentration results may be calculated after a linear fitting of the pairs of values (reading and standard concentration) available for each prepared and used standard solution. If the analytical systems shows curvature, a quadratic fitting may be tested.

All types of calculation, even those simply using equations

$$C = kR \tag{4}$$

$$C_1 = k_1R \tag{5}$$

where k and k_1 are constants calculated from the linear fitting, and C, C_1,

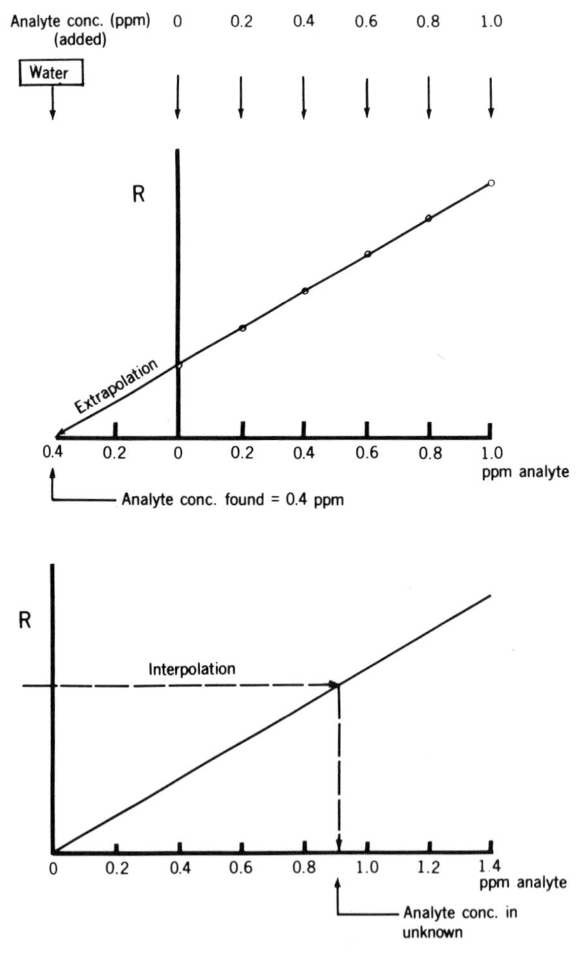

Fig. 4. Extrapolation procedure in addition method and interpolation procedure in the use of linear calibration curve. [Modified from *Flame Notes* **4,** No. 1, 12 (1969). Courtesy Beckman Instruments, Inc.]

and R have the values already mentioned for Eqs. (2) and (3), can be performed easily and rapidly with the use of computers. Ramírez-Muñoz *et al.* (1966) have described for the first time the application of computer techniques to flame analysis. Very simple programs can be prepared for work at terminals in time-shared computer systems; these cover most of the calculation needs of flame photometrists and allow them an easy and clear presentation of final results.

7 Applications

A review of the flame photometry literature published in the last decade leads one to believe that emission flame methods are used in analytical fields. There are some fields in which flame methods have been fully adopted, have remained the recognized methods for many years, and are still used daily. In other fields flame photometry has been used in isolated applications to solve a particular problem or as a confirmatory method.

High sensitivity and reasonable accuracy and precision, in addition to the low cost of instrumentation, maintenance, and operation, have been features which have brought about the widespread use of flame photometry in applied analysis.

Some publications, like the one by Burriel-Martí and Ramírez-Muñoz (1957), have described, in full, some of the major fields of applications. Interest in ecology and related fields has increased the use of flame methods in certain areas even more. Three main groups might be mentioned in today's applications:

(a) *Ecological applications,* including 1. air analysis, 2. potable water analysis, 3. industrial water analysis, 4. sewage analysis, 5. food and beverages analysis, 6. biochemical analysis, 7. clinical analysis, and 8. agricultural analysis.

(b) *Applications to natural mineral products.*

(c) *Applications to manufactured and related products.*

To the extent to which the general composition of the samples to be analyzed is known and the analytes of interest have been defined, the systematization of the analytical methodology follows very similar patterns in most applications: (a) preparation of the series of necessary standards and the blank; (b) preparation of a suitable subsample to be fed into the instrument; (c) readings of standards, blank, and samples in the same working session; and (d) evaluation of results. Step (b) is the only one which may differ among the various applications according to the nature of the original sample. Even taking into account possible differences, the preparation of the subsample always involves (1) conversion of the original sample into a soluble subsample, and (2) adjustment of the concentration of the analyte. However, two exceptions might be cited: (a) gaseous samples, mixed with one or both of the gases used to maintain the flame and fed directly into the flame; (b) solid samples which are not dissolved, but which are fed into the flame as a suspension.

The majority of applications involve liquid and solid samples. The liquid samples are prepared by dilution or physical concentration (evaporation of solvent), addition of a new solvent (miscible), or a transference of phase with the analyte (liquid/liquid extraction). Solid samples require dissolution (action of dilute acids, acid digestion, etc.), selective extraction (as in the preparation of soil extracts), or fusion. Fusion should be avoided as much as possible because of the undesirable amount of salt-type concomitants introduced in the subsample by the fusion process.

In most applications the adjustment of analyte concentration is necessary. It is not common to find cases in which the analyte is already present in the original sample at the best working concentration interval.

If the analyte is a macrocomponent, the original sample should be highly diluted in order to lower the relative concentration of the analyte. If the analyte is present in the original sample at concentrations lower than working intervals suitable for a reliable quantitative determination, the analyte should be preconcentrated.

In emission flame work *preconcentration* by extraction with organic solvents (simple physical transference of phase, or a combination of extraction and the action of specific or selective chelating agents) is the procedure selected as a first choice.

Preconcentration procedures should be immediately quantified by means of a *concentration factor,* calculated as a *concentration ratio* (final concentration/original concentration) and normally found in practice by a *volume ratio* (original volume/final volume). No problems are found in the preconcentration of large samples unless the analyte is present at such a low concentration that impractical factors are necessary (for instance, an element present at the concentration of 1 ppb would need a factor of 10,000 to be brought up to the concentration of 10 ppm, which might be an unrealistic goal). However, it is possible to combine physical concentration by evaporation with a second step involving liquid/liquid extraction. Real problems may be encountered when the flame photometrist has some sample-size limitation. If at least 2 ml of the final subsample are needed for an analysis, and a concentration factor of 5 is required, any sample less than 10 ml would be unusable.

Applications of flame photometry seem to be confined to minor components and components at trace level, at least taking into account most of the recommendations of many authors. This is justifiable in light of the method's outstanding sensitivity. Nevertheless, flame photometry can also be applied to the analytical determination of major components; these can be diluted to the most convenient range of concentration as already mentioned. Flame photometry cannot, however, in terms of accuracy, compete with classical analytic procedures for major components. But if the analyst can sacrifice some accuracy to achieve a reduction in analysis time and expense, he will do well to consider flame photometry.

References

Alkemade, C. T. J., and Milatz, J. M. W. (1955). *Appl. Sci. Res. Sect. B* **4**, 289; see also *J. Opt. Soc. Amer.* **45**, 583 (1955).

Barnes, R. B., Richardson, D., Berry, J. W., and Hood, R. L. (1945). *Ind. Eng. Chem.* (*Anal. Ed.*) **17**, 605.

Burriel-Martí, F., and Ramírez-Muñoz, J. (1955). "Fotometría de llama," Monografías de Ciencia Moderna, Vols. I and II. Consejo Superior de Investigaciones Científicas, Madrid.

Burriel-Martí, F., and Ramírez-Muñoz, J. (1957). "Flame Photometry, A Manual of Methods and Applications" (3rd. reprint, 1960). Elsevier, Amsterdam.

Dean, J. A. (1960). "Flame Photometry." McGraw-Hill, New York.

Gilbert, P. T. (1963). *Proc. Colloquium Spectroscopicum Internationale, 10th.* pp. 171–187. Spartan Books, Washington, D.C.

Lundegårdh, H. (1929, 1934). "The Quantitative Spectral Analysis of Elements," Vols. I and II. Gustav Fischer, Jena. (In German.)

Mavrodineanu, R., and Boiteux, H. (1965). "Flame Spectroscopy." Wiley, New York.

Mitchell, R. L. (1964). The Spectrochemical Analysis of Soils, Plants, and Related Materials. Tech. Comm. No. 44A. Commonwealth Bureau of Soils, Harpenden. Reprinted with Addendum, Commonwealth Agricultural Bureau, Farnham Royal, Bucks., England.

Ramírez-Muñoz, J. (1968). "Atomic-Absorption Spectroscopy" (reprinted in 1969). Elsevier, Amsterdam.

Ramírez-Muñoz, J. (1969). *In* "Flame Emission and Atomic Absorption Spectrometry" (J. A. Dean and T. C. Rains, eds.), Vol. I, pp. 25–48. Dekker, New York.

Ramírez-Muñoz, J. (1971). *In* "Spectrochemical Methods of Analysis" (J. D. Winefordner, ed.), pp. 127–188. Wiley (Interscience), New York.

Ramírez-Muñoz, J., Malakoff, J. L., and Aime, C. P. (1966). *Anal. Chim. Acta* **36**, 328; see also *ibid.* **42**, 515 (1968); **43**, 37 (1968).

Walsh, A. (1955). *Spectrochim. Acta* **7**, 108.

CHAPTER 23

Gas Analysis Techniques and Combustion Methods

R. V. Peterson

The Aerospace Corporation
El Segundo, California

Introduction

This chapter includes gas analysis methods not discussed in other chapters of this work and which are, for the most part, instrumental in character. These methods are covered in Section 1. Section 2 contains a discussion of the instrumental techniques and methods available for determining the elemental composition of a material by combustion gas analysis. Combustion gas analysis involves the burning of a sample and the measurement of the effluent gases from the combustion. Since the quantitative analysis of these gases is the subject of Section 1, combining these two subjects into one chapter seems justified.

1 Gas Analysis

1.1 General

Not long ago, gas analysis brought to mind a maze of glass bulbs and stopcocks representative of gravimetric and volumetric wet chemical techniques. Now the association that comes to mind is more likely an instrument such as a mass spectrometer, a gas chromatograph, or a gas monitor. The speed, accuracy, and range of gas analysis have greatly increased with the advent of these instruments. The great variety of methods available can be seen from Tables 1 and 2 in Chapter 1. One of the listings, gas analysis techniques, is the subject of this section, which presents methods for the quantitative analysis or monitoring of gaseous components in gas mixtures. These methods have been associated with the process industries and are generally not applicable to qualitative analysis; nevertheless, they are an important part of the total area of gas analysis. This is especially so at this time. The Occupational Safety and Health Administration has recently published maximum allowable concentrations in the atmospheres of laboratories and factories for over 400 different gases and vapors. Stringent new pollution standards have been set for automobiles and a large variety of stationary sources of gaseous emissions. A new national standard for atmospheric pollutants has been adopted by the Congress in the Clean Air Act of 1970. These recent activities place an unprecedented demand on analytical instruments for fast, reliable analysis of a variety of gases over a wide concentration range.

These new standards have prompted many studies of the available methods for pollution monitoring. One such study was performed by Snyder *et al.* (1972) for the Environmental Protection Agency (EPA). Several types of instruments for the analysis of nitrogen oxides were evaluated in a controlled laboratory test and in actual power plant monitoring studies. The new standards have also prompted study of new methods for pollution analysis. In a study by Gelbwachs *et al.* (1972), NO_2 was monitored on a real-time basis by use of a fluorescence technique.

Table 1 lists a selection of methods used for gas analysis and serves as a guide in the presentation of the following brief discussions on each technique. With few exceptions, the selection was based on instruments presently in use. The discussion that follows gives a brief summary of each of these methods along with the applications.

Note that many of the methods listed for gas analysis are used as detectors for gaseous effluents in gas chromatography (see Chapter 24).

1.2 Gas Sampling

There are two basic methods of sampling for gas analysis: static or dynamic. In static sampling, one obtains a "grab" sample by capturing a portion of a gaseous mixture and transferring this portion *in toto* or in part to the instrument for analysis.

The containers used to capture the sample may be a balloon, bag, syringe, flask, gas bottle, etc. The problems associated with this type of

TABLE 1

SELECTED METHODS FOR GAS ANALYSIS[a]

Measured physical/chemical characteristics of gas	Gas analysis method	Gas analyzed
Density	Gas density balance	Binary mixtures of most gases
Refractive index	Interferometry	Binary mixtures of most gases; see Chapter 15 on refractometry
Magnetic susceptibility	Magnetometry	O_2
Radioactivity	Scintillation or Geiger counter	Radioactive gases
Induced radioactivity	Neutron activation analysis	Hg and others that can be made radioactive; see Chapter 12
Sound propagation (velocity of sound)	Acoustic	Binary mixtures of most gases
Thermal conductivity	Thermal conductivity (kathorometer)	Binary mixtures of most gases
Absorption of electromagnetic radiation	UV-visible nondispersive photometry	O_3, Hg vapor, aromatic hydrocarbons SO_2, CS_2, Cl_2. Metal carbonyls
	Infrared nondispersive photometry	Gases that absorb in the infrared
	Optoacoustic spectrometry	Gases that absorb in the infrared
	Second-derivative spectrometry	Gases that absorb in the uv or visible
Emission of electromagnetic radiation	Flame photometry	SO_2, H_2S, SO_3, mercaptans, phosphorus gas; see Chapter 22
	Gas discharge	Selected gases
	Fluorescence	NO, NO_2, CO
Electrochemical properties	Conductance	O_2, H_2O, SO_2, and other gases that cause conductance change in electrolyte
	Coulometry	O_3, O_2, H_2, H_2O, SO_2, H_2S, NO, mercaptans
	Polarography	O_2, SO_2, NO_2, Cl_2, Br_2, CO
	Ion-selective electrodes	Dissolved gases

Table 1 continued

Measured physical/chemical characteristics of gas	Gas analysis method	Gas analyzed
Ionization	Flame ionization	Carbon compounds
	Electron-capture	Mostly halogen-containing compounds
	Cross-section ionization	Most gases
	Argon ionization	Gases with ionization potential below 14.7 eV
Chemical reaction	Absorption	Selected gases
	Heat of absorption	Selected gases
	Crystal oscillator	H_2O; selected gases
	Calorimetric	Selected gases
	Chemiluminescence	O_3, NO_2
	Colorimetric	Selected gases; see Chapter 18, uv-visible spectrophotometry
	Combustion	O_2, H_2, CO, hydrocarbons

[a] See also Chapter 1, Flow Charts 1 and 2.

sampling are obtaining a representative sample, reaction with the container wall, diffusion through the wall, gaseous reactions, condensation of the gas in the container, and the time delay between sampling and analysis. Furthermore, there may be difficulties associated with introduction of the sample into the instrument. These general considerations are also discussed in Chapter 1, along with methods for extracting gases from solids and liquids.

Much of the sampling for gas analysis, especially in the area of air pollution, employs dynamic techniques. These techniques use probes or tubes that continuously sample the gas environment, thereby giving a constant flow of gas to the instrument. These methods are especially useful for monitoring changing environments such as polluted atmospheres, gaseous reactions, and the like. The problems with these techniques are associated with obtaining a representative sample, but in addition, a gas pump or method of drawing the gas into the instrument may introduce its own problems, i.e., the possibilities of contamination and reaction with, or loss through, the pump and tubing walls. It is also likely that filters will have to be installed in flow systems to protect the instrument from dust and reactive gases. Although the sampling is generally done quickly in the flow systems, there is nevertheless a finite time delay between sample probe and instrument detection. The problems associated with dynamic sampling in combustion processes are discussed in detail by Tiné (1961). Gas sampling for air pollutants is discussed by Lodge and Pate (1971).

1.3 Methods for Gas Analysis

1.3.1 Density (Gas Density Balance)

The gas density balance is used to detect the density or molecular weight of a gas in binary mixtures. For detectors based on this principle, the best results are obtained when the two components have large differences in molecular weight or density.

There are three types of detectors that measure the density of gases. With the first type, the density difference is measured by a difference in the temperature (and the resistance) of a pair of matched hot wires or thermistors of the type used in thermal conductivity cells. The change in temperature of the detector is due to a change in flow rate in one leg of the reference gas stream.

The first type of gas density balance is shown in Fig. 1, where the reference gas stream (A) flows past the detectors (B_1) and (B_2) and finally recombines at the outlet (D). The sample gas enters the gas cell at (C) and tends to flow up or down depending on the density of the sample gas relative to the reference gas. Any gas that differs in density from the reference will, on entering the cell at (C), cause a disruption of the equilibrium flow of the reference gas in one or the other leg unless the sample gas

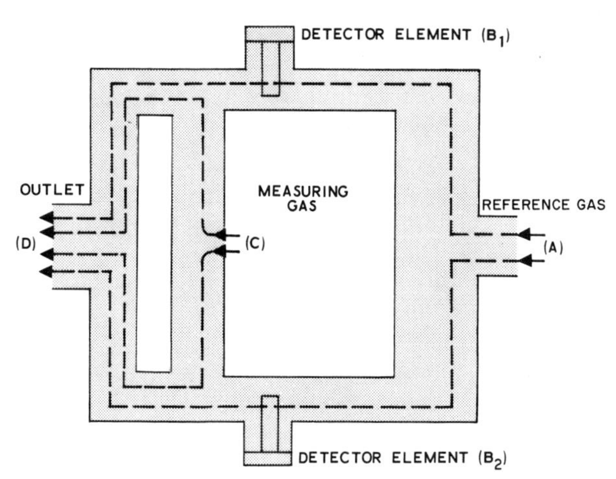

Fig. 1. Gas density balance for measurement of gas concentrations. [Courtesy of Gow-Mac Corporation.]

has the same density as the reference. A change in the equilibrium flow causes an imbalance in the temperature (resistance) of the two thermal-conductivity-type detectors. A Wheatstone bridge arrangement is used to monitor the change in resistance. This type of detector is discussed by Nerheim (1963).

The temperature of the detector cell must be controlled to $\pm 0.5°C$, and the sensitivity is a function of many factors including the design of the cell, the choice of reference and sample gases, flow rates, temperature, and the nature of the filaments.

The second type of gas density balance uses the buoyancy of a reference gas contained in a sample gas atmosphere to obtain a measure of the relative density of the two gases. In this instrument two balls are mounted in a dumbbell arrangement. One ball is punctured so that it is insensitive to changes in gas density. The other ball is sealed and reacts to changes in the density of gas atmosphere surrounding the ball. The position of the dumbbell is directly related to the density of the sample gas bathing the dumbbell arrangement. The displacement can be obtained from the potential needed to rebalance the dumbbell in an electrostatic field. The instrument can be calibrated to give density readings as a function of displacement or the potential applied to achieve equilibrium of the dumbbells.

The third type of gas density analyzer is the Ranerex specific gravity indicator, which uses an impeller and an impulse wheel. The impeller element spins a gas at high speed against the vanes of a companion impulse wheel. The torque on the comparison wheel is proportional to the density of the gas impinging upon it.

The gas density balance is used as a detector cell in gas chromatography and in certain monitoring devices. Since the filaments in the first type of density balance do not contact the sample gas, this balance offers several advantages over thermal conductivity detectors. For instance, reactive, oxidizing, or corrosive gases are not changed by the detector, and the sample gases themselves do not shorten the life of the detectors. Therefore, potentially explosive mixtures may be successfully run in the gas density balance.

1.3.2 Magnetic Susceptibility (Magnetometry)

Magnetometry is the measurement of the magnetic susceptibility of a gas. Most gases are slightly diamagnetic; that is, they are repelled out of a magnetic field. Oxygen, NO, and NO_2 (to a lesser degree) are paramagnetic or attracted into a magnetic field. The strong paramagnetic characteristic of oxygen relative to other gases has led to the development of instruments for the quantitative measurement of oxygen in gaseous samples.

There are two types of instruments that use magnetometry to measure oxygen concentration. In one, a small glass dumbbell is suspended on a quartz fiber in a nonuniform magnetic field. As the gas containing O_2 is introduced into the sample chamber, the magnetic force acting on the dumbbell is altered, causing a change in the position of the dumbbell. The scale may be read directly, or the voltage required to move the dumbbell to its rest or null position may be measured. Ideally, this instrument can measure oxygen concentration from 10 ppm to 100% oxygen by using several span ranges. The accuracy ranges from ± 0.5 to 1.0% of the scale reading, with a response time as low as 10 sec in a flowing or static gas sample. The accurate operation of this instrument requires that the gas pressure remain constant and the gas be free of solids.

The second type of instrument used for the measurement of O_2 combines the magnetic susceptibility measurement with a thermal conductimetric measurement. This instrument uses a reference and a sample cell. The only difference in these two cells is that the sample cell has a magnetic flux around a thermal conductivity filament. The gas sample is passed across the bottom of both gas cells, which contain heated thermal conductivity filaments. A strong magnetic field is directed across the sample filament and attracts the oxygen around this hot wire. Oxygen loses its magnetic susceptibility in inverse proportion to the square of the absolute temperature, and the heated gas is continually displaced by a countercurrent flow of the cooler gases in the opposite direction to the rising, heated O_2. A gas flow proportional to the amount of oxygen present is established around the filament detector, cooling the detector filament and causing a change in its resistance, which is detected by a bridge arrangement. Accuracy is $\pm 0.25\%$ O_2 up to 20% concentration and $\pm 2.5\%$ of range up to 100% concentration.

1.3.3 Radioactive Methods

Radioactive gases are analyzed by use of low-level radioactive counting techniques. Scintillation counters, Geiger tubes, and ionization-type counters have been used to measure the concentration of the radioactive species. (These may be static gas or flow counters.) For low concentrations, the radioactive gas may be concentrated by absorption on a proper absorbent. Moses *et al.* (1963) and Collinson and Haque (1953) discuss radioactive methods in relation to gas analysis.

Gudzinowicz and Smith (1963) discuss a novel method for gas analysis that employs a radioactivity detector. The source of radioactivity is a krypton-85 quinol clathrate that, on exposure to inorganic oxidants, releases radioactive krypton atoms, which are then detected with a Geiger tube. This method can determine 1 to 20-ppm fluorine in air. Hommel *et al.* (1961) have used this method for ozone. Br_2, NO_2, Cl_2, OF_2, NO_2F, and NO_2Cl also have high reactivities with the krypton-85 clathrate.

1.3.4 Sound Propagation (Acoustic)

The velocity of sound in a gas is a function of the molecular weight and the specific heat of the gas; this property can therefore be used as a method for gas analysis in binary mixtures. A review of the applications of sound velocity measurement to gas analysis is given by Crouthamel and Diehl (1948).

Acoustic detectors have been developed mainly for use in gas chromatographic analysis. In this application Testerman and McLeon (1962) used frequency changes to determine sound velocity, while Noble *et al.* (1964) measured the change in wavelength accompanying a velocity change by utilizing phase-change measurements. The velocity of sound in most gases ranges from 0.3 to 3 times the velocity in air. These methods can typically measure gas concentrations from 40 ppm to 100% in a cell volume of 5 to 50 μl.

1.3.5 Thermal Conductivity

Thermal conductance is the transport of energy in a material due to a gradient in temperature. This can be represented by

$$H = KA\ dT/dx \tag{1}$$

where H is the heat flux, A the cross-sectional area, and dT/dx is the temperature gradient. The quantity K is the coefficient of thermal conductivity, which is usually given in units of cal/sec cm deg. Thermal conductivities of gases are independent of pressure down to approximately 5 Torr.

Touloukian *et al.* (1970) give several methods for measuring thermal conductivity. However, since it is very difficult to obtain absolute measurement of thermal conductivity, these methods are not used as qualitative analytical techniques. On the other hand, quantitative analyses of binary mixtures are accomplished by use of one of the gases in the mixture as a reference gas. Used in this fashion, thermal conductivity methods are among the most important in gas analysis. Touloukian *et al.* (1970) also give comprehensive lists of thermal conductivity values for a large number of gases as well as binary mixtures.

Thermal conductivity techniques that may be used to analyze tertiary or higher mixtures of gases include

(1) selective absorption or separation of undesired components in the sample gas stream;

(2) measurement of the ratio between the reference and the sample gas at more than one temperature; and

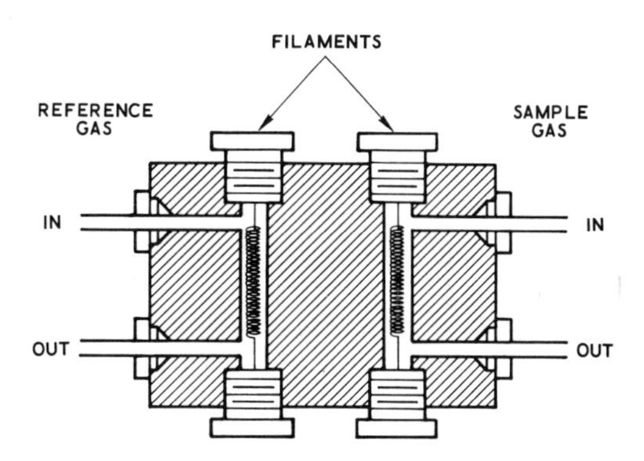

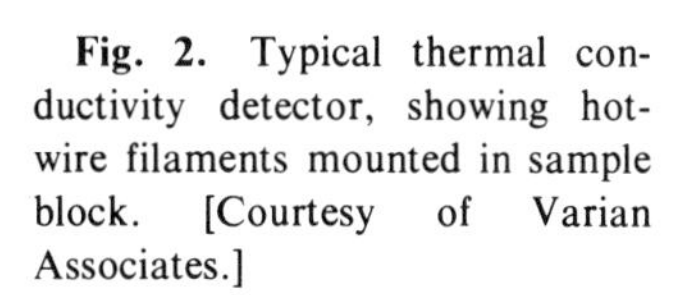
Fig. 2. Typical thermal conductivity detector, showing hot-wire filaments mounted in sample block. [Courtesy of Varian Associates.]

(3) measurement of a mixture in which the component of interest has a substantially different thermal conductivity than all the other components in the mixture.

Thermal conductivity analyzers are quite simple in design. The functional units are a flow system for maintaining a constant flow of sample gas, a detector, and a readout device. In some instruments the reference gas may be sealed in the detector cell, but usually both reference gas and sample gas will be adjusted to the same pressures and flow rates.

The katharometer or detector cell (Fig. 2) is composed of a reference and a sample cavity in a cell block, generally of brass or stainless steel. The heated elements in the reference and sample cavities are one or two matched pairs of filaments or thermistors that are heated by a current from a constant voltage supply. A sample with a different thermal conductivity than that of the reference will cause a change in the resistance of the sample filament. This change is detected by an arrangement of the heated elements in a bridge circuit. The detector cell must be controlled to $\pm 0.5°C$ for accurate results.

Instruments are available for laboratory analysis as well as for industrial monitoring and detection of gases.

The volumes of the detector cells range from 10 ml to less than 0.05 ml. The sample size in a flowing gas can be 0.01 ml or less at standard pressure and temperature (STP)* conditions with low-volume cells.

The time for analysis depends mainly on the speed of purging for the sample cell. This time is generally on the order of 30 to 60 sec and depends on the size of the conductivity cell and sampling line. The response time for the detector and electronics circuit is usually negligible compared to the purge times involved.

* STP: 25°C and 760 Torr.

The thermal conductivity method for analysis is a sensitive method that is limited in practice by the accuracy of the calibrating mixtures, the sample size, and the difference between sample and reference thermal conductivities at constant temperature. The sensitivity of the detector is approximately proprotional to the square of the applied voltage. Sensitivities on the order of 500-ppm air in helium have been quoted by some manufacturers.

The method is applicable to almost all gases, including inert gases, and is destructive only if the sample is diluted with reference gas. The sample analyzed may be recovered from the reference gas in some instances by trapping with absorbents or cold traps.

Thermal conductivity detectors are calibrated by use of known mixtures of gases over the concentration range of interest. Thus a plot of the response of the detector versus the concentration of the gas of interest is obtained. The number of points needed to define the curve depends on the linearity of the detector response to changes in concentration and on the precision needed. These plots should extrapolate to zero and normally have a linear increase in response as a function of an increase in concentration (linear dynamic range) of 1×10^4.

Kieselbach and Schmit (1972) and Weaver (1951) give very good review articles on thermal conductivity and its use in gas analysis.

1.3.6 Absorption of Electromagnetic Radiation

A large number of analytical techniques use the absorption of electromagnetic radiation as the means of detecting and monitoring concentrations of gaseous components. The methods discussed here are the nondispersive photometric methods; the infrared dispersive and uv-visible methods are discussed in Chapters 9 and 18, respectively.

The nondispersive methods lack a prism or grating to disperse or separate the electromagnic radiation into its components but have a radiation source, a detector, and a means of making the instrument selective for a given gaseous component. The methods are widely used in the quantitative analysis of samples where one component or type of component is to be monitored. They are simple in design, sensitive, and usually portable, and they can be quite specific in their responses to gases. The variety of instruments available makes it possible to analyze a diversity of gases ranging from ozone to hydrocarbons to mercury vapor.

a. Nondispersive Infrared Analyzer. The nondispersive infrared analyzers consist of an infrared source, sample and reference cells, a detector, and, in some cases, an optical chopper and filters. There are two ways of making nondispersive analyzers selective: (1) use of a broad-band source with a selective filter and/or detector and (2) use of a selective source in conjunction with a broad-band detector. Examples of broad-band sources are tungsten lamps, glowers, or glowbars, while an example of a selective source is a laser. Thermocouples and bolometers are examples of broad-band detectors, while the condenser microphone is considered a selective detector.

Figure 3 is a schematic of an infrared broad-band source with a selective detector. In this instrument, the radiation from two infrared sources is chopped; one beam is passed through a reference gas filled with a nonabsorbing gas, and the other beam passes through the sample gas. The sample gas (in the sample cell) absorbs some of the radiation, while the remaining energy is absorbed by the gas in the sample side of the detector. Since the reference cell contains a nonabsorbing gas, most of the initial energy falls on the reference side of the detector. The detector is a condenser microphone type. In this detector two compartments filled with the gas of interest are separated by a flexible metal diaphragm. One compartment has a window for the reference beam and the other a window for the sample beam. The difference in absorption between the reference and sample cells causes dissimilar heating and, consequently, unequal pressure

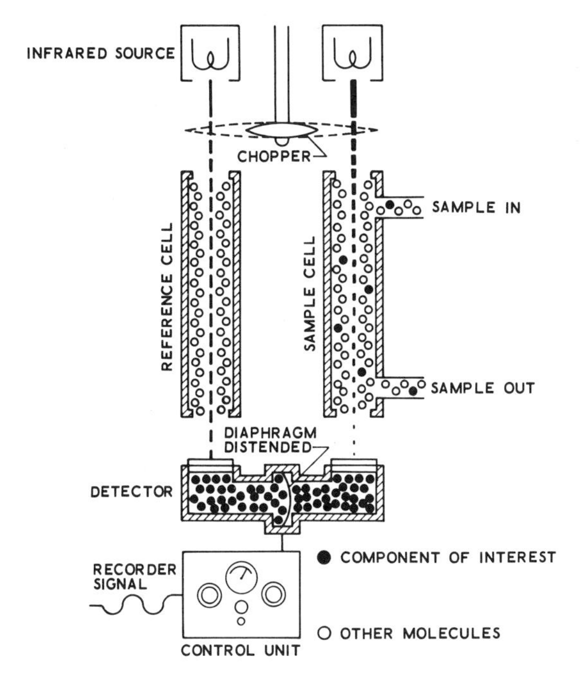

Fig. 3. A nondispersive infrared analyzer. Selectivity is obtained by filling the detector with the gaseous component of interest. [Courtesy of Beckman Instruments.]

between the compartments. As the beam is chopped, it causes the metal diaphragm to oscillate between equilibrium and distended states, creating a change in the electrical capacitance. The capacitance change modulates a radio-frequency signal from an oscillator, which is proportional to the absorption of the component of interest in the sample cell. Sensitivity and specificity of the method can be optimized by a judicious selection of optical filters to maximize absorption by the component to be analyzed.

Any contaminant that has an absorption band overlapping with the component of interest is an interference that can cause errors in the measurements. The instrument is sensitive down to the ppm region. The infrared nondispersive methods are especially suitable for CO and CO_2 and are also used for the analysis of hydrocarbons, water vapor, HCl, SO_2, NH_3, and freon.

A comprehensive review on nondispersive infrared instruments for gas analysis is given by Hill and Powell (1968). Fisher and Huls (1970) compare nondispersive infrared methods with chemical methods for the determination of oxides of nitrogen.

b. Nondispersive UV-Visible Analyzer. An example of an instrument that operates in the uv-visible region of the spectrum is shown in Fig. 4. Radiation from the source passes through the sample and onto a beam splitter (half-reflecting mirror). The reflected beam is directed to a phototube through a filter that passes only the wavelength band strongly absorbed by the constituent of interest in the sample. The transmitted beam

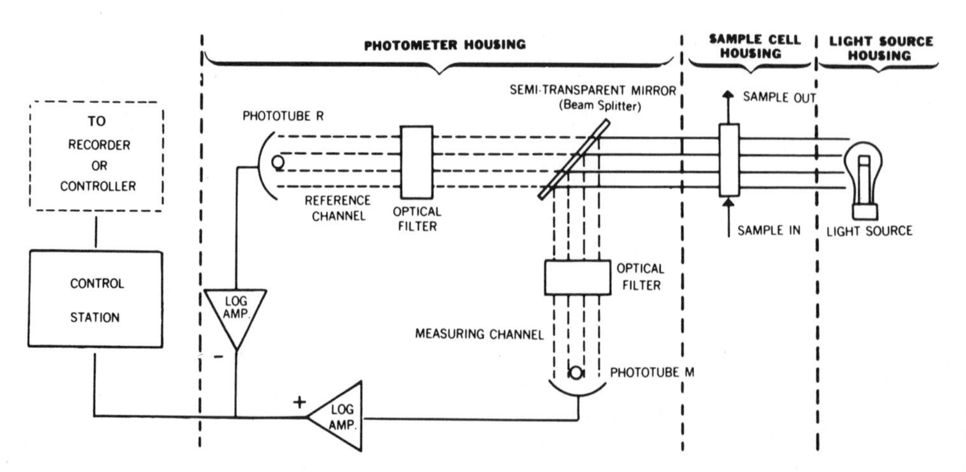

Fig. 4. A nondispersive uv-visible analyzer. Optical filters in the sample and reference beams increase sensitivity. [Courtesy of Du Pont Instruments.]

passes through an optical filter designed to pass the reference wavelength, which is not absorbed in the sample. The ratio of the electric currents generated by the phototubes is proportional to the concentration of the component of interest in the samples. Any errors that are not wavelength-dependent, resulting from dust particles in the sample or changes in the intensity of the light, are canceled out.

Photometric methods for the determination of NO_2 (and NO by oxidation to NO_2) are given by Schulze (1970) and by Keenan (1967) in a study of automobile exhausts.

c. OPTOACOUSTIC. Kreuzer (1971) describes a method that uses a tunable laser radiation source and microphone detector to identify and measure the infrared-absorbing components in a gas sample. A laser beam is directed into a l-cc cell; the wavelength of the laser is tuned to match the absorption of a gaseous component of interest in the sample. The amount of energy absorbed is in direct proportion to the concentration of the gas in the cell. The absorbed energy causes an increase in pressure, which is detected with a condenser microphone similar to that used in the nondispersive instrument. This information is transmitted to a computer, which matches these data with the absorption spectrum of known gases. This method can be used to detect concentrations as low as 10 ppb and is useful for gases that absorb in the infrared region of the spectrum. A commercial instrument is presently being developed.

d. SECOND-DERIVATIVE SPECTROMETRY. Second-derivative spectrometry measures the curvature of characteristic narrow molecular absorption bands that occur when molecules are irradiated with uv or visible light. The gas molecules that exhibit narrow-band absorption (0.1 to 5 nm) useful for measurement by second-derivative methods are most aromatic hydrocarbons, aldehydes, and some metals in the vapor state. In addition, several gases important in air pollution studies, such as NO, NO_2, SO_2 and O_3, can be analyzed. The derivative spectrum of a gas gives both qualitative and quantitative information, i.e., identity by the location of the peaks and quantity by the peak height. Overlapping absorption bands difficult to resolve in absorption spectrometry can frequently be analyzed by means of second-derivative techniques. The limits of detection are typically in the low ppb range.

The derivative spectrum in a second-derivative spectrometer is obtained by using wavelength modulation whereby the uv or visible monochromatic radiation is varied in time by lateral oscillation of the entrance slit position. The modulated radiation passes through a long path-length sample cell and is detected by a photomultiplier tube. Either a flow or static sample cell may be used. The derivative signal is obtained by demodulation of the photomultiplier signal. A more detailed discussion on this new technique for trace gas analysis is given by Hager (1973).

1.3.7 Emission of Electromagnetic Radiation

The characteristic emission or fluorescence spectrum of an excited gas is used as a qualitative and quantitative method of analysis for a gas. The analysis of the spectral emission of gases in the excited electronic state is discussed in Chapter 21 on emission spectrometry. The line and band emission spectra from a gas discharge comprise the subject of a book by Bochkova and Shreyder (1965). In addition, flame photometry and fluorescence photometry are methods applicable to gas analysis, and their use in this manner is described below.

a. FLAME PHOTOMETRY. Flame photometry is widely used to identify and determine the amount of a gas present by observing the spectral line or band emission in a flame. Each gas has its characteristic emission lines, and this information can be used for the qualitative and quantitative analyses of a material. In one type of flame-photometric detector analyzer, the emissions from sulfur compounds in a hydrogen-rich flame are measured. A narrow-band optical filter is used to select the emission band. Chapter 22 on flame photometry discusses gas analyzers using excitation of gases in flames.

b. FLUORESCENCE (PHOTON EXCITATION). The laser has been instrumental in stimulating applications of new (and old) methods for the analysis of materials. The ability of the laser to supply a high flux of monochromatic radiation had led to the investigation of gas fluorescence techniques. This method has been recently explored by Gelbwachs *et al.* (1972) for NO and NO_2. It shows promise of becoming an accepted instrumental technique, especially in air pollution analysis.

In this method a sample gas is drawn into a chamber where it is irradiated by a laser. The energy is absorbed by the molecules, raising them to excited energy states. The subsequent fluorescent emission from the excited molecules is filtered to exclude the laser radiation and is measured with a phototube. The output of the phototube is proportional to the gas concentration. The method is very selective, has a sensitivity of 1 ppb, and measures the gas on a real-time basis.

An infrared fluorescence technique has also recently been developed for carbon monoxide analysis. An instrument, described by Link *et al.* (1971), employing this technique uses the spectral lines created by the fluorescence generated in a cell filled with equal amounts of $C^{16}O$ and $C^{18}O$ isotopes. These gases are irradiated with a chopped infrared source. Two narrow bands of radiation transmitted by the two isotopes are guided through a filter disk chopper and on through the sample. The $C^{16}O$ radiation exactly matches the vibrational–rotational absorption lines of the sample CO, while the $C^{18}O$ radiation is used as a reference. The radiation from the cell containing the $C^{18}O$ passes through the sample essentially unabsorbed because of the low concentration (0.2%) of the naturally occurring isotope. The detector used is typically a solid-state or pyroelectric device. Source fluctuations and dust or contaminants in the sample gas will not affect the detected ratio of $C^{16}O$ to $C^{18}O$; therefore, the method is very selective. The instrument exhibits high sensitivity in that it can detect less than 1 ppm of CO. This technique could be developed for use with other gases as well.

1.3.8 Electrochemical Methods

Electrochemical methods are those methods that use an electrical property of a gas as a measure of its concentration in a sample. These methods use conductance, coulometry, and electrochemical transducers.

Electrochemical gas analyzers are generally of fairly simply design, sensitive, portable, rugged, and relatively inexpensive; these attributes account for their wide application as gas monitors and process analyzers in the fields of air pollution, stationary emission sources, and controlled room analysis.

The electrochemical analyzers typically use an electrochemical cell containing electrodes and a solid or liquid electrolyte. The current or other measured parameter of the cell is proportional to the gas concentration in the cell. The instrument can be designed and calibrated to read directly in gas concentration units using a meter or recorder output.

The specificity of these instruments is not inherent to the method; rather it is obtained by selecting gas scrubbers (absorbents or semipermeable membranes) to filter interfering gases. The lack of selectivity is a serious disadvantage to electrochemical methods, in general, and must be taken into account when they are used. Furthermore, electrochemical methods still retain the disadvantage of electrolyte and sensor deterioration with time, which can be remedied only with frequent calibration and replacement of the electrolyte or sensor. Another deficiency of these analyzers, in some cases, is slow response time. In many instruments the sample gas must be pumped or otherwise conveyed to the gas sensor.

a. CONDUCTIMETRIC ANALYZERS. The conductimetric analyzer measures the conductance of a solution after contact of the solution with the sample gas. The dissolved gas causes formation of ions, which increases the conductance of the solution. The observed increase in conductivity is proportional to the absorbed gas concentration. These instruments generally have high sensitivity, quick response, and are simple to operate. A major disadvantage is their nonspecificity in that they respond to a variety of gases; furthermore, the conductance is dependent upon temperature and sample gas flow rate.

Urone *et al.* (1965) describe a method for SO_2 whereby the gas is determined by measurement of the conductance of sulfurous acid or of the sulfuric acid formed after mild oxidation according to the following equations:

$$SO_2 + H_2O \rightarrow H_2SO_3$$
$$SO_2 + H_2O_2 \rightarrow H_2SO_4$$

b. Coulometric (Amperometric) Analyzers. In principle, the coulometric analyzers measure the generating current necessary to maintain a constant electrolyte concentration in an electrochemical cell when a sample gas has been introduced. In a typical cell a sample gas entering the reaction chamber disrupts the titrant ion concentration, producing a change in the sensor electrode signal and, in turn, causing current to flow to the generating electrodes. The generating electrodes restores titrant in the electrolyte to its original concentration. The measured current (or voltage) is proportional to the concentration of the reacting sample gas.

In one instrument, a neutral buffered potassium iodide (KI) electrolyte is used for NO_2 and total oxidants. It is also applicable to NO after oxidation to NO_2 and to ozone after reaction with NO to form NO_2. An air sample is drawn through the detector cell, which contains a buffered KI electrolyte, and the oxidants present in the sample oxidize the iodide ion present in the electrolyte. This action produces free iodine, which is immediately reduced back at the cathode to iodide ion, which generates a current proportional to the concentration of atmospheric oxidants in the sample gas.

The coulometric analyzer can measure gases such as O_3, NO_2, SO_2, ammonia, and halogens by the proper selection of scrubbers or electrode and electrolyte combinations. The major advantage to these methods is the minimal maintenance requirements. They are, however, nonspecific, and interfering gases must be removed from the sample by scrubbing.

Miller *et al.* (1971) have pretreated the atmospheric gases to obtain specificity for NO, NO_x, and SO_2 by means of a microcoulombic detector cell. These instruments are also useful in the ppm region (sensitivity $\sim$ 0.01 ppm) but have a somewhat slow response time of 2 min or greater.

A galvanic cell (Hersch, 1957) measures oxygen in a gas stream by monitoring the current produced by a cathode–anode reaction using an alkaline electrolyte. The oxygen reacts at a silver cathode by the reaction

$$O_2 + 2H_2O + 4e \rightarrow 4OH^-$$

with the resultant anode reaction of

$$Pb + 6OH^- \rightarrow 4e + 2HPbO_2^- + 2H_2O$$

The measured voltage (current) is proportional to the amount of oxygen in the cell. Kendall (1971) has used an improved version of this cell to monitor the O_2 concentration of gas atmospheres in nuclear reactor coolant systems.

c. ELECTROCHEMICAL TRANSDUCER ANALYZERS. The electrochemical transducer analyzers measure the current generated by the oxidation of the gas of interest at a sensing electrode. This sensor is a sealed amperometric device in which gas diffuses through a selectively permeable membrane, passes through a thin film of electrolyte as a dissolved gas, and is finally electrocatalytically oxidized at the electrode. The measured current between the sensing electrode and a counterelectrode is proportional to the sample gas concentration.

These analyzers are simple in operation and have low maintenance requirements, a fast response, and no reagents to change. A disadvantage is that the tranducers deteriorate gradually and require frequent calibration. Also the sensors are somewhat nonselective. Breiter (1968) describes an instrument in which CO is measured in an electrochemical cell, where the CO is electroxidized to CO_2 at a catalytically active electrode (e.g., platinum) in an aqueous electrolyte: $CO + H_2O \rightarrow CO_2 + 2H^+ + 2e$. The electrode current for the oxidation is proportional to the partial pressure of CO in the gas sample.

A monitor for oxygen is based on a polarographic sensor. The sensor consists of an anode and a cathode in an aqueous KCl solution. A thin, gas-permeable membrane confines the electrolyte while allowing the oxygen to diffuse into the sensor; this produces the following oxidation-reduction reaction:

Cathode reaction: $O_2 + 2H_2O + 4e \rightarrow 4OH^-$

Anode reaction: $4Ag + 4Cl^- \rightarrow AgCl + 4e$

Oxygen in a sample gas is reduced at the cathode upon application of a dc voltage across the electrodes; this causes the flow of a current whose magnitude is proportional to the partial pressure of oxygen present in a gaseous sample.

d. ION-SELECTIVE ELECTRODES. Ion-selective electrodes are used in the determination of the concentration of a large variety of ions in aqueous solution.

The method using ion-selective electrodes depends on the direct measurement of a membrane potential resulting from the selective transfer of an ion across the membrane, and it is especially useful for the determination of very low concentrations (ppm) of ions. As an example of its use in gas analysis, Lee (1969) has monitored the hydrogen chloride concentration in gaseous mixtures. A general review of ion-selective electrodes is given by Durst (1969).

1.3.9 Ionization Methods

Ionization methods for gas analysis are widely used as detectors in gas chromatography and are used to a lesser extent in gas monitors. Their use in gas chromatography is discussed in Chapter 24.

In these methods (Lovelock, 1961) a sample gas is ionized either in a flame or by bombardment with a high-energy radiation source, and the ionization current or the change in the equilibrium ionization current is detected. Ionization detectors are adversely affected both by gases that capture electrons and by contaminants that condense on the electrodes.

a. FLAME IONIZATION DETECTOR. In a flame ionization detector (Fig. 5) the sample gas is introduced into a hydrogen flame, and the current from this ionization of the sample gas is measured by electrodes in the vicinity of the flame. The detector responds only to substances that produce charged ions in the flame; however, this includes almost all organic compounds. By the same token it is relatively insensitive to inorganic gases whose ionization is comparatively inefficient.

The flame ionization detector has a useful sensitivity range of 1 ppb to about 1% of a component in a sample gas. The response of the detector is a function of the flow rate and nature of the sample gas; therefore this method requires regulated flow rates and may require calibration for a given gas type. Dust particles should be filtered from the detector, since they may perturb the ionization current.

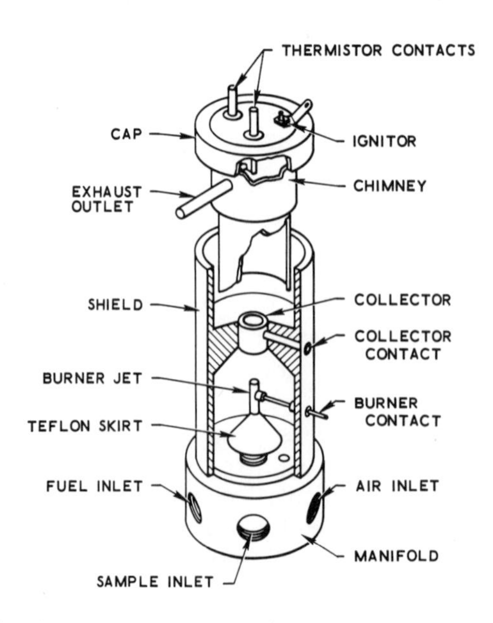

Fig. 5. Partial cutaway of a flame ionization detector of the type used in gas analysis. [Courtesy of Beckman Instruments.]

The detector is especially useful in hydrocarbon gas analyzers. These analyzers generally employ the flame ionization detector to measure the hydrocarbon content in automobile exhausts, atmospheric pollution, or stack and flue gases. Andreatch and Feinland (1960) have explored the use of these detectors as monitors for hydrocarbon pollution in the atmosphere.

b. CROSS-SECTION IONIZATION DETECTOR. The cross-section ionization detector can be used to monitor any gas or vapor in almost any type of carrier gas; however, the lighter gases, such as hydrogen and helium, give the greatest sensitivity. The gases are ionized in the detector cell by collision with particles emitted by a disintegrating radioisotope. The detector response depends on the absorption of ionizing radiation according to the atomic cross sections of the constituent elements in the gas.

c. ARGON DETECTOR. The argon detector is similar in design to the cross-section ionization detector but uses argon as the carrier gas and radium, strontium-90, or tritium as a source of beta rays for producing ionization in the argon (ionizing radiation). Ionization will take place in any gas sample that has an ionization potential equal to or less than 11.7 eV (the energy of metastable argon). The energy from the metastable argon is transferred to the sample gas; the rare gases and others with a high ionization potential do not respond to this detector. The response is generally related to the molecular weight of the sample gas; therefore, this detector can be used to measure the molecular weight of a gas. The detector is sensitive in the ppb region and has a response time on the order of milliseconds. Water vapor and air interfere with the detector performance even though they cannot be measured.

d. ELECTRON-CAPTURE IONIZATION DETECTOR. The electron-capture ionization detector is even more sensitive than the flame ionization detector, but it responds to only a few classes of compounds, mainly those that contain a halogen. The electron-capture detector has two electrodes placed in series in the sample gas stream. One of the electrodes is treated with a radioisotope that emits β particles, i.e., high-energy electrons. The high-energy electrons produce many low-energy secondary electrons in the sample gas, all of which are collected by the second, positively polarized, electrode, causing a steady-state current. Molecules that capture electrons as they pass between the electrodes reduce this current in proportion to the number of molecules present.

1.3.10 Chemical Reaction Methods

a. GAS ANALYSIS BY ABSORPTION. Gas analysis by absorption is simply the use of absorbents to selectively remove components in a gas mixture, followed by volumetric and gravimetric methods to measure these components quantitatively. The Orsat, Hempel, Bunte, and other similar apparatus, are refinements of volumetric techniques that use absorbents to remove a known gas. The amount of sample component combining with the reagent is determined by measurement of either the sample's loss in volume or decrease in pressure. Gas burettes are used to measure the gas volumes, while the absorptive reagents are contained in a series of absorption pipettes. A confining liquid (saline solution or mercury) is used to transfer the gas between burette and pipettes through a manifold arrangement equipped with stopcocks. In a gravimetric method, the gas is absorbed on a solid absorbent, and the increase in weight is proportional to the sample gas present. These methods are discussed in more detail by a number of authors, including Snell and Biffen (1964), Heron and Wilson (1959), and Hill (1939). An interesting account of the gas analysis methods used in the Manhattan Project is given by Rodden (1950). Note that these methods require that the reagents be applied in an orderly fashion. The selectivity for particular gases in many of the absorbing reagents is not great, and care must be exercised to minimize errors due to interfering species.

These methods suffer from several deficiencies, notably, (1) the reagents are specific not to individual gases but to certain classes or types of gases; (2) the accuracy of the results depends upon operator proficiency; (c) they require large sample volumes, usually over 1 ml; and (4) several minutes to several hours may be required for the analysis of relatively simple gas mixtures. These methods have almost universally been replaced by the superior gas-chromatographic, mass-spectrometric, and other instrumental methods.

Absorption techniques are used in some instruments as a quantitative analysis scheme for elemental analyzers. The combustion gases, CO_2, SO_2, and H_2O, are readily dissolved in a solid or liquid absorbent and measured gravimetrically or volumetrically. These instruments are discussed in Section 2 of this chapter. The absorbents that have been used for gas absorbent analysis are listed in Table 2. Miller (1949) discusses the theoretical aspects of the absorption of gases in solids.

b. HEAT OF ABSORPTION. Gases may also be analyzed by determination of the heat of absorption of a gas into a solid or liquid. In this method the rise in temperature from the absorption of the gas into the absorbent is measured with a temperature sensor; the temperature rise is proportional to the gas concentration. This method is, of course, limited in specificity by the absorbent.

c. PIEZOELECTRIC CRYSTAL OSCILLATOR. Piezoelectric crystals oscillate at a fixed frequency when placed in an alternating electrical field under constant conditions. The frequency varies inversely with the mass and is a very sensitive indicator of any mass change on the oscillating crystal. The detection limit of the crystal is estimated to be 10^{-12} g.

TABLE 2 Selected Absorbents For Gases

Absorbent	Gas	Comments and Refs.
Molecular sieve	H_2O; selected gases	Hersh (1961). Adsorbent used for separation of gases at reduced temperatures
Activated carbon	Most reactive gases	Strong adsorbent: used for air purification in gas scrubbers
Silica gel	H_2O; selected gases	Absorbs most reactive gases also; used for separation of gases
Calcium sulfate (Drierite)	H_2O	Hammond (1958)
Magnesium perchlorate	H_2O	
Phosphorus pentoxide	H_2O	
Dilute acid solutions	NH_3	Also removes other alkaline gases
Sodium hydroxide in asbestos (Ascarite)	CO_2	Also removes SO_2 and NO_2
KOH, NaOH, BaOH	CO_2	Absorbent may be solid or in aqueous solution
Ethylene diamine	CO_2	
Cuprous chloride (Cosorbent)	CO	
Copper sulfate and β-naphthol in H_2SO_4	CO	
Chromous chloride solution (Oxorbent)	O_2	Very rapid
Pyrogallol	O_2	Basic solution
Sodium hydrosulfite	O_2	
Yellow phosphorus	O_2	
Copper, hot reduced	O_2	
Manganous hydroxide	O_2	Iodometric titration
Potassium iodide	O_3	pH 7
$FeSO_4 \cdot 7H_2O$	O_3 and H_2O_2	Removes most oxidants
Colloidal palladium	H_2	
Palladiumized asbestos	H_2	
Ammoniacal cuprous chloride	Acetylene	
Concentrated sulfuric acid	Olefins	Alkaline constituents will also absorb
Bromine water	Olefins	
$H_2S_2O_7$	Olefins	
Maleic anhydride	Conjugated olefins	Molten
Water	Cl_2	
Potassium arsenite	Cl_2	
Potassium iodide	Cl_2	I_2 liberated
Lithium, magnesium	N_2	Forms nitrides at high temperatures
Calcium fluoride	HF	
Magnesium oxide	HF	
Silver wool or gauze, silver vanadate	Halogens and sulfur oxides	
Manganese dioxide	Nitrogen oxides	
Chromic oxide in sand	SO_2, H_2S, and mercaptans	

Therefore, if an absorbing coating is placed on the crystal, any gas absorbed by the coating will result in a reduced frequency of oscillation of the crystal. King (1964) and Karasek and Gibbons (1971) discuss these types of gas detectors.

An instrument using this technique measures the amount of moisture in an air sample by comparing changes in the frequency of two quartz crystal oscillators, one for a reference and the second for the sample. This latter oscillator, coated with a hygroscopic agent, vibrates at 9 MHz (million cycles per second) in a dry atmosphere. Sorption of moisture by the coated crystal lowers its frequency in proportion to the amount of water vapor present in a flowing gas sample. A ratemeter circuit is used to compare the frequency of the two crystals. This detector has a sensitivity of 0.1 to 25, 000 ppm with a response time of 3 min. The detector is adversely affected by gases which react with the absorbent coating. This method for gas analysis can be used for other gases as long as a suitable sorbent can be found; hydrocarbons have been detected down to 1 ppm by means of this technique.

d. COMBUSTION CALORIMETRY. Combustion methods can be used on any combustible or oxidizable gas. A combustion method simply provides for a contained combustion of a gas with an appropriate technique for measuring the temperature increase from the reaction. There are several approaches to combustion analysis, and they depend on the information desired from the analysis. Generally this method can be employed for analyzing or monitoring the total combustible gas or a binary gas mixture containing one oxidizable component in a gas sample or gas stream. In this case, the gas is catalytically combusted or ignited in an oxygen-rich atmosphere to complete oxidation. The heat generated is measured by a thermocouple, hot wire, or thermistor, in which a potential or resistance change is proportional to a change in temperature.

In one instrument, CO is monitored by oxidation in a catalyst bed to CO_2; the heat of reaction is measured by thermocouples. The incoming sample gas is filtered for particulate matter and chemically scrubbed for hydrocarbons and deleterious gases. Sturtevant (1971) reviews the general subject of calorimetry.

e. CHEMILUMINESCENCE. The chemiluminescence method for gas analysis measures the amount of emitted radiation from the chemical reaction of a sample gas with an appropriate reactant. The emitted light from the reaction is monitored with a phototube detector.

Steadman (1972) and Fontizn *et al.* (1970) have used a chemiluminescence instrument for measurement of ozone and nitric oxide in photochemical smog studies. The measurement is based on the reaction

$$O_3 + NO \rightarrow O_2 + NO_2 + h\nu$$

A similar chemiluminescence reaction in which ozone reacts with ethylene to give off light is discussed by Hodgeson *et al.* (1970). The chemiluminescence methods have generally selective response to the gas being measured and are sensitive into the ppb range with a response time of about 1 sec. Species that interfere with the method are those that quench the excited atoms before they can emit light or those that give competing chemiluminescent reactions. Filters can be installed in front of the phototube detector to minimize the unwanted radiation from competing chemiluminescence reactions. Chemiluminescent monitors are also feasible for SO_2, CO, H_2S, and NH_3 gases.

Patterson and Henein (1972) discuss chemiluminescence methods in relation to emissions from combustion engines.

f. COLORIMETRIC METHOD. The colorimetric method is also a chemical reaction method like chemiluminescence, but the reaction products are colored and, therefore, detected spectrophotometrically. In Chapter 18, which discusses uv-visible spectrophotometry, this subject is dealt with in some detail.

One visual colorimetric method that has been used for quite some time to crudely monitor the amount of CO in an atmosphere is the reaction with silicomolybdic acid, $H_8Si(MO_2O_7)_6$, impregnated on silica gel. The color of this reagent changes from bright yellow through green to blue as it is reduced by CO.

g. CONDENSATION NUCLEI METHOD. The condensation nuclei method for gas analysis (Van Luik and Rippere, 1962) produces submicroscopic particles from the gas molecules. Water vapor condenses on these particles, and the resulting fog is measured photometrically. This method is used only for low concentrations of gases. An example of this method is the monitoring of SO_2. An air sample is drawn through a filter and then converted into an H_2SO_4 aerosol. The H_2SO_4 aerosol, which consists of condensation nuclei, is drawn into a cell in the presence of H_2O vapor. The optical transmittance of the fog is an indication of the number of condensation nuclei, which is proportional to the SO_2 concentration in the sample gas.

2 Combustion Methods for Elemental Analysis

2.1 INTRODUCTION

Instrumental techniques for the analysis of gases have been discussed in Section 1. In this section instrumental combustion methods for the analysis of materials for selected elements are reviewed. These elements are carbon, hydrogen, nitrogen, oxygen, and sulfur. Virtually any inorganic or organic material that contains one or more of these elements and that pyrolyzes or oxidizes at temperatures up to 1100°C can be analyzed. The gaseous products of these reactions are analyzed quantitatively by one of the methods described in Section 1. From these analyses the amounts of the respective elements may be calculated.

2.2 Description of Methods

Combustion methods have been used for some time to determine selected elements present in a material. The classical Pregl and Dumas methods are combustion techniques for the determination of carbon, hydrogen, and nitrogen in organic materials. The Unterzaucher method was developed for oxygen analysis. These and other methods, which have been gradually modified and automated, provide the basis for most commercial elemental analyzers in use today. Completely automated methods have been developed for the elemental analysis of carbon, hydrogen, nitrogen, oxygen, and sulfur in organic and inorganic materials. The overall elemental reactions of interest that take place in the combustion of a material are

$$
\begin{aligned}
&C + O_2 \rightarrow CO_2\\
&H + O_2 \rightarrow H_2O\\
&N + O_2 \rightarrow N_xO_y \xrightarrow{\text{red}} N_2\\
&O + C \rightarrow CO \xrightarrow{\text{ox}} CO_2\\
&S + O_2 \rightarrow SO_2
\end{aligned}
$$

The reaction conditions are chosen to yield these final reaction products. It is imperative that the element of interest be completely converted to the single gas product for accurate calculation of the initial elemental composition.

The elemental information derived from the elemental analysis is used to determine or monitor the percent composition of the element in the material. For pure compounds this information can be used to calculate the empirical formula once the chemical constituents are known.

2.3 Sampling

The samples for elemental analysis may be in the form of a soild, a liquid, or with modification to the instrument, a gas. The solid sample can be in the form of a crystal, powder, or slab. Fast-burning samples may be diluted with an inert material, while the slow-combusting material can be intimately mixed with an oxidizing catalyst. Solid samples are weighed into an aluminum or nonreactive boat, which is then placed into the combustion tube.

A liquid sample is generally weighed into a capillary tube by means of a microliter syringe or injected directly into the packing of the combustion tube. In handling volatile samples extreme care must be taken to avoid losses due to evaporation. A microbalance (sensitive to 0.001 mg) is necessary for determining the weight of the samples. It is possible to introduce a gas directly into the combustion chamber via a syringe or as the effluent from a gas chromatograph. Liebman *et al.* (1972) have used this latter technique for analyzing chromatograph effluent gases.

Sample weight requirements are dependent upon sample composition; however, 0.5–50 mg of material are typically used in organic analyzers, while the inorganic instruments generally require 0.5–1 g of material, since the inorganic samples have, as a rule, less carbon, hydrogen, nitrogen, and sulfur present.

Calculation of meaningful empirical formulas for organic compounds may require special purification techniques. These techniques are outlined in Chapter 1, Volume I.

2.4 Instruments for Elemental Analysis

The essential components for all elemental analyzers are a source of gases for combustion and for purging, a combustion train with a sample tube (usually containing catalysts and reactants) in a heated zone, scrubbers for removing interfering gases, and finally a detector or means of quantitatively measuring the amount of gases from the combustion. An example of an organic elemental analyzer is shown schematically in Fig. 6. This instrument was developed from the original design of Simon and his co-workers (Clerc *et al.* 1963).

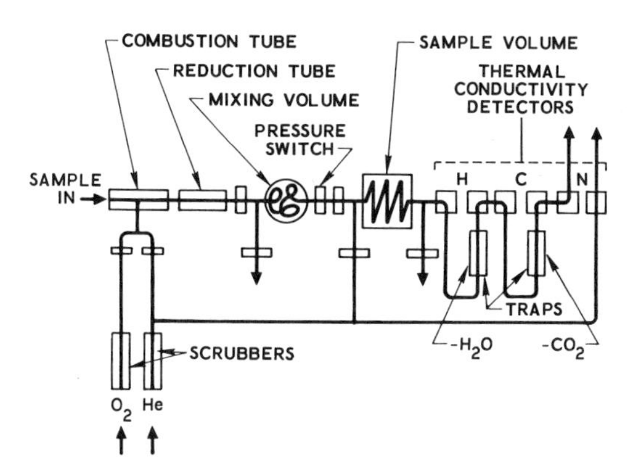

Fig. 6. A carbon–hydrogen–nitrogen elemental analyzer that combusts a material in an oxygen atmosphere and measures the amount of gaseous products with thermal conductivity detectors. [Courtesy of Perkin-Elmer Corporation.]

In this instrument a weighed organic sample is combusted in a static O_2 atmosphere in the presence of a catalyst. The combustion products (CO_2, H_2O, and N_xO_y) and the remaining O_2 are passed through a reduction chamber containing hot copper to reduce the O_2 to CuO and the N_xO_y to N_2. The remaining products (CO_2, H_2O, and N_2) are swept into a mixing chamber. An amount of this gas mixture is then measured in a sample volume and swept into the first set of thermal conductivity detectors. The initial detector measures the gas mixture with H_2O present, while the next detector measures the gas mixture after the H_2O has been removed by the H_2O trap or scrubber. The next set measures the CO_2 by difference in a similar manner, and the final set measures the nitrogen with the reference detector containing only pure helium. Kainz and Wachberger (1967) employ an absorption–desorption scheme with thermal conductivity detection for gas analysis. Many modifications to this scheme are used in other instruments and for different elements, as will become apparent in the following discussion.

The instruments available for the elemental analysis of organic materials are reviewed in an article by Fish (1969). Ma (1972) reviews the literature on methods for organic elemental analysis.

A method for the determination of organic carbon in water by pyrolysis oxidation-reduction is described by Dobbs *et al.* (1967). Lewis and Nordozzi (1966) describe an analyzer for determining carbon in steel.

Table 3 outlines the reaction schemes for the different elements in both organic and inorganic samples. This table in conjunction with the following discussion describes the function for each component in an elemental analyzer.

2.4.1 Sweep Gas

The sweep gas is used to purge the combustion train of any gaseous impurities before combustion and to transport the combustion products from the combustion tube through the gas scrubbers or reduction tube into the gas detectors. Helium is widely used in carbon, hydrogen, and nitrogen instruments that combust the sample in a static oxygen atmosphere and that use thermal conductivity detectors. Nitrogen is used to some extent in instruments for carbon and hydrogen analysis. Oxygen may be used as the sweep gas in carbon, hydrogen, and sulfur analyzers, where the combustion takes place in a flow-type reactor and where it does not interfere with the detection of the product gases. In one nitrogen analyzer, CO_2 is used as the sweep gas where the nitrogen is measured volumetrically after removal of the CO_2 sweep gas.

2.4.2 Combustion Train

If the combustion method in elemental analysis is to work effectively, the material must be completely converted to gaseous products in the combustion process. The factors influencing combustion are the temperature, catalyst, and the gas flow rate (in flow-type reactors). The proper choice of catalyst and temperature conditions minimizes too fast or too slow combustion, with their associated errors. The combustion catalysts used for organic materials are typically oxides of copper, manganese, and tungsten, while vanadium pentoxide is used extensively for inorganic materials. Dugan and Aluise (1969) developed a flash combustion technique for the determination of carbon, hydrogen, nitrogen, sulfur, and oxygen in organic materials. Vecera and Synek (1960) discuss Co_3O_4 as a catalyst in the microdetermination of nitrogen in organic compounds.

It should be noted from Table 3 that the combustion temperature is significantly different for organic and inorganic materials. Both inorganic and organic materials give the same gaseous end products, but the optimum combustion conditions for each type require different instruments. Resistance-type heaters are generally used for analysis of organic materials, while in many instances induction heaters are employed for inorganic materials. The induction heaters can supply a large amount of energy to the sample in a very short time, allowing a higher temperature with rapid, complete combustion. This flash combustion can be carried out without the aid of a catalyst. Too rapid heating sometimes allows volatile intermediates to escape the heated zone, resulting in incomplete combustion.

Elemental analyzers for the determination of oxygen in a sample employ a pyrolysis tube in place of a combustion tube. The oxygen in the sample reacts with a carbon catalyst or crucible in an inert atmosphere to form carbon monoxide. As in the carbon–hydrogen analyzers, the inorganic materials are pyrolyzed at a much higher temperature than the organic materials.

2.4.3 Reduction Train

Not every analyzer is equipped with a reduction train; however, the nitrogen analyzers must have some provision to completely convert the nitrogen oxides to nitrogen for accurate detection and calculation of nitrogen. The reduction train is used to convert nitrogen oxides to nitrogen, and oxygen to a nonvolatile oxide; the reducing agent is usually hot copper metal. Oxygen must be removed from the sweep gas before entering the thermal conductivity detectors, since, its presence would interfere with the thermal conductivity readings.

2.4.4 Mixing and Sample Volumes

Mixing and sample volumes are the tubes or containers used to collect and measure the combustion gases before their entrance into the detector portion of the analyzers. They are mainly used on instruments employing thermal conductivity detectors, since the detector response cannot be ac-

TABLE 3

ELEMENTAL ANALYZERS SCHEME

Element analyzed	Sweep gas	Sweep gas scrubber	Combustion or pyrolysis reaction	Gas[a] scrubber	Reduction[a] or oxidation reaction	Gas[a] scrubber	Detector
Carbon	O_2 for combustion. O_2, He, or N_2 for carrier	For removal of CO_2	Oxidation of organic carbon at 750–1050°C in presence of catalyst. Oxidation of inorganic carbon, metals, glasses, etc., at >1650°C in induction furnace. $2C + O_2 \rightarrow 2CO$ $2CO + O_2 \rightarrow 2CO_2$	For removal of sulfur oxides and halogens	Reduction of nitrogen oxides and oxygen at 400°C in presence of reducing catalyst $N_xO_y + Cu \rightarrow N_2$ $O_2 + 2Cu \rightarrow 2CuO$	For removal of N_xO_y and H_2O	Thermal conductivity or gravimetric
Hydrogen	Same as for C	For removal of H_2O	Combustion conditions the same as for carbon $4H + O_2 \rightarrow 2H_2O$	Same as for carbon	Same as for carbon	For removal of N_xO_y and CO_2	Thermal conductivity or gravimetric

Nitrogen	O_2 for combustion. CO_2, O_2 or He for carrier		Combustion conditions the same as for carbon $N + O_2 \longrightarrow N_2 + N_xO_y$	Same as for carbon	Same as for carbon	For removal of CO_2 and H_2O	Thermal conductivity or volumetric
Oxygen	He or N for carrier	For removal of H_2O and CO_2	Pyrolysis of organic oxygen at 750–1050°C in presence of platimized carbon catalyst. Pyrolysis of inorganic oxygen at 2100°C in graphite crucible in electrode furnace $O + C \longrightarrow CO$		Oxidation of CO at 500°C in presence of CuO $2CO + O_2 \longrightarrow 2CO_2$	For removal of H_2O	Thermal conductivity, volumetric, or gravimetric
Sulfur	O_2 for combustion. O_2 or He for carrier	For removal of H_2O and CO_2	Oxidation of inorganic sulfur at 1650°C in an induction furnace $S + O_2 \longrightarrow SO_2$		Removal of CO_2, H_2O, and N_xO_y	For removal of CO_2, H_2O N_xO_y	Chemical reaction

[a] May be absent in some instruments.

curately integrated over the time period needed for complete combustion to take place. In Fig. 6 the sample volume allows a measured amount of combustion gas to enter the detector region in a concentrated volume. The separation of gas in other instruments in accomplished with a chromatographic column or a series of traps or scrubbers. (See Chapter 24 on gas chromatography.)

2.4.5 Gas Scrubbers

The gas scrubbers selectively remove interfering gases in any part of an elemental analysis scheme. They may be used to purify the sweep gas; scrub out unwanted combustion gases, such as sulfur oxides and halogens; or selectively remove gases in the detector section of the analyzer. Table 2 lists materials that may be used as scrubbing agents.

The selection of absorbents and their location varies for different instruments; nevertheless, the accuracy and reproducibility of all elemental analyzers are dependent upon the performance of the scrubbers, and therefore their proper maintenance is mandatory. This maintenance may involve a regeneration by heating, etc., but more likely will involve replacement of the absorbent.

2.4.6 Detectors

The detectors or methods used to measure quantitatively the product gases in elemental analyzers vary from instrument to instrument. Most of the methods listed in Section 1 could be used; however, thermal conductivity and, to a lesser extent, gravimetric and volumetric methods are typically preferred. Bandi *et al.* (1966) report a coulometric titration method for the determination of sulfur in iron and steel. Thermal conductivity methods have the advantage that they use a direct electronic readout proportional to the change in thermal conductivity and, therefore, the concentration of a gas in a binary mixture. Operator error is minimized, and the method is fast, sensitive, and accurate. Gravimetric and volumetric methods have the advantages that they are selective, fairly accurate, and able to measure the gas over an extended period of time. These and other methods for gas analysis are discussed in more detail in Section 1.

2.5 Data Form and Calculations

The data form is, to a large extent, dependent upon the detector used in a particular elemental analyzer. Thus, the analyzers using gravimetric methods involve the weighing of the absorption trap or scrubber to determine the weight of the combustion gas; calculation from the appropriate stoichiometric equation of combustion will then yield the weight of the element present in the sample. Volumetric techniques give a measure of the volume of a combustion product. Calculation of the weight of the gas from its known volume permits determination of the percent composition, as in gravimetric methods. In the elemental analyzers employing thermal conductivity detectors, it is necessary to calibrate the response of the detector. It is a wise practice to check routinely the accuracy of elemental analyzers. These checks as well as instrument calibrations are accomplished by combusting in the instrument materials of known composition, such as those obtained from the National Bureau of Standards. Ideally, the calibrating standards should encompass a composition range similar to the unknown samples, as instruments do not always respond in a linear fashion over large ranges in sample compositions. In addition, a blank (no sample present or one devoid of the components of interest) should be run in the instrument as a check on its zero response.

2.6 Applications and Limitations

Combustion elemental analyzers are selective for the quantitative measurement of specific elements in materials, as shown in the previous discussions. A single instrument can usually perform carbon and hydrogen analysis and, in many cases, nitrogen analysis as well. The instrumental methods are, of course, destructive methods and are generally as accurate as the classical combustion methods. The accuracies quoted are generally better than 0.3% of the element present, with sensitivities on the order of 0.01%. Samples difficult to combust and those with appreciable fluorine content lead to greater than normal errors. Factors affecting precision of elemental analyzers include weighing the sample (and the absorption tubes, where applicable), readout response, detector operation, temperature control, and flow rate of the gas. The time required for analysis ranges from 8 to 30 min for organic materials and 1½ to 3 min for inorganic materials (with the use of induction-type heaters for combustion).

References

Gas Analysis

Andreatch, A. J., and Feinland, R. (1960). *Anal. Chem.* **32**, 1021.
Bochkova, O. P., and Shreyder, E. (1965). "Spectroscopic Analysis of Gas Mixtures." Academic Press, New York.
Breiter, M. W. (1968). *J. Phys. Chem.* **72**, 1305.
Collinson, A. J. L., and Haque, A. K. M. (1953). *J. Sci. Instrum.* **40**, 521.
Crouthamel, C. E., and Diehl, H. (1948). *Anal. Chem.* **20**, 515.
Durst, R. A. (ed.) (1969). NBS Spec. Publ. No. 314.
Fisher, G. E., and Huls, T. A. (1970). *J. Air Pollut. Contr. Ass.* **20** (10), 666.

Fontizn, A., Sabedell, A. J., and Ronco, R. J. (1970). *Anal. Chem.* **42,** 575.
Gelbwachs, J. A., Birnbaum, M., Tucker, A. W., and Fincher, C. L. (1972). *Opto-electron.* **4,** 155.
Gudzinowicz, B. J., and Smith, W. R. (1963). *Anal. Chem.* **35** (4), 465.
Hager, R. N., Jr. (1973). *Anal. Chem.* **45** (13), 1131A.
Hammond, W. A. (1958). Drierite and its Applications in the Drying of Solids, Liquids, and Gases, Hammond Drierite Company, Xeria, Ohio.
Heron, A. E., and Wilson, H. N. (1959). *In* "Comprehensive Analytical Chemistry" (C. L. Wilson and D. W. Wilson, eds.), Chap. III, Elsevier, Amsterdam.
Hersch, P. (1957). U.S. Patent 2,805, 191.
Hersh, C. K. (1961). "Molecular Sieves." Van Nostrand Reinhold, Princeton, New Jersey.
Hill, A. H. (1939). *In* "Scotts Standard Methods of Chemical Analysis" (N. H. Furman, ed.), 5th ed., p. 2336. Van Nostrand Reinhold, Princeton, New Jersey.
Hill, D. W., and Powell, T. (1968). "Non-Dispersive Infra-Red Gas Analysis in Science, Medicine and Industry." Plenum Press, New York.
Hodgeson, J. A., Krost, K. J., O'Keeffe, A. E., and Stevens, R. K. (1970). *Anal. Chem.* **42,** 1795.
Hommel, C. O., Chleck, D., and Brousaides, F. J. (1961). *Nucleonics* **19** (5), 94.
Karasek, F. W., and Gibbons, K. R. (1971). *J. Chromatogr. Sci.* **9,** 535.
Keenan, C. A. (1967). *Anal. Instrum.* **4,** 57.
Kendall, D. R. (1971). *Anal. Chem.* **43** (7), 944.
Kieselbach, R. R., and Schmit, J. A. (1972). *In* "Treatise on Analytical Chemistry" (I. M. Kolthoff, P. J. Elving, and E. B. Sandell, eds.), Pt. I, Vol. 10, p. 5939. Wiley (Interscience), New York.
King, W. H., Jr. (1964). *Anal. Chem.* **36,** 1735.
Kreuzer, L. B. (1971). *Anal. Chem.* **43** (11), 87A.
Lee, T. G. (1969). *Anal. Chem.* **41** (2), 391.
Link. W. T., McClatchie, E. A., Watson, D. A., and Compher, A. B. (1971). *Joint Conf. Sensing of Environmental Pollutants*, AIAA Paper No. 71-1047.
Lodge, J. P., Jr., and Pate, J. G. (1971). *In* "Treatise on Analytical Chemistry" (I. M. Kolthoff, P. J. Elving, and F. H. Stross, eds.), pp. 199–252. Wiley (Interscience), New York.
Lovelock, J. E. (1961). *Anal. Chem.* **33,** 162.
Miller, A. R. (1949). "The Adsorption of Gases on Solids." Cambridge Univ. Press, London and New York.
Miller, D. F., Wilson, W. E., Jr., and Kling, R. G. (1971). *J. Air Pollut. Contr. Ass.* **21** (7), 414.
Moses, H., Lucas, H. F., and Zerbe, G. A. (1963). *J. Air Pollu. Contr. Ass.* **13,** 12.
Nauman, R. V., West, P. W., Tron, F., and Gaeke, G. C., Jr. (1960). *Anal. Chem.* **32,** 1307.
Nerheim, A. G. (1963). *Anal. Chem.* **35,** 1640.
Noble, F. W., Abel, K., and Cook, P. W. (1964). *Anal. Chem.* **36** (8), 1421.
Patterson, D. J., and Henein, N. A. (1972). Emissions from Combustion Engines and their Control, pp. 305–334. Ann Arbor Science, Ann Arbor, Michigan.
Rodden, C. J. (Ed.-in-Chief) (1950). "Analytical Chemistry of the Manhatten Project," p. 644. McGraw Hill, New York.
Schulze, F. (1970). U.S. Patent 3, 512, 937.
Snell, F. D., and Biffen, F. M. (1964). "Commercial Methods of Analysis." p. 603. Chem. Publ., New York.
Snyder, A. D., Eimutis, E. C., Konicek, M. G., Parts, L. D., and Sherman, P. L. (1972). Instruments for the Determination of Nitrogen Oxides Content of Stationary Sources Emissions, Vol. II, APTD-0942. Monsanto Res. Corp., Dayton, Ohio.

Steadman, D. H. (1972). *J. Air Pollut. Contr. Ass.* **22** (4), 260.
Sturtevant, J. M. (1971). *In* "Techniques of Chemistry" (A. Weissberger and B. W. Rossiter, eds.), Vol. I, Pt. V, p. 347. Wiley (Interscience), New York.
Testerman, M. K., and McLeon, P. C. (1962). *In* "Gas Chromatography" (N. Brenner, J. E. Cullen, and M. D. Weiss, eds.) p. 183. Academic Press, New York.
Tiné, G. (1961). "Gas Sampling and Chemical Analysis in Combustion Processes," pp. 4–56. Pergamon, Oxford.
Touloukian, Y. S., Liley, P. E., and Saxena, S. C. (1970). "Thermal Conductivity, Nonmetallic liquids and Gases," Vol. 3 of "Thermophysical Properties of Matter." Plenum Press, New York.
Urone, P., Evans, J. B., and Noyes, C. M. (1965). *Anal. Chem.* **37,** 1104.
Van Luik, F. W., Jr., and Rippere, R. E. (1962). *Anal. Chem.* **34** (12), 1617.
Weaver, E. R. (1951). *In* "Physical Methods in Chemical Analysis" (W. G. Berl, ed.), pp. 387–437. Academic Press, New York.

Combustion Analysis

Bandi, W. R., Buyok, E. G., and Straub, W. A. (1966). *Anal. Chem.* **38** (11), 1485.
Clerc, J. T., Dohner, R., Santer, W., and Simon, W. (1963). *Helv. Chem. Acta* **46,** 2369.
Dobbs, R. A., Wise, R. H., and Dean, R. B. (1967). *Anal. Chem.* **39,** 1255.
Dugan, G., and Aluise, V. A. (1969). *Anal. Chem.* **41** (3), 495.
Fish, V. B. (1969). *J. Chem. Ed.* **46** (5), A323.
Kainz, G., and Wachberger, E. (1967). *Microchem. J.* **12,** 584.
Lewis, L. L., and Nordozzi, M. J. (1966). *Anal. Chem.* **38** (9), 1215.
Liebman, S. A., Ahlstron, D. H., Creighton, T. C., Pruder, G. D., Averitt, R., and Levy, E. J. (1972). *Anal. Chem.* **44** (8), 1411.
Ma, T. S., and Gutterson, M. (1972). *Anal. Chem.* **44** (5), 445A.
Vecera, M., and Synek, L. (1960). *Mikrochim Acta.* 208–219.

Bibliography

Cook, G. A. (1961). "Argon, Helium and the Rare Gases," Wiley (Interscience), New York.
Coordinating Research Council, Inc. (1965). *Rep. CRC Symp. Exhaust Gas Analysis.*
Gannoe, R. E. *et al.* (1971). Gas Monitoring Equipment Investigation, IV00014-69-C-0352. Battelle Memorial Institute, Columbus, Ohio.
Gollob, F. (1971). *Amer. Lab.* 3 (3, 8). Industrial gas analysis review.
Hauber, W. P. (1971). Hydrocarbons in Auto Exhaust by Various Instrumental Methods. ISA Anal. Instrum. Soc. P. D-3.
Hobbs, A. P. (1966). *Anal. Chem.* **38** (5), 166. Review of analytical methods for specific gases.
Hollowell, C. D., Gee, G. Y., and McLaughlin, R. D. (1973). *Anal. Chem.* **45** (1), 63A. Instrumental methods for continuous monitoring of SO_2.
Lawrence Berkeley Lab. (1972). Instrumentation for Environmental Monitoring, AIR. LBL-1, Vol. 1.
Lewis, C. D., (1972). *In* "Treatise on Analytical Chemistry" (I. M. Kolthoff, P. J. Elving, E. G. Sandell, eds.), Pt. 1, Vol. 10, pp. 6293–6404. Wiley (Interscience), New York.
Mueller, P. K., Kothny, E. L., Pierce, L. B. Belsky, T., Imada, M., and Moore, H. (1971). *Anal. Chem. Rev.* **43** (5), IR. Review of gas analysis methods in air pollution.
Mullen, P. W. (1955). "Modern Gas Analysis." Wiley (Interscience), New York.

Norton, A. C. (1970). Survey of Commercial Laboratory Instruments, Vol. 2 of Commercial Instrumentation for Space Station Application. NASA-CR-102941.

Ruch, W. E. (1966). "Chemical Detection of Gaseous Pollutants," Ann Arbor Science, Ann Arbor, Michigan. An annotated bibliography.

Siggia, S., (1959). "Continuous Analysis of Chemical Process Systems." Wiley, New York.

CHAPTER 24

Gas Chromatography

Gerald R. Shoemake

Texas Engineering and Science Consultants, Inc.
Houston, Texas

Introduction

Gas chromatography is an analytical technique whereby the sample to be analyzed is volatilized, injected into a flowing carrier gas stream; and passed through a column, where the heterogeneous sample is separated into its individual constituents; the individual constituents are then detected and measured by a detector as they emerge from the column. The time required for the detection of each sample component, under a controlled set of conditions, provides the qualitative information. The magnitude of the detector response may be used to obtain a quantitative interpretation.

The technique of gas chromatography is one of the most rapidly developed methods to become available to the analyst in modern times. Since its introduction by James and Martin (1952) it has been expanded to include the qualitative and quantitative analyses of both organic and inorganic samples, which may be either liquids, solids, or gases, in their initial states. References to the use of this method are found in practically all fields of endeavor throughout the spectrum of the physical and biological sciences. The instrumentation, applications, and user modifications have become so numerous and varied that separate volumes dealing only with a single topic or application have been published. It is not the purpose of this undertaking to cover in detail each topic, but rather to attempt to provide a general understanding of the gas-chromatographic technique and its potential uses. Included are references that will allow the reader to pursue, in detail, any area discussed. An attempt has been made to emphasize the application of gas chromatography to the systematic analysis of materials.

1 Theory

1.1 General

Chromatography is a physical means of separation whereby the substances to be separated are transported in a moving or mobile phase through the system (column) and allowed to interact with a fixed or stationary phase. This interaction causes a distribution of the sample components between the two phases that ultimately results in separation of the sample into its various constituents.

The stationary phase may be a solid material and the mobile phase either a liquid (liquid–solid chromatography) or a gas (gas–solid chromatography). Alternatively, the stationary phase may be a liquid and the mobile phase either a liquid (liquid–liquid chromatography) or a gas (gas–liquid chromatography). Gas chromatography includes both gas–liquid and gas–solid systems.

For either gas–liquid or gas–solid chromatography there are several techniques available to effect a separation. These are elution development, frontal analysis, and displacement development. The majority of all gas-chromatographic applications involve the use of elution development. In this technique the sample is placed in contact with the stationary phase, then eluted by passage of the mobile phase (carrier gas) through the column. As already noted, it is the distribution of the sample components between the mobile and stationary phases that brings about a separation of the components of the sample.

The methods of detection and measurement of sample components as they emerge from the column can be divided into two categories. These are called integral and differential methods. In integral detection some property of the effluent from the chromatography column is measured in an additive manner. An example of this method is the recording of the volume of titrant used in the automatic titration of a reacting species in the column effluent. Integral methods are seldom used in gas chromatography today, although they are still of academic and historical interest. In differential detection some transient property of the column effluent is measured. An example of this method is the measurement of the change in thermal conductivity of the column effluent as the sample components emerge. In this chapter treatment of the subject of gas chromatography will be restricted to differential detection methods.

To summarize, the discussion of gas chromatography that follows will be limited to either gas–liquid or gas–solid systems using elution development techniques and differential detection methods, since these are the forms utilized almost exclusively in current gas-chromatographic applications.

The theory of gas chromatography is essentially the theory of the separation that occurs in the chromatographic columns. An ideal case is illustrated in Fig. 1. The carrier gas, G, is flowing through a column, in the

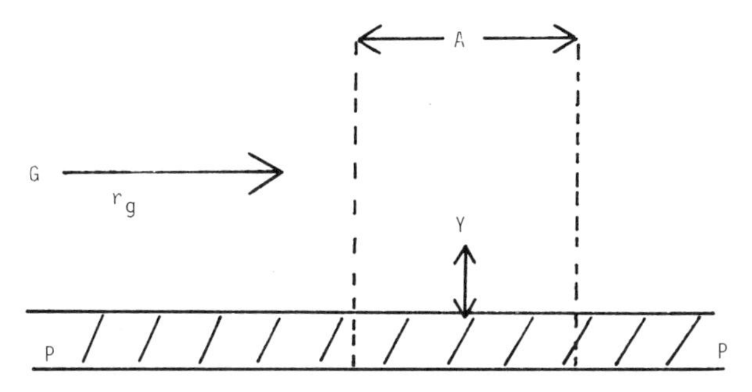

Fig. 1. Ideal gas–liquid chromatography: G, carrier gas; r_g, carrier gas velocity; Y, sample component; A, sample bandwidth; P, liquid stationary phase.

direction indicated by the arrow, with a uniform velocity r_g. A single component sample, Y, has been placed in the column in a narrow band having width A. The sample is being transported through the column in the same direction as the carrier-gas flow. As the sample moves through the column, it interacts with the stationary phase, P. If we make the assumptions that the carrier-gas velocity is uniform along all of the column dimensions, that the transport of sample across the carrier gas–stationary phase interface is instantaneous and equilibrium is maintained, and that there is no diffusion of the sample in either the carrier gas or stationary phase, we have an ideal situation. Under these circumstances the transport rate of the sample along the length of the column is determined by the carrier-gas velocity and the distribution of the sample between the mobile and stationary phases. If we consider a cross section of the column of width A, the distribution coefficient k is given by

$$k = \frac{\text{concentration of sample in the stationary phase}}{\text{concentration of sample in the carrier gas}} = \frac{C_{sp}}{C_{cg}} \tag{1}$$

If V_{sp} and V_{cg} are the volumes of the stationary phase and carrier gas, respectively, in the cross section being considered, the fraction of sample in the gas phase, R_f, is given by

$$R_f = \frac{C_{cg}V_{cg}}{C_{cg}V_{cg} + C_{sp}V_{sp}} = \frac{1}{1 + k'} \tag{2}$$

where $k' = kV_{sp}/V_{cg}$. If we make the further assumption that the distribution coefficient k is independent of sample concentration, the width of the sample band does not change as it passes through the column. It follows that the rate of movement of sample passing any column cross section equal to A is $R_f \times r_g$.

If we now consider a multicomponent sample, each component having a different valve of R_f, the rate of transport of each component through the column, at a constant-carrier-gas velocity will be different, and it would be predicted that they would emerge from the column at different points in time. This is what is observed in the practice of gas chromatography.

The foregoing theoretical treatment is oversimplified but serves the purpose of outlining the main sample transport mechanism through the column. A more detailed and comprehensive treatment of the rate theory may be found in many of the standard reference books on gas chromatography (see Selected Additional Reading).

1.2 Gas–Liquid Chromatography

In gas–liquid chromatography the mobile phase is the carrier gas, and the stationary phase is a liquid coated as a thin film on an inert solid support, which may either consist of particles (packed column) or the walls of a narrow-bore open tube (capillary column). The sample is distributed (partitioned) between the liquid and the gas phase. In the previous section we considered an ideal situation. In the practice of gas–liquid chromatography there are many factors which cause the observed separations to deviate from the ideal behavior.

1. Owing to column characteristics the flow of carrier gas is not uniform. In a packed column all of the flow is not directed along the axis of the column because of deflection caused by flow of the carrier gas around the particles. In addition, part of the gas stream will take a more tortuous path. This causes band spreading because portions of the sample will take longer paths and lag behind the average, while other portions will travel shorter paths and arrive ahead of the average (eddy diffusion). Because of frictional forces and curvature of the column, there will also be a gradient of the gas velocity from the walls toward the center of the column. Moreover, because of the compressibility of the carrier gas, its linear velocity increases from the column inlet to the column outlet.
2. In the gas phase the sample diffuses from a region of higher to one of lower concentration. This leads directly to broadening of the sample band.
3. At the carrier-gas velocities encountered in gas chromatography, the transfer of sample across the carrier gas–liquid stationary phase interface is perturbed, and equilibrium is seldom, if ever, obtained. There is a resistance to mass transfer across this interface. As a result some portions of the sample remain too long in the gas phase and move ahead of the sample band, while other portions of the sample are slow to diffuse out of the liquid stationary phase and will trail the average.
4. Owing to diffusion of the sample through the thin film of the liquid stationary phase and to the nonuniformity of liquid thickness, interaction occurs between the sample and the solid support. Thus, adsorption may occur in addition to partition.

Fortunately, except for the adsorption effects, these deviations from ideal behavior are of a statistical nature and give rise to a symmetrical broadening of the sample band.

A typical record (chromatogram) showing the elution of a sample component from the column, as measured with a differential detector, is shown in Fig. 2. As with distillation and other similar processes the quality of the column can be measured. This measurement is denoted by n, which is referred to as the number of theoretical plates. The value of n can be calculated from the chromatogram shown in Fig. 2 according to the relationship

$$n = (OA/DE)^2 \times 16 \tag{3}$$

In practice it is found that the value of n, is a function of the column length L under a given set of experimental conditions. Thus it is more meaningful

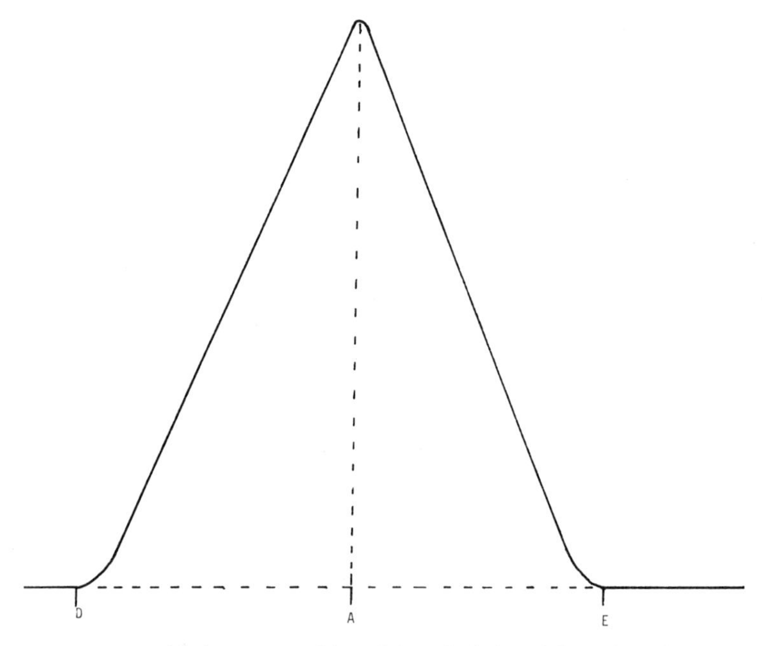

Fig. 2. Model chromatographic peak for calculation of theoretical plates.

to derive a new term as the measure of the quality of the column. This term is height equivalent to a theoretical plate (HETP), usually denoted H. It is calculated from the relationship

$$H = L/n \tag{4}$$

The quantity H, can be related to the deviations from ideal behavior previously discussed. The equation describing this relationship has been given by van Deemter *et al.* (1956), Gluekauf (1955), and Klinkenberg and Sjenitzer (1956) and is

$$H = 2\,\lambda d_{\mathrm{p}} + \frac{2\,\mu g D_{\mathrm{g}}}{u} + \frac{8}{\pi^2} \cdot \frac{k'}{(1 + k')^2} \cdot \frac{d_f^2}{D_{\mathrm{L}}} \cdot u \tag{5}$$

where

H = the HETP;
λ = correction factor ($\sim$0.5–0.2);
d_{p} = average particle diameter of the solid support;
μg = correction factor ($\sim$0.7);
u = carrier gas velocity;
D_{g} = diffusivity of the sample in the carrier gas;
k' = $kV_{\mathrm{sp}}/V_{\mathrm{cg}}$;
d_f = average film thickness of the liquid stationary phase; and
D_{L} = diffusivity of the sample in the liquid stationary phase.

The equation is often written in the simpler form

$$H = A + (B/u) + Cu \tag{6}$$

where A accounts for eddy diffusions; B/u describes the effect of molecular diffusion; and Cu, is related to the establishment of equilibrium or coefficient of mass transfer.

Other authors (Jones, 1961; Kieselback, 1960; Giddings, *et al.*, 1960) have proposed separating the term C into two terms C_L and C_g for the coefficient of mass transfer in the liquid stationary phase and carrier gas, respectively. The equation in this form is

$$H = A + (B/u) + C_L + C_g u \tag{7}$$

A more detailed account of the theory of separations using gas–liquid chromatography has been given by Giddings (1966), Tranchant (1969), and Littlewood (1970). A recent review of theoretical studies has been presented by Juvet and Cram (1970) and Juvet and Dal Nogare (1968).

1.3 Gas–Solid Chromatography

In gas–solid chromatography the mobile phase is still the carrier gas, and the stationary phase is a solid material such as an adsorbent or solid chemical agent. The solid may be packed directly into a column as particles, coated on an inert support material, or coated on the walls of a narrow-bore open tube. In gas–solid chromatography the sample is distributed between the solid and the gas phase. Rather than with the solubility of the sample in the stationary phase, as in gas–liquid chromatography, gas–solid chromatography is concerned with the direct interaction of the sample with the solid surface. This interaction may involve direct chemical reaction (bonding), electronic attraction, ionic attraction, or geometric relationships between the sample molecules and the solid. All of these factors are discussed in a general way under the heading of adsorption. The theory of gas–solid chromatography is not as well understood and developed as that of gas–liquid chromatography. Suffice it to say that in gas–solid chromatography the same deviations from ideal behavior appear as in gas–liquid chromatography (i.e., nonuniformity of carrier-gas velocity, sample diffusion, and resistance to mass transfer) and affect the quality of the chromatogram. In addition, there is the added difficulty of accounting for adsorption effects. This latter factor is not statistical, as in gas–liquid chromatography, but depends upon the adsorption isotherm for the sample on a given solid. The net result consists of chromatographic peaks that are, as a rule, not symmetrical.

Giddings (1964) has shown that the HETP for gas–solid columns may be calculated provided that a factor C_k, related to adsorption/desorption effects are substituted in the Van Deemter equation in place of the terms for coefficient of mass transfer in the liquid stationary phase. Also, instead of writing an equation for the mean value of H, he integrates H at a point in the column. The resulting equation is given by:

$$H = \left[\frac{1}{A_i} + \frac{1}{C_{gi}} u\right]^{-1} + \frac{B}{u} + C_k u \tag{8}$$

where i is the simple plate being considered.

The dynamics of gas–solid chromatography is treated in detail by Giddings (1966). Kiselev and Yashin (1969) discuss both the theory and practice of gas–solid chromatography.

2 Gas Chromatography Apparatus

2.1 Basic System

The choice of components for a gas chromatograph is large and depends primarily upon the particular analysis being performed. *Science* (1971–1972) lists approximately 50 manufacturers of commercial gas chromatographs marketed in the United States. Compared to that of many analytical instruments, the construction of gas chromatographs is simple and inexpensive. Regardless of origin or manufacturer, all gas chromatographs contain the same basic elements. Figure 3 shows a schematic block diagram of a gas-chromatograph instrument. It contains a source of carrier gas that serves to transport the sample through the instrument. The flow of carrier gas is controlled by a suitable gas regulator. The sample inlet is that part of the instrument that accepts the sample, converts it to a gas or vapor, and injects it into the flowing carrier-gas stream at the inlet to the column. The column performs the separation of the sample into individual components as it it is carried along the length of the column by the flowing carrier-gas stream. The detector monitors the concentration of the sample components as they

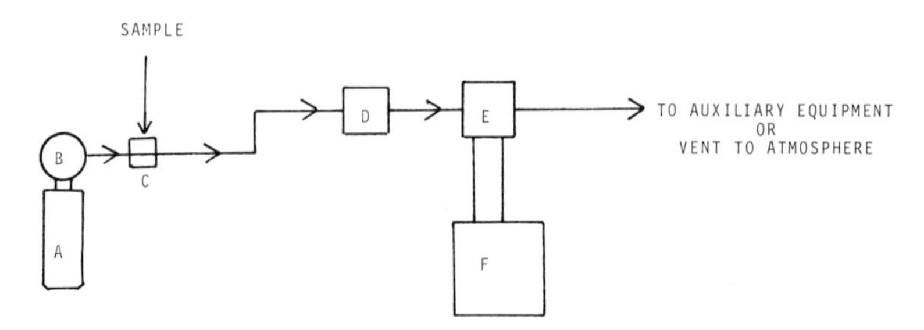

Fig. 3. Schematic diagram of gas-chromatograph instrument: A, carrier gas source; B, carrier gas regulator; C, sample inlet; D, chromatograph column; E, detector; F, data system.

emerge from the column. The data system provides a record of the response or signal from the detector. After exiting the detector, the sample components may be vented to the atmosphere, collected for further analysis, or presented directly to complementary analytical instrumentation, such as an infrared spectrophotometer or mass spectrometer. The sample inlet, column, and detector are generally housed in an oven to allow vaporization of the sample in the inlet, eliminate sample condensation in the column and detector, and provide optimum column temperature for the desired separation.

2.2 Carrier Gas

The choice and flow rate of the carrier gas is the primary factor in the determination of the retention times (analysis times) of sample components with any given chromatographic column. The proper selection is dictated by several considerations. Foremost among the considerations is which detector is being used in the system. Thermal-conductivity detectors require the use of the lighter gases H_2 or He for optimum sensitivity and performance, although N_2 is also used satisfactorily. Ionization detectors, exclusive of the flame ionization detector, require one of the rare gases, He, Ne, Ar, Kr, for proper operation. The flame ionization detector may be used with H_2 as the carrier gas.

The common carrier gases are H_2, He, N_2, Ar, CO_2, and various combinations of these. Other gases have been used for specialized operations. Nelson and Lysyj (1969) have reported the use of steam as a carrier gas with a hydrogen flame ionization detector. Tsuda *et al.* (1969) have used the vapors of organic solvents as the carrier gas.

Carrier gases that react with the sample should be avoided for general applications. However, in certain instances a reactive carrier gas may be used to advantage. Zlatkis *et al.* (1960) employed hydrogen as the carrier gas to reduce aldehydes to methane and water, thus simplifying calibration.

In gas–solid chromatography care must be taken not to select a carrier gas with a polarity that allows competition with the sample components for the active sites of the adsorbent. Hoffman and Evans (1966) have studied the effect of carrier gas on retention time, peak width, peak height, and column efficiency in gas–solid chromatography.

Other factors to be regarded in the selection of carrier gas are the resistance to mass transfer and diffusion of the sample in the gas phase. The ligher gases allow greater diffusion of the sample in the gas phase, thus reducing column efficiency. To reduce this undesirable effect, the heavier gases are preferred. Perrett (1965) has shown that diffusion in the liquid stationary phase and resistance to mass transfer are relatively independent of the carrier gas used.

In general applications the carrier gas chosen should be dry and free from impurities. Prescott and Wise (1966) discuss the need for clean gases in gas-chromatographic systems. Depending upon the detection sensitivity required, the allowable impurities range from a fraction of 1% to a fraction of a part per billion for the most critical applications. In addition to other undesirable effects, water in the carrier-gas stream often leads to a rapid degradation of the chromatograph column, particularly at elevated temperatures. To achieve high carrier-gas purity, desiccants, cryogenic traps, precolumn reactors, and catalytic chambers are sometimes employed in the flowing carrier-gas stream.

In most gas-chromatograph instruments the carrier gas is supplied from high-pressure cylinders. Pressure regulation is provided by diaphragm valves operating in the pressure regime desired. The most stable gas-chromatographic systems employ very precise pressure controllers in addition to the cylinder regulators. These devices are placed between the high-pressure cylinder regulator and the sample inlet. In instances where H_2 is used as a carrier gas, commercial electrochemical hydrogen generators are available for supplying regulated quantities of electrolytically pure H_2. These devices have found wide general acceptance.

The flow rate of the carrier gas through the column affects the degree of separation of sample components and the analysis time. For each analysis there is an optimum flow rate, determined by factors previously discussed (see Section 1). The measurement of the flow rate of carrier gas is made at the column or detector outlet. The apparatus for measurement of the flow rate may be a rotometer: a simple manometer, where the measured pressure differential is a function of flow rate; a soap bubble flow meter, which measures the movement of a soap film in a calibrated volume; or other electromechanical devices that measure some physical property of the carrier gas, which in turn is related to flow rate.

Schwartz (1967) and Krugers (1968) have reviewed the effects on column performance of the various carrier gases.

2.3 Inlet Systems

2.3.1 General

The ideal sample inlet system for a gas-chromatograph instrument should allow placement of the sample, in its natural form, onto the first few plates of the gas-chromatograph column, where it is instantaneously vaporized or volatilized. Unfortunately, such an ideal system with universal application has yet to be devised. Thus the analyst must select an inlet system based upon the closest approximation to the ideal situation. The sample type and desired analysis predetermine, in most instances, the sample inlet system used.

It is common practice to operate the inlet system at temperatures slightly above the boiling point of the least volatile component in the sample mixture to assure rapid and efficient vaporization of the materials to be analyzed. Caution must be taken, however, not to exceed the temperature that would cause thermal degradation of the sample.

In the selection or design of an inlet system, large volumes and nonuniform flow paths should be avoided between the point of sample admission and the gas-chromatograph column. These factors contribute to exponential dilution of the sample with carrier gas and diffusion of the sample band, thereby degrading the system performance. Another factor often overlooked is that the flow of carrier gas through the sample inlet system must allow for a very rapid mass transfer of sample from the inlet system to the chromatographic column. Thus, if the sample inlet were uniform in path and contained a volume of only 100 μl, a flow of carrier gas of 100 μl/min would require a minimum of 1 min to transfer the sample to the chromatograph column. Therefore, the minimum peak width for a sample component would be 1 min, which is unsatisfactory. Transfer times of no more than several seconds are required.

An inlet system should be constructed of materials that minimize sample sorption and do not react chemically with the sample.

Because of the increasing sensitivity of detectors and the use of small-diameter packed and capillary columns, with their limited sample capacities, there is often a requirement to inject ever smaller quantities of sample into the instrument. For samples of less than $\frac{1}{10}$ to $\frac{1}{100}$ μl, it is required that the sample be subdivided in the inlet system, prior to being admitted into the column. In this mode only a small fraction of the sample placed in the sample inlet is allowed to proceed to the column, the remainder being vented to the atmosphere or collected for subsequent use. Halasz *et al.* (1970) have recently described a stream splitter which compensates for mass fractionation of the sample components. Dubsky (1970) has reported on a disk-type splitter with a split ratio of up to 10^3:1 and reproducible to 1%. Ettre and Averill (1961) have also described systems capable of providing this splitting of the sample.

Figure 4 shows a schematic diagram of the salient features of a standard inlet system that utilizes syringe injection. The sample is admitted through a resealable plastic septum where it is vaporized (if the sample is a liquid or solid) and some portion of it swept into the chromatographic column.

Ashley (1967) describes the relationship between sample input profile and peak shape. Cramers (1967) has discussed the influence of sample injection on the accuracy and resolution of the chromatographic system. The practical aspects of sampling and inlet systems have been reviewed by Condon and Ettre (1968) and Hamilton (1968). A good review of sample inlet requirements is also given by Byrnes (1972).

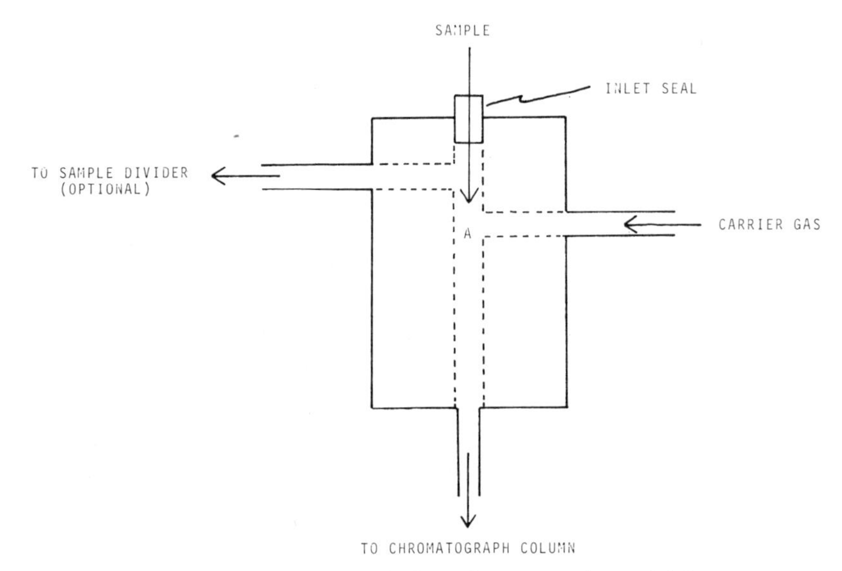

Fig. 4. Schematic diagram of standard gas-chromatograph sample inlet system.

2.3.2 Techniques

a. Syringe. The most widely used method of admitting either a liquid or gas sample is by the use of a special hypodermic-type syringe. The syringe must be inert toward the sample and capable of storage and injection of the sample without leakage. Such syringes are available from a number of commercial suppliers. The syringe needle penetrates the inlet seal (refer to Fig. 4), and the sample is injected into the flowing carrier-gas stream. Syringe injection can also be used for solid samples provided that the sample is first dissolved in a solvent that will not interfere with the analysis being transformed. The inlet system may be designed to allow placement of the tip of the syringe needle directly onto the chromatograph column. Liquid syringes are available with total capacities of 1 μl or greater. Gas-tight syringes with a volume of 0.05 ml or greater may be obtained commercially. With a little practice syringe injection can be performed with a precision of about 1%. Factors affecting the precision of syringe injections have been discussed by Pitt (1969). The problems associated with injection ports using polymer septums are reviewed by McKinney and Sheppard (1968), Tucknott and Williams (1969), and Callery (1970).

b. Solids Probe. An analogous technique to syringe injection for solid samples is the solids probe. The solid sample is melted and the probe inserted into the melt. The probe fills by capillary action. When withdrawn from the melt, the sample in the probe again solidifies. The probe is then inserted into the inlet system in a manner similar to syringe injection. The inlet temperature must be sufficiently high to vaporize or sublime the sample into the flowing carrier-gas stream. Yannone (1968) has described a solids injector for gas chromatography. Barakat (1967) and Roberts (1968) have also described solid samplers, pointing out their usefulness for samples that leave a solid residue.

c. Sample Valve. Sample valves are continuously being developed for use in gas chromatography. They are used for both liquid and gas samples, although their primary application is with the latter. They have the potential of very high precision and have proved to be extremely versatile. As with other sample inlet systems, the sample valve must be chemically inert toward the sample and leak tight.

The two basic sample valve designs are the rotating disk and the sliding plug. A typical rotating-disk-type sample valve is shown schematically in Fig. 5. In position (a) the carrier gas is directed from the carrier-gas source

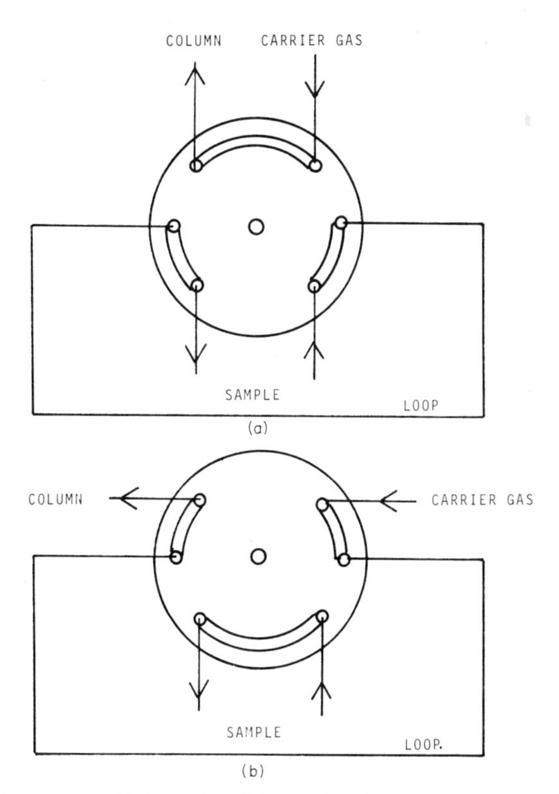

Fig. 5. Diagram of typical rotating disk sample valve: (a) normal operation; (b) sample inject position.

directly to the chromatographic column. Simultaneously the sample loop is filled with the sample to be analyzed. In position (b) the carrier gas is diverted through the sample loop containing the sample. The flowing carrier-gas stream displaces the sample from the loop and transfers it onto the chromatographic column. From this simple beginning valves that contain up to 16 separate ports for performing functions such as sample injection, column switching, and backflushing of the chromatograph column have now become commercially available.

Harris (1968) has described a sliding-plug-type sample valve, whereas Karas (1967) has been granted a patent on a rotating-disk-type sample valve. Both practical and quantitative aspects of valve sampling have been given by Meyers and Rosman (1969), Dodd (1971), and Todd and Courneya (1968). Almost all commercial gas-chromatograph manufacturers supply a complete line of sample valves that operate from subambient pressures to pressures of several atmospheres and at temperatures up to approximately 250°C. Oberholtzer and Rogers (1969) have critically evaluated the performance of a number of commercially available sample valves.

d. OTHER SAMPLE INLETS. In addition to the most common types of sample inlets discussed above, others have been reported for special applications.

When a detector that is insensitive to changes in carrier-gas flow rate is used, it is possible to interrupt the carrier-gas flow, open the system, and place a liquid or solid sample directly on the column. Although employed in the past, this technique is little used today because of the inconvenience and loss of system performance that are usually associated with it.

Another means of introducing liquid samples, once quite popular but again no longer widely used, is the pipette. In this sample inlet system the liquid sample is taken into a capillary. The capillary is placed in a gas lock that is part of the inlet, and the internal volume of the gas lock is swept with carrier gas. The gas lock is then opened to the injector block and the capillary inserted into the heated zone of the injector block. The sample is forced out of the capillary by a combination of its vaporization and a pressure differential between the lock and the injector system. As Amy and Baitenger (1963) have pointed out, this system is somewhat awkward to use and is no more precise than syringe injection.

Another injection technique of growing use is sample encapsulation. In this technique the sample to be analyzed is sealed in either a fusible metal or glass enclosure. The sealed enclosure is placed in the flowing carrier-gas stream, and the system is allowed to equilibrate. The sample is released by melting or breaking the encapsulating material. Kim *et al.* (1970) and Dunlop and Pollard (1971) have described the use of samples encapsulated with a fusible metal in gas chromatography. Back *et al.* (1969) discuss a simple device for injecting a sample from a sealed glass tube into a gas chromatograph.

A precolumn has also been found to be a convenient means of introducing samples into a gas chromatograph. In this technique the sample is placed directly on a short section of packed column either by direct means, by adsorption of the sample on a suitable material, or by trapping the sample cryogenically on the precolumn. The precolumn is then placed in series with the chromatograph column and the sample displaced, usually by heating the precolumn. Cramers and Van Kessel (1968) have described such a technique for use with high-resolution capillary columns. Other applications have been given by Herout *et al.* (1970), Aue and Teli (1971), Llewellyn (1970), and Darbre and Islam (1970).

A technique analogous to the precolumn inlet is condensing the sample in a suitable trap. The trap is placed in the flowing carrier-gas stream and heated to release the sample rapidly into the chromatographic column. Gronendyk and Van Kemenade (1969) and Willis (1968) have described this technique.

Another sample inlet system that has found particular application in the characterization of polymers is the pyrolysis apparatus. By means of this apparatus the nature of the sample is deduced from the products of its thermal degradation. In the pyrolysis sample inlet the sample is either coated on a filament or placed in a small receptacle, which is then placed into the flowing carrier-gas stream before it enters the chromatograph column. The sample is then decomposed by heating, either directly by using resistance heating of the filament or receptacle or indirectly by means of an auxiliary heater. The thermal fragments are moved from the sample inlet onto the chromatograph column. The principles, techniques, and applications of pyrolysis have been reviewed by Wolf *et al.* (1971). Walker and Wolf (1970a,b) reported on a comparative study of three different pyrolysis techniques. In addition to thermal decomposition, pyrolysis inlet systems that use laser energy as well as photolysis are being developed. Barker and Purnell (1970) discuss photolysis, while Folmer (1971) and Guran *et al.* (1970) describe work with laser pyrolysis systems.

In addition to the foregoing, there are many more sample inlet systems that are unique and offer solutions to specific sample inlet problems. Novak *et al.* (1971) discusses sampling from high-pressure systems, while Goltz and Moffat (1970) are concerned with sampling from vacuum systems. Much work has also been done on sampling of head-space gases, on-stream sampling, and automatic, repeating sample inlet systems.

2.4 Columns

As previously noted, the theory of gas chromatography is essentially the theory of the operation of the gas-chromatograph column. Indeed, the chromatograph column is the very heart of any gas-chromatograph system or instrument. Although other elements of the system affect its performance, it is the column that ultimately determines the success or failure of the analysis. There are generally two types of chromatograph columns: packed and open tubular.

2.4.1 Packed Columns

A suitable packed column may be constructed of metal, glass, or plastic tubing. Common materials of construction are stainless steel; copper; aluminum; glass; Teflon; and aluminum, the inner surface of which has been coated with Teflon. The plastic column materials are of limited use because of their generally lower maximum operating temperatures and the problem of air diffusion through the plastic. The latter can contribute to deactivation or decomposition of the column and may create a particular problem when used with very sensitive detectors. The column tubing must obviously be stable at the conditions under which the analysis is to be performed. The column tubing should be clean and contain, other than the packing, no foreign material that would interact with the sample. Examples of impurities are oils and other chemicals used in the manufacture and extrusion of metal tubing as well as oxide films that result from exposure of the column tubing to the atmosphere. As with the sample inlet, the tubing selected for column construction should be inert chemically toward the sample and have no catalytic effect on any of the sample components. Levins and Ottenstein (1967) have noted adsorption losses of samples of polyols and vanillins when aluminum or stainless steel tubing were used for their columns. When an attempt is made to analyze very polar compounds such as alcohols, amines, and many compounds of biological origin, it becomes necessary to use a special deactivation procedure to make the inner column tube surface passive toward the sample. An example is steroid analysis using glass columns, in which the inner glass surface has been deactivated by replacing the free hydroxyl groups by silyl groups. This process is discussed in detail by Prevot (1969, p. 99) under the topic of treatment of solid supports. A good general rule to follow is that the more chemically active the sample constituents, the more inert the column tubing must be.

The length of a packed column may vary between several inches and tens of feet. The more common lengths fall between 3 and 12 ft. As the column becomes longer, the pressure drop along its length becomes significant causing a nonuniformity of carrier-gas velocity, which leads to a decrease in column performance.

The internal diameter of a packed column may vary from about 0.01 in. for micropacked columns to greater than 0.5 in. for preparative scale columns. The most widely used internal diameters seems to range from 0.06 to 0.12 in., with a tubing wall thickness averaging 0.02 in.

The column tubing is fabricated in a geometry suitable for use in the particular system employed, and its shape is largely dictated by the innermost dimensions of the column oven and placement of the carrier gas source, sample inlet, and detector. Typical configurations are a coiled helix, U shape, or W shape. A practical rule is that the shape of the column has little effect upon its performance, provided that the radius of any bend is at least ten times the internal diameter of the tubing. Thus a 0.0625-in. column can be coiled in a helix with a radius of 1.25 in. with no notable degradation in performance.

The finished chromatograph column is equipped with gas-tight fittings that allow it to be connected between the sample inlet and detector.

Packed columns may be subdivided into two types; adsorption or partition, depending upon whether the sample is separated into its constituents by an adsorption interaction with the solid packing or by a partition process with a liquid film contained on the solid packing.

a. Packed Adsorption Columns. For this discussion packed adsorption columns will include common adsorbents, porous polymers, and any solid support material in which the separation of the sample components is primarily a function of the solid packing. The placement of porous polymers in this category is questionable, since the separations that they afford may well be a combination of adsorption and partition, which would place them in a special class. However, they are included in this section as a matter of convenience and because their use and preparation more nearly resemble the packed adsorption columns than the packed partition columns.

Adsorption columns find their major application in the analysis of gases and low molecular weight or highly volatile samples. A schematic cross-sectional view of an adsorption column is shown in Fig. 6a.

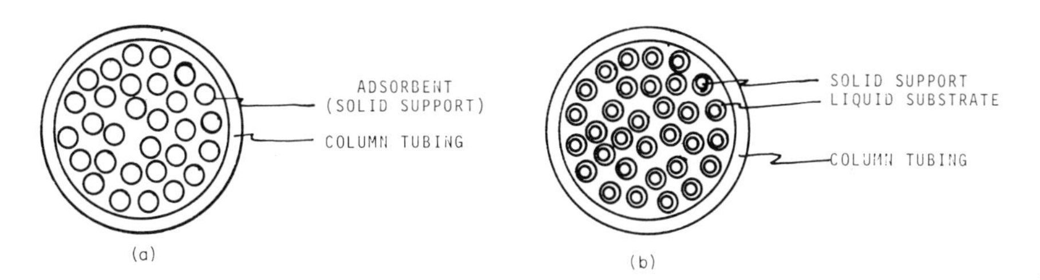

Fig. 6. Schematic cross-sectional view of packed chromatograph column: (a) packed adsorption column; (b) packed partition column.

The choice of adsorbents for use in gas–solid chromatography is varied, the selection depending upon the particular analysis desired. A list of adsorbents that have been used includes alumina, silica gel or silica beads, inorganic salts, molecular sieves (zeolites), porous glass, metal oxides, charcoal, carbon black, and carbon molecular sieves. In addition to adsorbents that are used alone, there is an increasing use of adsorbents whose surfaces have been modified especially to affect their adsorptive properties. Such modification is usually accomplished by addition of a material to the adsorbent to bind the more active sites. In addition, adsorbents have been coated on the surface of solid supports in a manner very similar to the preparation and use of partition columns.

List *et al.* (1967) have reported on the use of activated alumina for the separation of cis/trans isomers of olefins. Samarkina and Vereschagina (1970) have used modified alumina columns for the determination of C_1–C_5 hydrocarbons. Brookman and Sawyer (1968d) have published a series of articles dealing with modified alumina columns.

Grob and McGaugh (1971) and Onuska and Janak (1968) have described separations using inorganic salts as the adsorbent.

Brookman and Sawyer (1968b, 1968c) have reported on salt-modified silica beads as a gas–solid chromatographic column material.

Brunnock and Luke (1968), Dietz (1968), Parkinson and Wilson (1968), and many others have reported on the use of molecular sieve adsorbents. Brunnock and Luke (1969) have given details of the separation of napthenes and paraffins using both untreated molecular sieve and molecular sieve treated with alkali.

Frolov (1968) has used porous glass as an adsorbent for the separation of light hydrocarbons.

A discussion of the influence of surface chemistry on the separation obtained with silica gel has been given by Lebedeva *et al.* (1971), while Seide *et al.* (1970) discuss the use of modified silica gels.

Carbon molecular sieves are relatively new. Their characteristics and use are given by Kaiser (1970).

DiCorcia and Bruner (1970), and Kouznetsov and Scherbakova (1970) have reported on graphitized carbon black as an adsorbent for gas chromatography.

An in-depth review of the use of adsorbents for gas chromatography has been presented by Kiselov (1970). Kiselev and Yashin (1969) have prepared a complete volume covering the theory and use of gas adsorption chromatography.

One of the most widely used packing materials in the adsorbent category are the porous polymer beads. Their unique characteristics make them adaptable to the solution of many analytical problems. There are approximately 20 different varieties of porous polymer beads available, each having a slightly different property that allows a wide range of chromatographic separations to be performed. Supina and Rose (1969) and Dave (1969) have given reviews of the properties and use of the porous polymer beads.

A new adsorbent material, similar in nature to the porous polymer beads, is a polyurethane foam formed *in situ* in the column as reported by Ross and Jefferson (1970).

b. PACKED PARTITION COLUMNS. Packed partition columns include those in which the separation of the sample components is provided by a thin liquid film coated on an inert solid support. Fig. 6b shows a schematic cross-sectional view of a packed partition column. It consists of the column tube, the inert solid support, and the liquid stationary phase.

i. Solid Support. There are a very large number of materials available for use as the solid support in gas chromatography. The choice should be guided by the following criteria.

The solid support should (a) be mechanically and thermally stable, (b) have a high surface area and porosity, (c) not have strong adsorptive properties, and (d) be chemically and physically inert.

Some of the more common solid supports are the diatomaceous earths, glass beads, and Teflon. These materials and their modifications are marketed under a variety of trade names.

The diatomaceous earth materials are prepared from sedimentary formations of silicous shells of diatoms. The natural material is mined, crushed, compressed, and calcined (either with or without a flux). The type of flux used and the time and temperature of the calcining process determine the physical properties of the final material. The mechanical stability of this type of material is moderate. It is thermally stable up to the maximum operating temperature of most gas-chromatographic separations (400–500°C). The specific surface area for the diatomaceous supports varies between 1–10 m^2/g. The diatomaceous supports are not completely inert, and most exhibit adsorptive properties that may limit their application for many analyses.

The diatomaceous supports may also contain trace elements that interfere with certain analyses. To overcome this problem and that of surface adsorption, special treatment procedures have been developed to compensate partially for this deficiency. Trace elements can be removed from the support surface by washing the material with acid. To maintain chemical inertness, the solid support may be treated with bases to eliminate residual acidic properties from the acid treatment. Normally the packing material is washed with water and dried to eliminate residual base.

To eliminate or reduce the adsorptive properties of the diatomaceous support, it has become common practice to treat the material in order to bind surface adsorptive sites. It appears that the major causes of adsorption are surface hydroxyl groups on the silicous material. These are removed by treating the material with dimethyl dichlorosilane (DMCS) or hexamethyl disilazane (HMDS). Such treatment replaces the hydroxyl groups with silyl or trimethyl silyl groups which do not exhibit the adsorptive properties. Diatomaceous solid supports have also been coated with stable organic chemicals such as Teflon and polyvinyl pyrrolidone to cover surface adsorptive sites.

Depending upon the type of material used, from 1 to 30% of the weight of diatomaceous solid supports may be coated with a liquid substrate.

The thermal and mechanical stability of glass beads is good. Their specific surface area is less than 1 m^2/g because of their limited porosity. Recently porous glass beads have become available with increased surface area while still retaining thermal and mechanical stability. Glass beads are inherently more inert than the diatomaceous supports. However, their residual adsorptive properties may be treated as above with DMCS or HMDS to reduce this characteristic further. Because of their generally lower surface area, the amount of liquid substrate that may be coated on glass beads is less than 1% by weight of the solid support. The new porous glass beads may accept up to 10% of their weight in liquid substrate.

Teflon is one of the most inert solid supports available. Its mechanical stability is good, but because of static electrical charge on its surface, it has a tendency to agglomerate and become difficult to use. The maximum safe operating temperature for Teflon solid support is 200°C. Its specific surface area is 0.5–1.0 m^2/g. Up to approximately 10% by weight of liquid substrate may be coated on a Teflon solid support.

Other materials that have been used as solid supports for packed partition columns are stainless steel powder, metal fibers, quartz, dendritic salts, vermiculite, alumina, silica, organoclays, metal helices, ceramics, and powdered detergents.

When used as a solid support for gas chromatography, the materials are ground and then screened to yield a relatively uniform mesh size. The narrower the range of particle size of the solid support, the higher is the performance of the column owing to more uniform liquid substrate distribution and uniformity of carrier-gas velocity in the column. Common mesh sizes employed are 45–60 mesh, 60–80 mesh, 80–100 mesh, and 100–120 mesh. For specialized applications other mesh sizes may be used.

A review of the choice of solid supports has been given by Palframan and Walker (1967). Bryzgalova *et al.* (1968) have reported on a study of physical properties of several support materials. Various washing, heat treatment, and silanization methods have been discussed by Viska *et al.* (1970) and Blandenet and Robin (1966). A very good practical discussion on the selection and treatment of solid supports is also given by Tranchant (1969).

ii. Liquid Phase. The choice of liquid phases for use in gas chromatography is immense. A survey of one manufacturer's catalog has revealed over 200 choices of liquid phases, and these are only the ones in routine use. The choice of liquid substrates should be guided by the following criteria.

The stationary phase should (a) not react with the components of the sample, (b) coat easily and uniformily on the solid support, (c) be chemically pure since minor impurities may also react with the sample constituents, (d) be stable at the conditions of analysis, and (e) be essentially nonvolatile at the maximum temperature of the analysis.

The selection of stationary phase is based upon the analysis to be performed. In general, the more polar the sample constituents, the more polar the substrate should be. This is analogous to the old chemical rule of "like dissolves like" in the selection of solvents. Unfortunately, the selection of stationary phase is still made on empirical basis in many instances.

Several suggestions have been made for classifying the various common liquid substrates according to their polarity. Rohrschneider (1959) has proposed a classification based upon increments of Kovats' indices. Chovin and Lebbe (1961) and Brown (1962) have suggested different classifications for the liquid substrates. Each of the classification schemes is based upon the exhibited polarity of the liquid phase.

As mentioned above, the liquid substrate should not have a significant vapor pressure at the maximum temperature of operation. Similarly, a substrate should be a thin liquid film during the analysis. Therefore its lower temperature of operation should be above its crystallization point. Most manufacturers' catalogs give the normal operating temperatures, as well as the solvents, for the liquid substrates that they provide.

In the event that no single liquid phase is suitable for a particular analysis, it is possible to mix two or more phases to achieve the desired analytical separation. Many workers in the field have used and reported on this technique.

Preston (1970) has called attention to the use of a wide variety of liquid phases and suggests the selection of a limited number of preferred substances for specific analyses.

c. Preparation. A detailed discussion of the various methods of preparing packing chromatographic columns is beyond the scope of this text. For adsorption columns the procedure consists of

(1) selection of column length and diameter;
(2) selection of adsorbent;

(3) grinding and screening of adsorbent, if required;
(4) packing the column tube with the adsorbent;
(5) fabricating the column into the proper configuration for use;
(6) installation of column fittings;
(7) installation of the column in the instrument, and
(8) activation or conditioning of the column for analytical use.

For packed partition columns the procedure includes

(1) selection of column length and diameter;
(2) selection of solid support;
(3) grinding and screening of solid support, if required;
(4) selection of type and percentage of liquid substrate;
(5) preparing the packing by making a slurry of the solvent, solid support, and liquid substrate;
(6) removal of the solvent from the prepared packing;
(7) packing the column tube;
(8) configuring the column for use;
(9) installation of the column fittings;
(10) installation of the column into the instrument; and
(11) conditioning of the column prior to analytical use.

Smith *et al.* (1964) have performed a statistical evaluation of some important parameters in the preparation of gas-chromatographic packings. Many authors have reported techniques and apparatus for packing column tubes. More detailed discussions of the steps in the preparation of packed columns is given by Supina (1963) and Prevot (1969, p. 86).

Another new development worthy of note is the chemical bonding of the liquid substrate to the solid support. This technique has been reported by Waters Associates (1972) and Kirkland and DeStefano (1970).

d. PERFORMANCE. Factors such as quality of the column, sample size, and conditions of operation have a definite effect upon the performance of a packed gas-chromatograph column. For a specific analysis, the column will have an optimum set of operating conditions and an ideal sample size. These conditions may be estimated from theoretical considerations but are more commonly determined experimentally. The range of sample size for packed columns is a maximum of several milligrams for the small-diameter packed columns up to several grams for the large-diameter, preparative scale columns. The efficiency of packed columns may range from 1000 to 2000 theoretical plates for preparative scale columns to 50,000 theoretical plates and higher for the narrow-diameter packed columns. Any standard text on gas chromatography will include a discussion of the effect of sample size and operating conditions upon column performance.

2.4.2 Open Tubular Columns

An open tubular (capillary) column may be constructed of metal, glass, or plastic. The most common materials of construction are stainless steel and glass. The same selection criteria apply to open tubular column materials as apply to packed columns, i.e., chemical inertness, stability, etc. In fact, these criteria become even more important in the selection of tubing for an open tubular column.

The length of a capillary column may vary from several feet to several hundred feet. The more common lengths fall between 50 and 200 ft.

The internal diameter of a capillary column may vary from slightly less than 0.01 in. to 0.05 in. with a wall thickness of 0.006 to 0.02 in.

In use, the column tubing is almost always configured in a coil, the radius of which is determined by the dimensions of the column oven in the particular instrument being used.

As with packed columns, the finished open tubular column is equipped with gas-tight fittings that allow it to be installed into the instrument. In the selection of fittings for capillary columns, care must be taken to minimize the volume of the fitting. Since the capillary has a small internal volume, a large fitting volume would lead to sample diffusion and a serious degradation of the column performance.

Open tubular columns may be subdivided into two categories: (a) adsorbent and (b) partition columns.

a. OPEN TUBULAR ADSORPTION COLUMNS. Open tubular or capillary adsorption columns consist of two basic types: those in which the inner surface of the tubing has been modified to form an adsorptive surface, and those in which an adsorbent has been coated on the inner surface of the capillary tube.

Mohnke and Saffert (1962) have reported on an etching technique for glass capillaries that leaves an exposed silica surface. Such a column was used to separate hydrogen isotopes and other isomers. Petitjean and Leftault (1963) found that by treating the inner surface of a metal capillary, an oxide film was formed that was useful as a gas–solid adsorbent for capillary chromatography.

There are many examples of capillary columns that have had their inner wall surface coated with adsorbent. Schwartz *et al.* (1963) successfully coated the inner surface of metal and plastic capillaries with silica from a colloidal suspension. Halasz and Horvath (1963) prepared capillary columns containing thin wall coatings of carbon black, alumina, silicon carbide, and diatomaceous earth materials.

b. OPEN TUBULAR PARTITION COLUMNS. Capillary partition columns are of three basic types: (1) wall-coated capillary columns, i.e., those in which the liquid phase is coated directly on the clean capillary inner surface; (2) surface-modified capillary columns, i.e., those in which the inner surface of the capillary is modified in some manner prior to being coated with the liquid phase; and (3) support-coated open tubular (SCOT) or porous layer open tubular (PLOT) columns, i.e., those in which the inner surface of the capillary tube is coated with a solid support and the solid support coated with the liquid stationary phase.

Golay (1958) first suggested the use of open tubular columns after considering the theory of chromatographic separations. When the suggestion was reduced to practice, gas chromatography received one of the greatest improvements in its history. The literature is rich with references to the use of capillary columns for solving the most difficult analytical problems. Most of the applications reported use wall-coated capillary columns.

One problem associated with capillary columns was the limited amount of liquid substrate that could be coated on the inner capillary wall. The first attempt at solving this problem was to treat the inner capillary surface in order to increase its surface area prior to application of the liquid phase. Bruner and Cartoni (1964) etched glass capillaries with alkali to increase the internal surface area and studied the effect of increased amounts of liquid phase on column performance. Acid etching of the column surface has been used on both glass and metal capillary columns. Zlatkis and Walker (1963) used a dichromate treatment to modify the inner surface of capillary columns prior to coating.

The modifications that eventually overcame the limited liquid phase that could be applied to capillary columns were support-coated open tubular columns (SCOT) and porous layer open tubular columns (PLOT). These techniques were originally described by Golay (1960) and reduced to practice by Halasz and Horvath (1963). The SCOT and PLOT columns seem to offer the best attributes of both packed and wall-coated capillary columns.

A recent innovation in capillary columns is the "sandwich" capillary column reported by Liberti *et al.* (1968). In this type of column the fractionating medium is a graphite thread inside the capillary tube.

c. Preparation. Wall-coated open tubular columns are commonly pre-prepared by passing a solution of the liquid phase through the clean capillary tubing, using gas pressure as the driving force. Levy *et al.* (1968) have described a technique that allows for uniform velocity of the solution during the coating process. Novotny *et al.* (1969) have studied the effects of column radius and coating rate on liquid film thickness. A technique for cleaning open tubular columns has been described by Lavoue (1968).

Bruner and Cartoni (1964) and Zlatkis and Walker (1963) discuss the preparation of surface-modified capillary columns.

Kaiser (1968) has reported a simple procedure for the preparation of SCOT columns, while Cronin (1970) discusses the preparation of PLOT columns.

d. PERFORMANCE. The performance of an open tubular gas-chromatograph column depends upon the quality of the column, its length, sample size, and conditions of operation, i.e., flow rate and temperature. Wall-coated capillary columns will accommodate sample sizes up to several milligrams under optimum conditions. SCOT and PLOT columns have sample capacities of from 3 to 5 times greater than this. The efficiency of open tubular columns is greater than packed columns. It is customary to have capillary columns demonstrate an efficiency of 50,000 to 100,000 theoretical plates, with values of 1,000,000 theoretical plates having been reported.

For a complete review of open tubular columns, the reader is referred to the excellent reviews by Ettre (1965) and Desty (1967).

2.4.3 Combination Columns

Often it is an advantage to utilize the most desirable characteristics of more than one type of column in a single analysis. It is possible to combine two or more packed columns in a series or parallel arrangement, as well as to combine packed and capillary columns in a similar manner to perform the required analysis. The use of multiple columns and sample switching has found particular favor in process analyses, where speed of analysis is a requirement. Taylor and Shoemake (1972), Teranishi (1967), and Walker and Wolf (1970a) have described the use of combination packed and capillary columns. A review of multiple columns with column switching has been given by Chizhkov (1970).

2.4.4 Column Operation

For both packed and capillary columns there are several modes of operation, depending upon the nature of the sample and the speed required for the analysis. These modes of operation are (a) isothermal and isobaric; (b) isothermal with pressure (flow) programming; (c) isobaric with temperature programming, and (d) temperature and pressure (flow) programming.

Isothermal (constant temperature) and isobaric (constant pressure) operations of the chromatographic column are employed when the sample constituents are similar in nature, their boiling point range is narrow, and/or the speed of analysis is not an important factor. Under these circumstances it is possible to select one set of column operating conditions that is near optimum for the analysis of each component in the sample. This method of column operation is the one chosen in most analytical applications because the required instrumentation is less complex and interpretation of the resulting data is simplified.

For samples with a wide boiling range, it is possible to operate the chromatograph column at a predetermined temperature and to change the linear velocity of the carrier gas during the analysis by increasing the carrier-gas pressure at the column inlet. Thus for low-boiling components of the sample, the carrier-gas flow rate is slow to completely resolve the compounds of interest, whereas the carrier-gas flow rate is increased during the analysis to displace higher boiling sample components in a reasonable time and with greater efficiency. Ettre (1969) has reviewed the advantages of pressure programming.

Temperature programming is the most widely used method for the analysis of samples whose components exhibit a wide boiling range. In this technique the sample is injected into the chromatograph and the column temperature increased during the analysis in either a linear or nonlinear manner. The lower initial operating temperature allows adequate separation of the low-molecular-weight sample components, while the increased temperatures allow the higher molecular-weight components to be eluted from the column efficiently and within a reasonable time. A review of the theory of temperature programming is given by Myakishev (1967). A practical review of temperature programming is given by Tranchant (1969) and Harris and Habgood (1966).

A new innovation in programmed temperature column operation has been reported by Coudert *et al.* (1971). This technique involves establishing a temperature gradient along the column length in combination with temperature programming of the column during analysis.

Simultaneous pressure and temperature programming has not received wide application in analytical operation, although it is accepted practice to use a pneumatic controller for the carrier gas when performing temperature programming to keep the carrier-gas velocity constant through the column during the analysis.

2.5 Detectors

2.5.1 General

The function of the gas-chromatograph detector is to sense the compounds as they emerge from the chromatograph column and provide a signal that is a measure of the amount of each compound present. The characteristics of a good gas-chromatograph detector are the following.

(a) It should provide a rapid response so that its signal is an instantaneous profile of the components emerging from the column.

(b) It should be sensitive to the compounds being analyzed.

(c) Its response should be linear to facilitate quantitative interpretation of the data.

(d) It should be stable.

(e) It should be compatible with other system components and their operation; i.e., its operating temperature must be applicable to the analysis being performed, it should function properly with the carrier gas and gas flow rate selected for the analysis, etc.

Since almost any physical or chemical property of the sample components may be measured, there exists a large number of possible detectors applicable to gas chromatography. However, the majority of gas chromatographs utilize one of three detectors: the thermal-conductivity detector, the flame ionization detector, and the electron-capture detector.

2.5.2 Thermal-Conductivity Detector

A thermal-conductivity detector consists of a sensing element placed in a cavity of the detector body through which the carrier gas and column effluent flow. This sensing element may be a fine wire or a thermistor. An electric current is passed through the sensing element so that at normal operating conditions, with only carrier gas flowing through the cell, it will attain a certain temperature and therefore a certain electrical resistance. When a sample component from the column enters the detector cell, the heat-carrying capacity (thermal conductivity) of the gas surrounding the sensing element is changed. This causes a change in the temperature and consequently in the electrical resistance of the sensing element. During the elution of a sample component from the column, the resistance of the sensing element is changing continuously with time until the component has passed completely through the detector and only pure carrier gas is again flowing through the detector cell. At such time the resistance of the sensing element has returned to its original value. A measure of the time-based change in electrical resistance of the sensing element will correspond to the profile of the sample component as it was eluted from the column.

Normally a thermal-conductivity detector consists of two chambers with sensing elements. One sensing element is exposed only to the carrier gas and serves as a reference, whereas the other sensing element monitors the column effluent containing the sample components. With only pure carrier gas flowing through both chambers, a steady-state equilibrium will be established between the two sensing elements. When a sample component emerges from the column and enters the measuring chamber of the detector, this equilibrium will be displaced, and a measure of the difference in the electrical resistance between the two sensing elements, with respect to time, describes the sample component profile. The advantage in this type of system is that minor changes in operating parameters, i.e., temperature and carrier-gas flow rate, are compensated for, since they affect each sensing element approximately equally. This leads to a much more stable detector system.

In practical use of the thermal-conductivity detector, the two sensing elements are made a part of a Wheatstone-bridge system. When no sample component is present in the monitoring chamber, the Wheatstone-bridge circuit is balanced and in equilibrium. However, when sample enters the measuring cell, the bridge will no longer be in equilibrium and balanced. The voltage required to balance the Wheatstone bridge is measured and provides the signal related to the sample profile as it emerges from the chromatograph column.

The thermal-conductivity-detector response is based upon the difference in thermal conductivity between the carrier gas and sample components. For this reason greater sensitivity is attained when hydrogen or helium is used as the carrier gas because their values of thermal conductivity are approximately 10 times greater than those of most sample constituents. Nitrogen is also used as a carrier gas in instances where there is a restriction on the use of hydrogen or helium and/or sensitivity is not of prime consideration.

Thermal-conductivity detectors are not selective in that they will respond to any component that has a thermal conductivity different than the carrier gas. Thus, they have application in a broad range of organic and inorganic analyses.

For precise quantitative analysis using a thermal-conductivity detector, it is necessary to calibrate for each sample constituent. The response time of the detector is normally less than 1 sec at regular carrier-gas flow rates. Its linear dynamic range is approximately 10^5 and it has a sensitivity of 10^{-8} to 10^{-9} g/sec.

A review of thermal-conductivity detectors in gas chromatography has been given by Lawson and Miller (1966).

2.5.3 Flame Ionization Detector

A flame ionization detector is formed by measuring the ion current in a burning hydrogen flame. The carrier gas, containing the sample, and a source of hydrogen and air are brought together in a specially designed jet. The hydrogen gas is ignited. An electrical potential is impressed between the flame jet and an electrode located above the cone of the hydrogen flame. Sample components entering the hydrogen flame are burned and ionized as a result of the intense thermal energy of the hydrogen flame. The ions formed are collected on the electrode before there is a chance of their recombination. The current flow between the flame body and the collection electrode is indicative of the amount of sample present. Prescott and Wise

(1969) have described a very good design for a flame detector, as have several other authors.

The flame ionization detector will only respond to those compounds which are ionized in the burning hydrogen flame. Fortunately, this includes most of the organic species. Compounds which do not normally respond are the permanent gasses, i.e., N_2, O_2, CO, CO_2, etc.; water; and most inorganic compounds. For precise quantitative analysis this detector should be calibrated for each compound of interest.

The sensitivity of the flame ionization detector is 10^{-10} to 10^{-12} g/sec. Its linear dynamic range usually exceeds 10^6. Its response time is less than 1 sec. The flame ionization detector is the most widely used of the sensitive ionization-type detectors.

A great deal of work has been reported in the literature regarding the effect of electrode geometry and operating parameters upon the response of the flame detector. Response factors for many compounds have also been published. A review of the theoretical and practical considerations of using a flame ionization detector has been given by Garzo (1967). Guichard and Buzon (1969) also review important factors in the design and use of a flame ionization detector.

2.5.4 Electron-Capture Detector

The electron-capture detector is potentially the most sensitive of all detectors used in gas chromatography. However, its response is specific for certain types of compounds.

The electron-capture detector consists of a small-volume ionization chamber irradiated with low-energy beta particles from either a tritium or nickel-63 radiation source inside the detector cell. Normally the detector cell contains two parallel electrodes at opposite ends of a cylindrical ionization chamber. The carrier gas is argon containing small amounts of carbon dioxide or methane. The purpose of the second gas is to absorb much of the energy of the primary and secondary electrons formed in the irradiation process and give rise to a population of free electrons in the detector chamber of approximately thermal energy. When an electrical potential is applied to the two electrodes in the detector cell, all of the free electrons will be collected and an electrical current measured. The applied voltage may be either dc or pulsed dc voltage. There is evidence to suggest that the latter is the preferred mode of operation for this detector. When a sample component with an affinity for free electrons enters the detector cell, two reactions are possible:

$$AB + e^- \rightarrow (AB)^- \pm \text{energy}$$
$$AB + e^- \rightarrow A + B^- \pm \text{energy}$$

Either of these reactions leads to a reduction of the free-electron population in the detector cell and thus to a decrease in the measured electrical current in the cell. It is this reduction in electrical current that is measured and is indicative of the amount of sample present in the detector cell.

As mentioned, the electron-capture detector responds only to those compounds that have an affinity for free electrons. This class of compound includes most halogenated compounds and certain compounds containing conjugated unsaturated centers. In practice, the range of applicable compounds is extended by the preparation of chemical derivatives of the sample that have an affinity for free electrons.

The sensitivity of an electron-capture detector may be as high as 10^{-15} g/sec. However, the response for any given compound is dependent upon that compound's degree of affinity for electrons. This degree of affinity may vary as much as 10^6 among different compounds. For this reason calibration of the detector is essential for any quantitative application. The response time for an electron-capture detector is less than 1 sec.

The normal linear dynamic range of an electron-capture detector is small, being approximately 10^2 to 10^3. However, special linearizing circuits have been described by Fenimore *et al.* (1968) and others; these extend the linear dynamic range to approximately 10^5–10^6.

Lovelock (1961) describes the construction and operation of the electron-capture detector. Simmonds *et al.* (1967) has shown the design of a high-temperature electron-capture detector using nickel-63 as the source of ionizing radiation. Devaux and Guiochon (1970) have reported on a study of the determination of optimum operating conditions for this detector. There are many literature references on the application of this detector to specific analytical problems.

2.5.5 Other Detectors

In addition to the three most commonly used gas-chromatography detectors, there are many more that are used for specialized applications. Many have features which would make them the choice for a specific qualitative or quantitative analysis.

Flame photometric detectors which are specific to certain classes of compounds have been described. These detectors measure the intensity of a specific emission line of the sample as it is burned in a hydrogen flame. Bowman and Beroza (1968) describe such a system for the detection of phosphorous and sulfur-containing compounds. Nowak (1968) describes a slightly modified version that is specific for halogen-containing compounds. Yet another version specific for nitrogen compounds has been reported by Hartman (1969).

Lovelock *et al.* (1964) describes the design and operation of an ionization cross-section detector for use in gas chromatography. This device is of moderate sensitivity, but has the advantage that its response may be related to a molecular parameter that enables the analyst to obtain absolute quantitative information without calibration.

One of the first detectors employed in gas chromatography and one receiving a great deal of renewed interest is the gas-density balance. This detector is of moderate sensitivity, but its response is absolute insofar as precise quantitative information may be obtained without calibration. Conversely, if the sample size is known, then the precise molecular weight of an unknown sample may be determined. Construction details and measurement of response for a gas-density balance have been given by Creitz (1969).

Lovelock (1961) has described a family of argon ionization detectors for use in gas chromatography. These detectors offer high sensitivity for the detection of organic compounds but generally have not been as widely used as the flame ionization detector.

Other detectors designed primarily for gas chromatography include conductometric, coulometric, helium ionization, photoionization, plasma, radioactivity, and mass detectors; glow discharge, polarographic, ultrasonic, and capacitance devices.

In addition to gas-chromatographic detectors other analytical instruments have been used as detectors for gas-chromatographic apparatus. These include mass spectrometers and infrared and ultraviolet spectrophotometers.

An excellent review on the choice of detectors for use in gas chromatography has been given by Gaugh and Walker (1970).

2.6 Data System

There are two forms of data system used in gas chromatography—analog and digital. Both are often used simultaneously to maximize the amount of information obtained for a single analysis. Digital data provide a convenient form for further data use, but analog formats are more informative of what is actually occurring in the chromatographic process.

Analog data systems are no more than strip-chart recorders coupled to suitable amplification circuits for the measurement of detector response on a time base. Digital data are obtained by converting the basic analog data.

A more detailed discussion on the data system is contained in Section 3.

3 Data Form

If the chromatograph column is the heart of the gas-chromatograph system, then the detector is the brain. It is the output signal provided by the detector, in response to sample components, that provides the data upon which subsequent analytical interpretation is based. Without exception, the untreated signal from a differential detector is in analog form, i.e., it is a direct and continuous measure of some physical or chemical parameter of the column effluent with respect to a convenient time base. By proper manipulation of the basic analog information, digital data may be obtained for reasons of convenience, accuracy, precision, or ease of interpretation.

3.1 Analog Data

The basis for analog information is the measurement being performed by the chromatographic detector. The detector response is converted to an electronic signal and measured directly or amplified and measured for subsequent recording. The recording device may be a strip-chart potentiometric recorder, a recording oscillograph, or any other convenient means of allowing the analyst to visualize the information. To be acceptable, the electronic conversion and recording of the detector signal must be essentially instantaneous with respect to what is occurring in the detector cell at any point in time. An example of analog data is shown in Fig. 7, which is a typical chromatogram of the major components of air. The basic

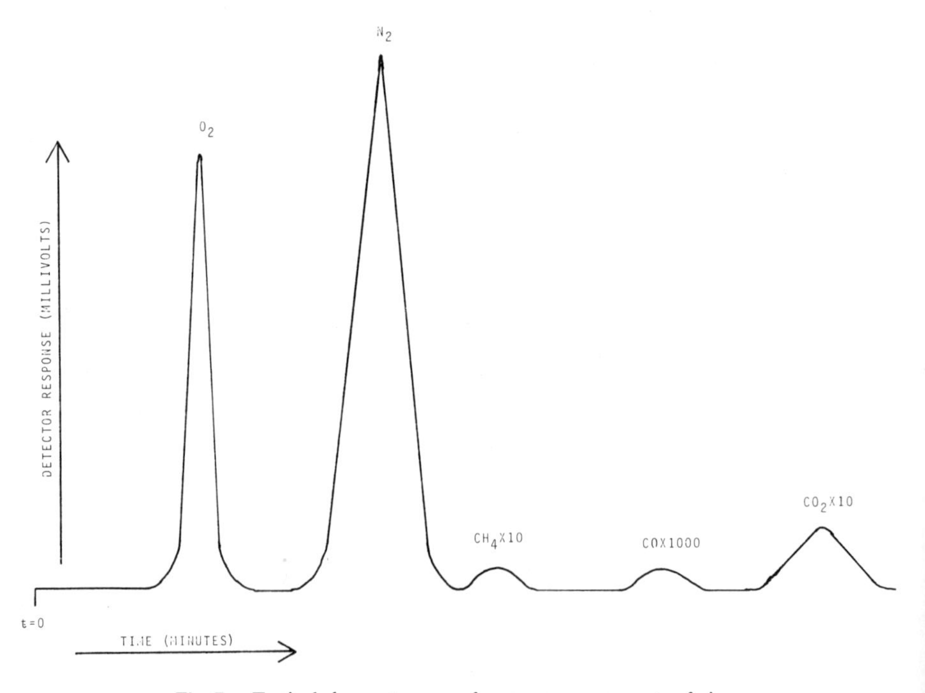

Fig. 7. Typical chromatogram of common components of air.

analytical information is contained in the response of the detector versus time. For a given set of operational conditions, the time of peak emergence is a qualitative measurement. The area beneath each peak, which is dependent upon the peak magnitude and duration, is a quantitative measure. Analog information is always desirable because it is a picture of what has occurred during the chromatographic separation. An experienced analyst can derive a great deal of information about the chromatographic process, and therefore about the analysis, from both the shape of the individual peaks and the relationship between peaks. This information might be obscured if only digital data were available. The interpretation of analog information will be treated in the sections on quantitative and qualitative analyses.

Szonntagh (1968) has reported on a comparison of various types of recorders and associated equipment for the measurement of analog data.

3.2 Digital Data

While analog data are a graphic representation of the analysis, the reduction of these data to obtain qualitative and quantitative information is tedious, time consuming, and subject on the part of the operator to measurement errors that adversely affect the accuracy and precision of the analysis. For this reason it is now common practice to convert the analog signal to a digital form and either display these converted data directly on a printed tape or as is done more often, apply computer techniques to manipulate the digital output in order to obtain detailed analytical information, such as the retention time of each peak; the area beneath each peak (total response for each sample component); the response for each sample component, corrected by the appropriate detector response factor; and the percentage of each sample component present.

An example of direct presentation of digital data for the analysis shown in Fig. 7 might be a printed tape containing the following information:

1.02	25741
2.06	80526
2.97	00215
3.15	00002
6.29	01099

The qualitative information would be obtained from a knowledge of the retention time for each peak, which is shown in the first column. Each number in this column corresponds to the retention time, in minutes, for a single sample component. Thus we would deduce that the sample contained 5 components, each with the retention time indicated. From prior calibration of the gas-chromatographic system, we would have ascertained that

TABLE 1

Peak #	Retention time	Area	Response factor	Corrected area	%
1	1.02	25741	0.85	21880	21.88
2	2.06	80526	0.97	78110	78.11
3	2.97	00215	0.93	00200	2×10^{-3}
4	3.15	00002	0.90	00002	2×10^{-6}
5	6.29	01099	0.91	01000	0.01

the components with these retention times were O_2, N_2, CH_4, CO, and CO_2. The quantitative information for this particular analysis would be obtained from the second column which is the integrated area beneath each peak. The analyst would multiply each area by the appropriate detector response factor, obtained from prior calibration. Finally, the amount of each component present in the sample would be determined.

These above-outlined steps for the handling of digital data may be performed by a computer. A typical computer output for the same data might appear as shown in Table 1.

Powell (1969) has discussed the design of electronic digital integrators for gas chromatography. An automatic digital integrator for the determination of peak areas has also been given by Watkin *et al.* (1969).

The use of computer techniques to assist in gas-chromatographic data presentation and reduction has been the subject of many reports. Pattee *et al.* (1970) describe methods of preparing the data and include the computer programs. Gladney *et al.* (1969) provide a general discussion on computer-assisted gas chromatography. The reports by Gill and Perone (1969), Oberholtzer (1969), and Tochner *et al.* (1969) cover various aspects of computer data systems for gas chromatography.

4 Applications and Limitations

4.1 Qualitative Analysis

4.1.1 General

The prime consideration in qualitative gas chromatography is that the system provide a finite and measurable response to all the components of a sample. Therefore, consideration of column efficiency and resolution becomes important. Also, all sample components that enter the gas chromatograph must pass through the system and be measured by the detector. Thus, irreversible loss of sample components in the system and lack of detector response to certain compounds must be considered.

The most serious shortcoming of gas-chromatographic analysis is that it does not provide for unequivocal qualitative information. This aspect has been recognized, and recently much work has been devoted to improving the qualitative aspects of gas chromatography. Now available are a number of techniques that enable sufficient qualitative information to be obtained to characterize the sample components.

a. IRREVERSIBLE LOSSES AND DETECTOR RESPONSE. The irreversible loss of all or a portion of the sample components is caused primarily by the following factors: (1) adsorption in the system; (2) decomposition due to thermal or catalytic activity in the system; and (3) chemical reaction with the system components. In the first case certain components of the original sample will not appear at the detector, and their presence will remain unknown. In the cases (2) and (3) certain sample components may not reach the detector, or equally important, compounds not present in the original sample may be detected. In all these cases erroneous qualitative information is obtained. Discussion of gas-chromatographic apparatus has attempted to call attention to these possibilities, and techniques and procedures are recommended to minimize their occurrence.

If the gas-chromatograph detector is insensitive to or has no response for any of the sample components, neither qualitative or quantitative analysis will be possible. The choice of detector must assure that a response will be obtained for each component present in the sample.

b. EFFICIENCY AND RESOLUTION. The efficiency of a gas-chromatographic system is primarily the efficiency of the column, although there are other contributing factors. Efficiency is one measure of the ability of the chromatographic apparatus to separate the sample into its individual constituents, which is necessary to obtain a qualitative analysis.

A model chromatogram is shown in Fig. 8. The efficiency n for the first peak is calculated from the relationship

$$n = 16(OA/BC)^2 \tag{9}$$

where OA is the distance from the time of injection to the appearance of the chromatograph peak maximum, and BC is the width of the base of the chromatograph peak. The efficiency for the other peaks in the chromatogram would be calculated in a similar manner.

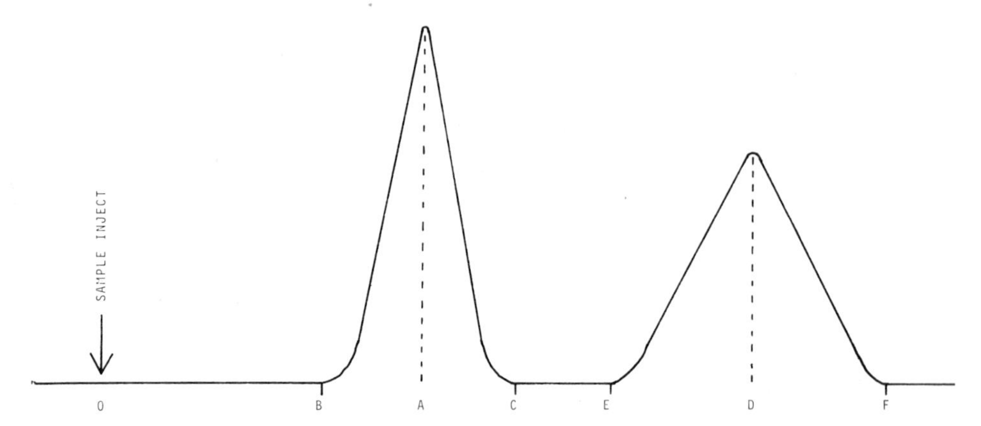

Fig. 8. Model chromatogram for calculation of efficiency and resolution.

Resolution is a measure of the ability of the system to resolve two neighboring peaks. Resolution may be calculated by either of two methods (refer to Fig. 8):

$$R_1 = 2\,(OD - OA)/(EF + BC) \tag{10}$$

or

$$R_2 = 2\,(OD - OA)^2/[(EF)^2 + (BC)^2] \tag{11}$$

Equation (11) has the slight advantage in that it is proportional to the length of the column.

It must be remembered that in practical gas chromatography the resolution of sample components is primarily a function of the type of column employed and of the partitioning liquid or adsorbent used. It may not be possible to perform the desired separation of sample components on a certain column with an efficiency of 100,000 theoretical plates, whereas the desired separation may be performed on a different column with an efficiency of only 5000 theoretical plates.

4.1.2 Techniques and Interpretation

a. Comparative Technique. In gas chromatography the majority of all qualitative analyses are performed by comparing the retention time of the unknown sample component to that of a known compound. In practice, pure compounds, corresponding to those whose presence is suspected in the unknown sample, are injected into the chromatograph under the same operating conditions used to analyze the unknown mixture. If the known compound is found to have the same retention time as one of the sample constituents, it is assumed that the unknown sample constituent is identical to the standard compound. For a single analysis this is positive, but certainly not conclusive, evidence upon which to base a qualitative identification. The probability of a successful qualitative analysis may be increased by performing the same comparative analysis using two or more chromatographic columns with different characteristics. For example, if an unknown sample component and a known compound have precisely the same retention time on a nonpolar column, which separates according to molecular weight of the sample, and on a polar column, which separates according to the functional groups present in the sample, then the presumptive evidence upon which to base a qualitative identification is strong.

A modification of the comparative technique is the use of relative retention values for qualitative identification. In this technique the ratios of the retention times of a series of known, pure compounds to the retention time of a preselected standard compound are determined, under a given set of operating conditions, using a single column. In qualitative analysis the standard compound is added to the unknown sample prior to its injection into the gas chromatograph. The ratios of retention times for the sample constituents to the retention time of the standard compound are measured and compared to those previously determined. If the ratio of an unknown sample compound is the same as one of the predetermined ratios, then qualitative identification is normally made. Provided that the retention times of the standard and unknown compounds are closely related, this method is reasonably accurate. To improve the probability of success, it is desirable to use more than one standard compound and determine several ratios for each component of the sample.

This technique offers several advantages. The ratios generally remain valid for two columns of the same material. Thus, data can be accumulated and exchanged among different analysts. Also, minor variations in operating conditions during the analysis do not effect the ratios to any great degree. A third advantage is that the standard compounds added to the sample may be used to assist in quantitative analysis.

b. Retention Data Plots. It has been shown that for a given chromatographic column and a specific set of operating conditions, the logarithm of the specific retention volume V_r (retention time measured from the air peak) is linearly related to the number of carbon atoms in a given homologous series of compounds. Figure 9 illustrates this relationship. A, B, and C represent different series of compounds, e.g., straight-chain saturated hydrocarbons, straight-chain unsaturated hydrocarbons with a single double bond, and branched saturated hydrocarbons. As can be

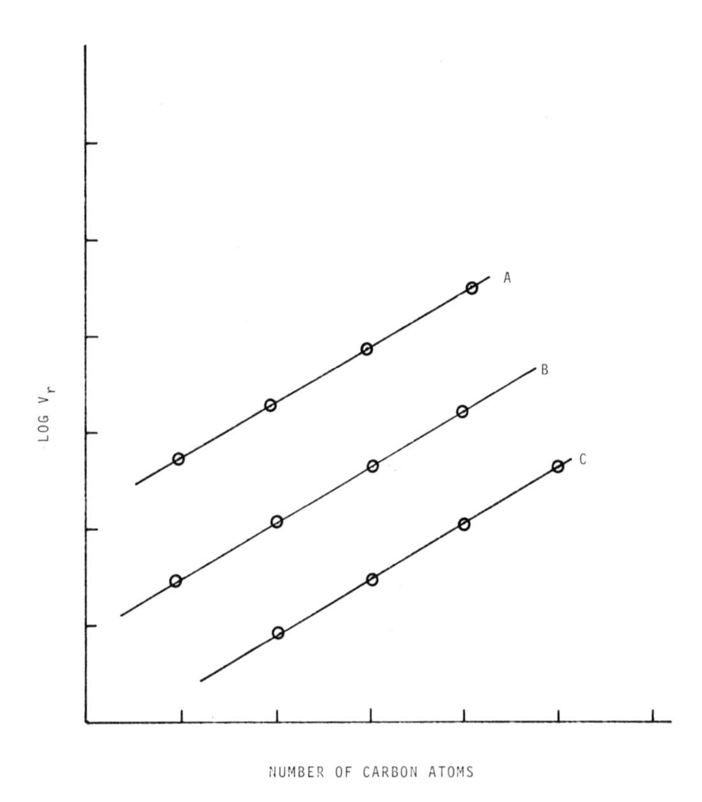

Fig. 9. Data presentation showing linear relationship between number of carbon atoms in the sample compound and log V_r. A, B, C: different homologous series.

seen, without prior knowledge of the nature of the sample, qualitative identification would not be possible.However, such a relationship may be useful in identifying which specific member of a known homologous series is present in an unknown sample. This relationship is not commonly used as a qualitative tool. However it does form the basis for the several useful qualitative techniques that follow.

c. RETENTION INDICES. Kovats (1958) has established a general identifitification scheme based upon the linear relationship between specific retention volume and the number of carbon atoms for a given homologous series.The relationship defined by Kovats is constant for a particular column, under a given set of operating conditions. In this technique the compound to be identified is related to two normal paraffins. By definition normal paraffins are assigned an index equal to 100 times their respective number of carbon atoms. The defined relationship is

$$I = 200 \frac{\log V_{g(x)} - \log V_{g(p_z)}}{\log V_{g(p_{z+2})} - \log V_{g(p_z)}} + 100z \qquad (12)$$

where

x is the sample component whose index is to be measured;
$V_{g(\ )}$ is the retention volume;
p_z is a normal paraffin with z carbon atoms, eluted from the column before the sample component;
p_{z+2} is a normal paraffin with $z + 2$ carbon atoms, eluted from the column after the sample component.

In practice it is determined experimentally which two normal paraffins satisfy the requirements of the calculation. These two paraffins are then mixed with the unknown sample. The Kovats retention indices for many compounds may be found in the literature, thus facilitating qualitative identification.

Maume (1965) has recommended a similar index based upon reference to normal primary alcohols instead of normal paraffins.

A plot of retention indices versus boiling point is linear for a given homologous series. In addition, a family of parallel curves are obtained for the same homologous series on different stationary phases. This factor has prompted Rohrschneider (1965) to propose the relationship

$$I = I_0 + ax + by + cz + du + es \qquad (13)$$

where

I is the corrected retention index;
I_0 is the retention index of the sample component on a squalane column;
x, y, z are factors related to the polarity of the stationary phase;
a, b, c are factors related to the polarity of the sample component; and
du, es are correction factors.

Like the Kovats retention indices, Rohrschneider constants for many common substrates have been published in the literature.

The effect of the polarity of the stationary phase on the retention index has led to another approach to qualitative information—that of the increment of index ΔI. The stated relationship is

$$\Delta I = I_p - I_a \qquad (14)$$

where

ΔI is the increment of index;

I_p the retention index of the sample component on a polar stationary phase; and

I_a the retention index of the sample component on a nonpolar stationary phase.

Kovats has shown that the increment of index is a function of the type of compound present and may be used for identification purposes.

A comprehensive review of the use of retention indices for qualitative analysis has been given by Kovats (1966), Kaiser (1969), and Takacs and Kralik (1970).

d. CHEMICAL CONVERSION. *i. Pyrolysis.* Pyrolysis gas chromatography has become a useful analytical technique, particularly for the identification of high-molecular-weight samples. The unknown sample is thermally degraded in the inlet portion of the chromatograph, and the fragments formed are chromatographed in the usual manner. Under a specific and regulated set of experimental conditions, this fragmentation process is reproducible for a given type of sample. The chromatogram obtained is a "fingerprint" of the sample and may be used for its identification.

This type of analysis has found particular application in the identification of polymers. Wolf *et al.*(1971) have discussed the principles, techniques, and applications of pyrolysis gas chromatography in a general review. Juvet *et al.* (1971) and Levy *et al.*(1971) have discussed factors affecting the reproducibility of this technique and the problem of reproducing the data in different laboratories.

The list of all applictions of pyrolysis gas chromatography to the qualitative analysis of materials is too extensive for inclusion in this text. However, specific examples are analyses of pure high-molecular-weight organic compounds, rubbers, and polymers; characterization of human hair and other materials of biological origin; identification of microbiological species; and determination of the organic content of meteorites and other geological samples (including samples from the lunar surface).

ii. Carbon Skeleton. In the carbon skeleton technique the sample to be analyzed is subjected to a catalytic reaction at the inlet to the gas-chromatograph column. As a result of the reaction, only the saturated hydrocarbon skeleton remains and is chromatographed. A knowledge of the carbon skeleton and of the products formed from the reaction often allows the original structure of the sample to be deduced.

Carbon skeleton analysis is obviously limited to carbon compounds. Care must be exercised in the control of experimental conditions to minimize modification from catalytic reaction of the carbon skeleton structure.

Beroza and Sarmiento (1964) used a hydrogenation reaction to investigate high-molecular-weight carbon compounds. Thompson *et al.* (1964) used deoxygenation, desulfurization, denitrification, and dehalogenation reactions to obtain the carbon skeleton. Franc and Kolowskova (1965) have also reviewed the details of this type of analysis.

iii. Elemental Analysis. Gas chromatography used in conjunction with catalytic oxidation of the sample can yield an elemental analysis. In this method a small quantity of the unknown sample is weighed into a reaction tube containing an appropriate catalyst and reacted with oxygen. The sample carbon is converted to CO_2; the sample hydrogen is converted to H_2O; the sample nitrogen is converted to NO_2; etc. All other elements are converted to their basic oxides. The products of the oxidation reaction are injected into a gas chromatograph and identified. With proper calibration the original elemental constitution of the sample can be calculated for samples containing only those elements for which the analysis has been performed.

Berezkin and Tatarisnky (1971) have described a system for the determination of C:H:N:O ratios. Mamaril and Meloan (1965) describe a technique for the elemental analysis of halogen-containing compounds. Schussler (1969) discusses the determination of sulfur in organic compounds. The determination of phosphorus in organic compounds has been described by Handy and MacDonald (1970).

iv. Other Techniques. In addition to pyrolysis, carbon skeleton, and elemental analysis, other chemical conversion techniques that aid in the qualitative interpretation of gas-chromatographic data have been reported.

A chemical conversion of the sample may be performed to facilitate qualitative analysis by allowing more sensitive or specific detection methods to be utilized. Such is the case when certain classes of chemical compounds are reacted to yield volatile halogen derivatives that are detected by an electron-capture detector after chromatographic separation. This type of chemical conversion is also exemplified by the reaction of water with calcium or aluminum carbide to form acetylene or methane, which are easily detected by a flame ionization detector.

Reaction of the sample with reagents specific for a particular type or class of compound is often performed. Thus the presence of the appropriate chemical derivatives in the column effluent is indicative of the original nature of the sample. Examples of this type of chemical conversion include methanolysis, methylene insertion reactions, esterification, and hydrogenation.

The foregoing chemical conversion techniques are normally employed before injection of the sample into the chromatograph, or in the inlet system prior to entry of the sample onto the gas-chromatograph column. However, chemical conversion of the sample may also be accomplished on the gas-chromatograph column. In this technique the chemical reagent is incorporated as part of the column, and reaction with the sample occurs during the chromatographic process.

Chemical conversion of the sample as a qualitative aid may also be accomplished after the chromatographic separation of the sample components but prior to detection. In practice, however, this technique has been limited to application as a quantitative aid for eliminating variations in detector response to different compounds. An example is the conversion of each sample component to CO_2 after separation on the column, and the detection of CO_2 by the detector.

Finally, chemical conversion of the sample may occur after primary gas-chromatographic detection. This technique allows reaction of the sample components with reagents specific to a given type or class of compound. Each sample component may be exposed to the reagent separately as it elutes from the detector. The qualitative information is gained by observing a color change or some other measurable response of the sample component with the reagent. A modification of this technique allows pyrolysis or elemental analysis to be performed on each sample component as it emerges from the first detector, with the resultant products being passed into a second gas-chromatographic system. In this manner pyrolysis patterns or elemental analysis may be obtained for each component of an unknown heterogenous sample.

Berezkin (1968), Ettre (1969), and Littlewood (1968a) have reviewed the techniques and applications of reaction gas chromatography.

e. Peak Removal and Addition. In peak-removal qualitative techniques the sample must be chromatographed more than once. The entire sample is analyzed, and the number and position of each peak corresponding to a single sample component are noted. The sample is then treated chemically or physically to remove certain types or classes of compounds. The sample is then analyzed again chromatographically, under identical experimental conditions as employed initially. The disappearance of certain peaks on the chromatogram indicates both that the sample components represented by these peaks were of a certain type and where they were present in the original sample. Typical examples of this qualitative technique are the removal of normal paraffins on a precolumn of molecular sieves during the second analysis, and the irreversible chemical complexing of olefins prior to the second analysis.

An alternative approach is to add the pure suspected component to the original sample. These suspected components are added to the original sample one at a time and after each addition a new analysis performed under the same experimental conditions. An increase in the relative detector response (peak size) for any one of the peaks would be supporting evidence that the added compound was present in the original sample.

f. SPECIAL DETECTORS. Certain gas-chromatograph detectors respond only to a particular type of compound. Thus, if a detector response is obtained for a given chromatographic analysis, the presence of a particular type of compound in the original sample is confirmed. Examples of this method are the use of electron-capture detectors to note the presence of compounds with an affinity for thermal electrons; the use of a flame-photometric detector to detect the presence of sulfur, phosphorous, and nitrogen compounds; and the use of a thermionic detector that responds to halogen-containing compounds.

g. MULTIPLE DETECTORS. The effluent from the gas-chromatograph column may be split and a portion diverted to each of several different detectors. If the response factor for the compounds of interest are significantly different in each detector, qualitative identification may be possible by means of response ratios. An example of this technique has been described by Zielenski *et al.*(1967), who used relative response data from a flame ionization and an electron-capture detector to obtain qualitative information from the analysis of difficult-to-separate isomers.

h. TRAPPING AND SUBSEQUENT IDENTIFICATION. The most common method of obtaining qualitative information from gas-chromatographic analysis is to trap the individual components from the separtion, after they have been seen by the detector, and subject them to separate analytical procedures in order to determine their exact character. After the sample constituents have been separated chromatographically and trapped, they may be identified by almost any convenient analytical method. Typical qualitative methods include mass spectrometry; infrared, visible, and ultraviolet spectrophotometry; and nuclear magnetic resonance spectrometry.

The trapping of sample components for qualitative determination imposes several criteria upon the trapping system. It should allow exclusive and complete removal of the sample components from the carrier-gas stream after detection. The nature of the trap should be such that no chemical or physical modification of the sample components occurs while they are contained in the trap. The sample components should be completely removable from the trap. Finally, the trap should allow convenient presentation of the trapped components for the subsequent qualitative analysis.

The type of trap depends upon the characteristics and operation of the gas chromatograph, the nature of the sample, and the ultimate requirement for further analysis. Examples of the types of traps that have been used include thin-walled, narrow-bore glass or metal tubing; infrared windows and gas cells; short packed columns, containing a suitable adsorbent or chemical agent; disposable pipettes; ordinary condensers; syringes; fritted filters; and glass wool.

i. Other Qualitative Techniques. In addition to trapping and subsequent qualitative identification of sample components, certain qualitative analytical devices may be directly joined to the gas chromatograph. These techniques are discussed separately in Section 6 on auxiliary techniques.

4.1.3 Selectivity of Qualitative Analysis

As previously mentioned, gas chromatography is not an absolute qualitative technique. It is generally applicable to volatile organic and inorganic substances or their volatile derivatives. However, the technique must be modified or derived calculations applied in order to obtain presumptive qualitative evidence from gas-chromatographic data alone. The selectivity is dependent upon the type of sample, the technique employed, and the gas-chromatographic system used.

4.1.4 Sensitivity of Qualitative Analysis

Because of the variations in types of available gas-chromatographic systems and their applications, no finite sensitivity may be assigned. Qualitative sensitivity extends throughout a range of 1 ppb (or less) to 100%, again depending upon the nature of the sample and the specific gas-chromatographic analysis applied. The optimum range for qualitative analysis is dependent upon these same factors.

4.1.5 Sample Requirements for Qualitative Analysis

The only sample requirement for qualitative analysis using gas chromatography is that the sample or a known derivative of the sample be injected into the gas-chromatograph instrument and there volatilized for passage through the system. It is often desirable to transform the sample prior to or during the gas-chromatographic analysis to enable the analyst to work with a more stable, more or less volatile, less polar, less adsorbable, more easily separable, or more readily detectable compound than that originally present in the sample. Caution must be exercised in the utilization of such transformations to assure that the resulting product can be related in a qualitative manner to the constituents of the original sample.

4.1.6 Nondestructive or Destructive Analysis

In qualitative gas-chromatographic procedures utilizing detectors or sample transformations that consume the sample, the analysis is destructive. An example is the use of a flame ionization detector to measure the sample components as they emerge from the column. Another example is the irreversible chemical reaction of the sample components with a specific chemical reagent after elution from the detector. In instances where destructive techniques are employed, partial retention of the sample may be accomplished by splitting the column effluent and directing a portion to a suitable trap, the remainder being consumed in the destructive technique.

Gas-chromatographic procedures using the unmodified sample and a detector that measures some physical or chemical property of the sample components are nondestructive. One example is the use of a thermal-conductivity detector to measure an untreated sample. In these cases the entire sample may be recovered, if desired, by trapping the sample components as they emerge from the detector.

4.1.7 Analysis Time

The time required for most gas-chromatographic analyses does not exceed one hour, exclusive of any time for sample pretreatment. Complete analyses accomplished in several seconds have been reported, whereas other analyses requiring 24 hr have been performed. Again, the exact total analysis time depends upon the sample and the specific gas-chromatographic apparatus used.

4.2 QUANTITATIVE ANALYSIS

4.2.1 General

It is in quantitative analysis that gas chromatography finds its primary application and makes its major contribution as an analytical technique. Gas chromatography is unsurpassed in its ability to separate the components of a sample mixture and provide a recorded response for each compound present. It is the determination of this response and its relation to the amount of each component present in the original sample that comprise quantitative analysis using the gas-chromatographic technique. The factors affecting the accuracy and precision of quantitative analysis are the response of the detector, the accuracy and precision of the recording device, and the accuracy and precision of the measurement of the area beneath each chromatographic peak. The weight of each sample component from a gas-chromatographic analysis is generally defined as

$$W_c = K_c - A_c \qquad (15)$$

where W_c is the weight of sample component; A_c is the area of the chromatographic peak representing the sample component; and K_c is a constant, dependent upon the gas-chromatographic system employed and the detector response.

4.2.2 Techniques and Interpretation

a. PEAK AREA MEASUREMENT. *i. Planimeter.* Simple mechanical integrators, like those normally used in mechanical drafting, can be used to measure the area of a gas-chromatographic peak. They are inconvenient and time consuming to use and have found little favor with the analyst. The use of such devices has no real advantage over other techniques.

ii. Weighing. One method of determining the areas for peaks on a chromatogram is to cut them out of the record (chromatogram) and weigh each peak individually. Again, this procedure is inconvenient and time consuming to use, particularly when there are a large number of peaks whose areas are to be determined. In addition, this method would seem to contain more potential error sources than other available techniques. The accuracy and precision of such a measurement is highly dependent upon the proficiency of the analyst as well as certain uncontrollable factors such as the uniformity of recorder paper, humidity, and the accuracy and sensitivity of the weighing device.

iii. Geometric Calculations. (a) *Triangulation: one-half base times altitude.* An ideal gas-chromatographic peak should be Gaussian in shape. Therefore it can be shown mathematically that a close approximation to actual peak area can be obtained by a method of triangulation; i.e., an isosceles triangle is superimposed on the peak whose base is the base line of the chromatogram and whose sides are tangent to the peak. This approximation is illustrated in Fig. 10. The area of the chromatographic peak is approximated by the triangle ABC. The area of the triangle ABC is given by

$$\text{area} = \tfrac{1}{2}(AB)(CD) \tag{16}$$

Mathematical comparison of this triangulated area to the actual area of the peak shows that the triangulated area is in reality approximately 0.97 of the value of the actual area.

(b) *Triangulation: height times width at one-half maximum peak height.* An alternative method is estimation of area based upon triangulation as shown in Fig. 11. The area of the triangle GHI is given by

$$\text{area} = \tfrac{1}{2}(GH)(JI) \tag{17}$$

or

$$\text{area} = (KL)(JI) \tag{18}$$

Again, mathematical comparison shows the calculated area to be approximately 0.94 of the actual peak area. In most instances where

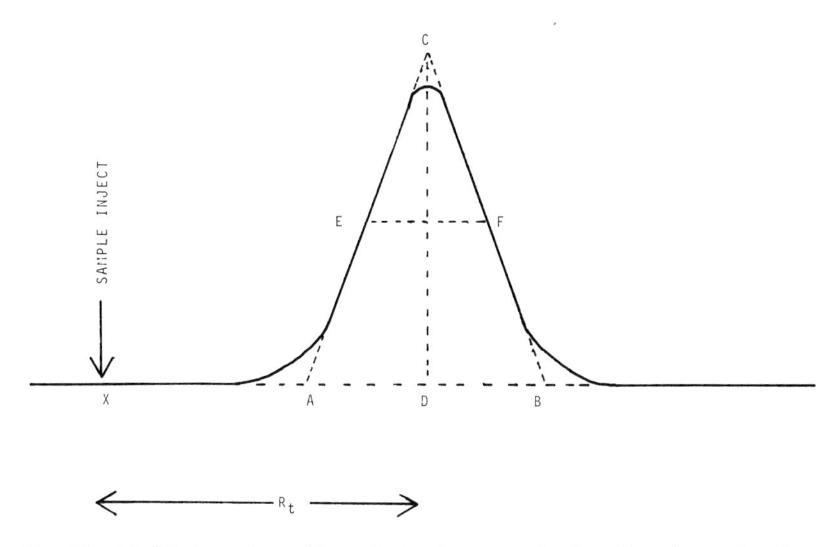

Fig. 10. Model chromatographic peak showing approximation of peak area by triangulation method: one-half base times altitude.

chromatograms are measured by the triangulation method, it is the measurement of peak height times the peak width at one-half the maximum peak height that is used.

(c) *Peak height times retention time.* An alternative method of estimating relative peak areas is based upon geometric considerations and the efficiency of the gas-chromatographic separation. If the experimental conditions are carefully controlled, then an estimate of relative peak areas may be obtained by a measurement of peak height and retention time of the

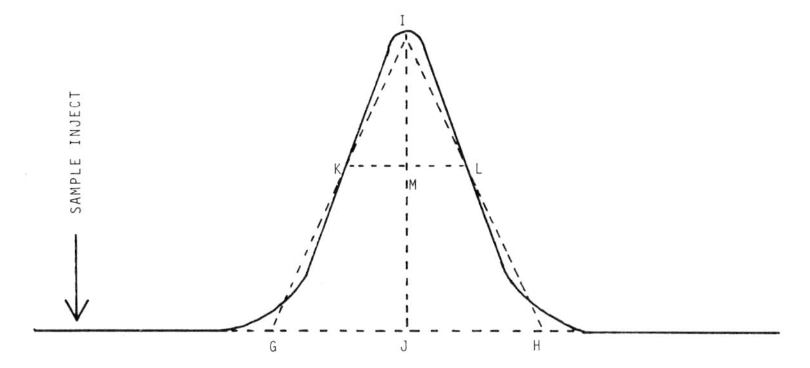

Fig. 11. Model chromatographic peak showing approximation of peak area by triangulation method: peak height times peak width at one-half the maximum peak height.

peak. Again, with reference to Fig. 10, the relative area would be given by

$$\text{area} = (CD)(XD) \tag{19}$$

and the weight of the compound represented by this peak could be estimated from the relationship

$$W_c = K\,(CD)(XD) \tag{20}$$

where K is an instrument constant.

It should be noted that in all cases of geometric calculations of peak areas, the peaks are assumed to be a perfect Gaussian shape. Even with this assumption only a reasonable approximation of the actual peak area is obtained. The more the chromatograph peak becomes skewed, the greater will be the error in estimating peak areas by this technique. In the literature there may be found other references that suggest different ways of geometrically measuring peak areas to compensate for the deviation from ideal peak shape. However, it is the belief of this author that if more accurate determinations are required, it is worthwhile considering an alternative automatic integration system that measures the area of the peak directly.

iv. Integration. There are two basic types of automatic integrators used in gas chromatography—electromechanical and electronic.

The electromechanical integrator attaches directly to the potentiometric recorder that is used to record the detector response to a sample component. This unit measures and records peak areas directly on the recorder chart by a series of oscillations which are proportional to the areas of the chromatographic peaks. The potentiometric recorder functions by measuring and recording the voltage required to maintain the recorder in balance (i.e., there is a constant input signal from the detector when no sample is present). As the detector signal changes owing to the presence of the sample component, the recorder compensates for this change by changing its internal electrical resistance, thereby remaining in the balanced position. If this variable resistance is used to control a power supply that drives a variable speed motor, the speed of the motor will be related to the detector signal at any point in time. The integrator is simply a revolution counter for the variable speed motor and records the total number of motor revolutions for each peak. In spite of certain difficulties encountered with the use of electromechanical integrators, they are inexpensive and generally reliable. They are widely used in gas chromatography.

The electronic integrator may be connected to the detector directly or in parallel with the potentiometric recorder. It electronically integrates (or sums) the detector signal during the elution of a chromatographic peak.

This integrated value is printed automatically. The electronic integrator begins its measurement as soon as the detector signal varies from the base-line conditions, and it ceases its measurement and prints the integrated value when the detector signal returns to the base-line value. This device also measure the rate of change in the slope of the detector signal as a peak is eluted, and prints a time measurement at the peak maximum, which is the retention time of the compound being eluted from the column.

The major disadvantage in the use of electronic integrators is their cost. Often their cost can exceed that of the basic gas-chromatographic apparatus.

The advantage of automatic integrators is that they have a vary rapid response and are generally the most accurate method of measuring and recording peak areas and retention times. Modern versions can compensate for a drift in detector base line and yield relatively accurate integrated values for two peaks that are not completely resolved by the gas chromatograph. When combined with a computer system, the electronic integrators can automatically perform the complete reduction of gas-chromatographic data.

Watkin *et al.* (1969) have described an automatic integrator for gas chromatography. Powell (1969) has also discussed automatic electronic integrators. There are a number of commercial suppliers of automatic integrators for use with gas-chromatographic apparatus, and their literature is descriptive of the present capabilities of such systems.

b. OTHER OPERATING TECHNIQUES. The foregoing discussion has been centered around the measurement of the area of a single gas-chromatographic peak. Once this measurement is performed, the question that remains is how this information can be related to the amount of a particular component present in the original sample.

i. Absolute. There are two absolute quantitative techniques that may be employed: (1) those in which the detector response is a measure of some sample component parameter that may be directly related to the amount of component present, and (2) those where the peak height or peak area of the sample component are compared with calibration curves prepared from pure samples. Precise quantitative analyses are usually performed using the latter technique.

The two gas-chromatographic detectors that provide an absolute response are the ionization cross-section detector and the gas-density balance.

The ionization cross-section detector response is a function of the ionization cross section of the component molecule, which may be approximated from the atomic cross sections of its constituent atoms. This relationship and the formula for determining absolute quantitative values have been given by Lovelock *et al.* (1964).

The response of the gas-density balance is a linear function of the concentration of the sample component in the carrier gas. The response factor for the sample component depends only upon the density difference between the sample component and the carrier gas. If the sensitivity is

$$S \propto (M - m)/M \tag{21}$$

where S is the sensitivity factor, M the molecular weight of the sample component, and m the molecular weight of the carrier gas, then the weight of sample component is related to the peak area by

$$\text{area} = W_c (M - m)k/M \tag{22}$$

where W_c is the weight of the sample component, k is the detector constant, or more conveniently,

$$W_c = \text{area } M/(M - m)k \tag{23}$$

The absolute nature of this detector allows the determination of molecular weights to better than 1%. Littlewood (1962) describes in detail the use of the gas-density balance in an absolute manner.

Calibration curves may be prepared by prior analysis of a series of known concentrations of pure compounds. The same amounts of standard sample is injected for each data point; only the concentration of pure compound present in the standard is varied. In this manner a calibration curve is obtained. A typical example is shown in Fig. 12. A separate curve (A,B,C,etc.) is necessary for each component present in the sample. For this technique to be effective, great care must be exercised in the preparation and use of the calibration curves. In addition, the injection of sample into the chromatograph must be very reproducible. Precise control must also be maintained on all operating variables such as carrier-gas flow rate, column and detector temperature, sample injection, etc. Another factor to be aware of is that the sample matrix can sometimes affect the response obtained for a given sample component. The validity of such calibration curves should be periodically verified to account for and correct deviations due to long-term changes in the performance of the gas-chromatographic system, particularly the gas-chromatograph column. With proper precaution and application this technique can yield quantitative information to better than 1%.

ii. Internal Standard. Many of the problems of operating-condition variations, which lead to errors in absolute techniques, can be minimized or avoided by the use of an internal standard technique. In this method a standard compound is added, in a known amount, to the sample to be analyzed.

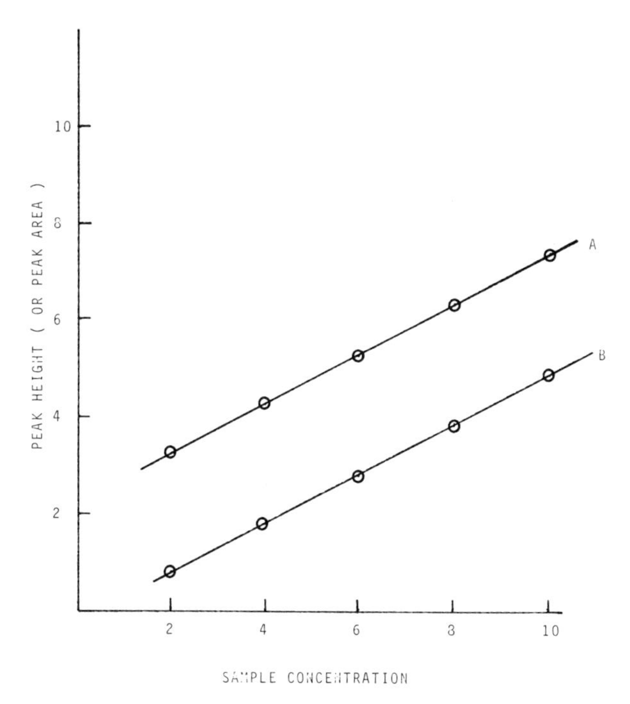

Fig. 12. Direct calibration curve for quantitative analysis. A, B: different sample components.

The peak height or peak area of the sample compound is measured and related to the peak height or peak area of the internal standard. In this manner minor variations in operating parameters are compensated for, since they should affect the sample components and standard compound equally. Also, with this method the sample size is not a governing criterion, within the linear range of the system. As before, calibration curves are prepared. These curves are prepared by analysis of a known concentration of the pure compound of interest to which the internal standard has been added. A typical example is shown in Fig. 13. Again, a separate curve (A,B,C,D,etc.) is obtained for each component in the sample.

Harvey and Chalkey (1955) point out that a suitable internal standard should meet the following requirements: (1) it should yield a completely resolved peak; (2) it should have a retention time near the sample component; and (3) the ratio of its peak height to that of the sample component should be close to unity. In addition, the detector response factors for the sample component and internal standard should be similar. Obviously the internal standard selected should not be present in the original sample. Proper ap-

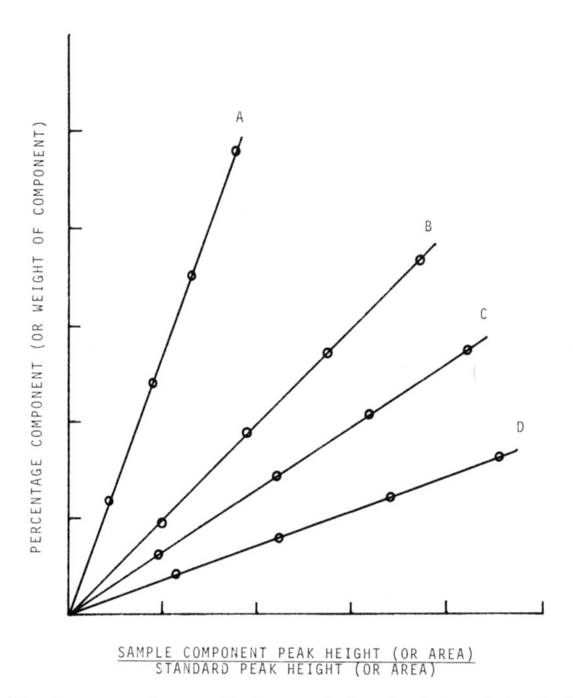

Fig. 13. Calibration curve for quantitative analysis using internal standard method. A, B, C, D: different sample components.

plication of this technique can provide quantitative data with a precision and accuracy better than ±1%.

iii. Area Normalization. Area normalization may be used when semi-quantitative information is all that is required. This technique is based upon the assumption that the ratio of the area of the chromatographic peak representing the sample component to that of the total area appearing beneath all peaks in the chromatogram is related to the amount of that sample component in the original sample. This may be better understood by reference to Fig. 14. The percentage of the sample component C_1, present in the original sample, would be calculated:

$$\%C_1 = A_1/(A_1 + A_2 + A_3 + A_4) \tag{24}$$

For a more general case:

$$\%C_i = A_i/(\sum A_i + \cdots + A_n) \tag{25}$$

where C_i is the concentration of sample component of interest, and A_i the area beneath each chromatographic peak.

For this technique to be valid a distinct and finite peak must be obtained for each component present in the sample. Any sample component that is

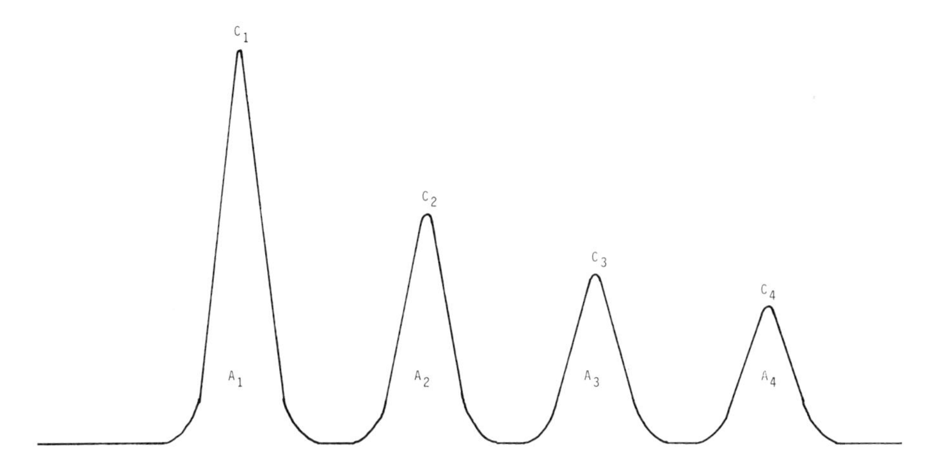

Fig. 14. Model chromatogram for calculation of percentage of each sample component using area normalization method.

not eluted from the column or that does not give a detector response would not appear in the calculations and thus would invalidate this approach.

The preceding calculations can be refined to account for differences in detector response factors for different sample compounds according to

$$\%C_i = f_i A_i / (\sum f_i A_i + \cdots + f_n A_n) \tag{26}$$

where f_i is the detector response factor.

Usual application of this technique yields quantitative values in the range ± 5–10% of the absolute value.

4.2.3 *Selectivity of Quantitative Analysis*

The selectivity of quantitative analysis depends primarily upon the type of sample and detector used. A quantitative analysis is possible for any sample whose components can be separated and detected by gas chromatography.

4.2.4 *Sensitivity of Quantitative Analysis*

Sensitivity of quantitative analyses using gas chromatography extends throughout the range of 1 ppb (or less) to 100%, depending upon the nature of the sample and the specific gas-chromatographic analysis applied.

4.2.5 *Accuracy of Quantitative Analysis*

With proper care and judicious selection of the appropriate quantitative gas-chromatographic technique accuracies of $\pm 1\%$ or better can be routinely accomplished.

4.2.6 Precision of Quantitative Analysis

Precision of quantitative gas-chromatographic analysis is better than ±1%, when the method is properly applied.

4.2.7 Sample Requirements for Quantitative Analysis

As with qualitative analysis, the only sample requirement for quantitative analysis using gas chromatography is that the sample, or a known derivative of the sample, be injected into the system, volatilized, separated chromatographically, and the components detected as they are eluted from the column. Sample dissolution or derivatization may be employed to make the analysis more convenient or to improve its quality.

4.2.8 Nondestructive or Destructive Analysis

Quantitative analysis using gas chromatography may be either nondestructive or destructive depending upon the technique and detector employed. The general discussion applied to qualitative analysis is also true for quantitative analysis.

4.2.9 Analysis Time

The time required to perform most quantitative gas-chromatographic analyses is less than 1 hr. The exact analysis time depends upon the sample and the specific gas-chromatographic apparatus and technique used.

4.2.10 Examples of Applications

Because of the many thousands of analyses to which gas chromatography has been applied, it is impossible within the scope of this text to discuss them all. What is presented is a brief discussion of a group of applications that are indicative of the potential of this technique.

a. PERMANENT AND INORGANIC GASES. Yu *et al.* (1970) describes a technique for the determination of moisture in air. Humidity determined in this manner is accurate to better than ±1°C dew point over the range 18–21°C.

Mindrup and Taylor (1970) discuss the analysis of liquid ammonia for traces of O_2, N_2, CO, CO_2, CH_4, and H_2O. A detection limit of better than 3 ppm is stated. Ayers (1969) describes the determination of ammonia in tobacco and smoke by means of a polyethyleneimine column with helium carrier gas and a thermal-conductivity detector. Jones (1967) has published a method using two columns of Porapak Q operated at different temperatures and a flow switching procedure for the analysis of NH_3, H_2O, H_2, O_2, N_2, CO, CO_2, H_2S, and the C_1–C_5 saturated hydrocarbons in refinery gases.

Graham and Stevenson (1970) have reported on the separation of argon, carbon dioxide, and phosgene. Dubansky (1968) has used a gas-chromatography technique to determine the amount of radiogenic argon in geological samples. Down to 10^{-9} mole of argon in 1 ml of gas sample can be determined with an accuracy of 1 to 5%. The solubilities of argon, helium, hydrogen, oxygen, methane, sulfur hexafluoride, and neopentane in aqueous solutions of KOH have been reported by Shoor *et al.* (1969). The saturated solutions were stripped with carrier gas and analyzed by gas chromatography.

Castello and Munari (1970) have reported on the use of a helium ionization detector for the determination of gaseous atmospheric pollutants.

Helzel (1970) describes a gas-chromatographic technique for the determination of gaseous inclusions in glass. The compounds analyzed for include CO_2, H_2, N_2, CO, O_2, and methanol. Walker (1968) uses multidiameter columns for the efficient determination of CO_2 in natural gas.

Carbon monoxide in blood and tissue has been determined by Blackmore (1970). Gas chromatography coupled with a special stripping technique has been used by Swinnerton *et al.* (1968) to measure CO in sea water.

Mallik and Khurana (1971) describe an analysis of automotive engine exhaust gases. Gas chromatography of gases emanating from a soil atmosphere is the subject of a report by Van Cleemput (1969).

The separation of ortho and para hydrogen has been outlined by Christofferson (1967). There have been many reports on the separation and determination of hydrogen and other occluded gases in steel.

Applebury and Schaer (1970) describe the analysis of pulp mill gases for H_2S, SO_2, and CH_3SH. Nebbia and Bellotti (1970) have discussed the determination of sulfur trioxide in air and sulfur dioxide.

Purer *et al.* (1969) use a cryogenic gas-chromatographic system for the separation of neon isotopes. Bruner and Di Corcia (1969) reported on the separation of nitrogen-14 and nitrogen-15 isotopes.

Lawson and McAdie (1970) describe a technique for the determination of oxides of nitrogen in air.

b. HYDROCARBONS. Specific retention volumes of the *n*-paraffins C_1–C_{10} and selected aromatic compounds have been determined using a hexadecane column; these have been reported by Snyder and Thomas (1968). Much work has been reported on the hydrocarbons C_1–C_4, both saturated and unsaturated. The hydrocarbons C_{14}–C_{32} were separated from fish products and analyzed by Gershbein and Singh (1969). Stuckey (1969) used an open tubular column for the analysis of C_6–C_{10} aromatics and C_1–C_{11} saturated hydrocarbons.

The analysis of C_5–C_8 cycloparaffins is the subject of a report by Leveque (1967). Kouznetsov and Scherbakova (1970) use graphitized carbon black to determine the configuration of the C_6–C_{12} cyclic hydrocarbons. Silver nitrate columns were used by Schmitt and Jonassen (1970) for the separation of high-boiling cyclic diolefins. Steranes, isoprenoids, and terpene hydrocarbons have also been analyzed by means of gas-chromatographic techniques.

Louis (1969) has studied the separating efficiency of 24 liquid phases for a homologous series of aromatic hydrocarbons. Impurities in benzene, toluene, and xylene products have been discussed by Terrada (1971). Brookman and Sawyer (1968a,b,c) have reported on several studies using salt-modified silica columns for the separation of aromatic hydrocarbons. Using a flame ionization detector, Sanchez (1968) provides the response range for biphenyl–terphenyl mixtures. Programmed temperature column operation for the quantitative analysis of biphenyls, terphenyls, and polyphenyls is discussed by Onuska and Janak (1968). Olson *et al.* (1967) describe the organic analysis of materials for aromatic hydrocarbons.

c. OXYGENATED COMPOUNDS. Gas chromatography has been used extensively used for the quantitative analysis of organic acids and esters. Nichikova *et al.* (1970) describe the separation of C_1–C_4 aliphatic acids in aqueous solution. Mahadevan and Zieve (1969) discuss the determination of volatile free fatty acids in human blood. Steam was used as the carrier gas by Nonaka (1970) for the analysis of C_2–C_{18} aliphatic carboxylic acids. Both dicarboxylic and tricarboxylic acids have been analyzed by gas chromatography. Many literature references describing the analysis of fatty acids after their conversion to a volatile ester are available. The applications include human physiological fluids, food materials, botanical specimens, microbiological and biological samples, and milk products. Many naturally occurring esters are routinely analyzed by gas chromatography in a variety of natural and synthetic products.

Doelle (1969) reported on the separation of C_1–C_7 alcohols by means of a polyethylene glycol liquid substrate. The analysis of the alcohol content of tequila and mescal is reported by Moreno and Llama (1970). Van Ling (1969) converted polyhydric alcohols to silyl ethers prior to analysis. Singliar and Dykyj (1970) give retention times and separation techniques for the mono-, di-, and tri-ethylene glycols. Other hydroxyl compounds analyzed by gas chromatography include the saccharides, glycerol, terpene alcohols, hydroxyl fatty acids, and phenols.

Other oxygenated compounds that are subject to quantitative analysis by gas chromatography include aldehydes, ketones, ethers, and steroids. The latter comprise a major field of application for the gas-chromatographic technique.

d. NITROGEN COMPOUNDS. Smith and Waddington (1969) describe the analysis of aliphatic amines. The same authors (1968) describe the use of porous polymers for the separation of aliphatic amines and aliphatic diamines. The method of determining primary amines of polynuclear aromatic hydrocarbons is given by Masuda and Hoffman (1969).

The amines of aliphatic carboxylic acids up to C_4 have been analyzed by Yasuda and Nakashima (1970). Dardenne *et al.* (1970) discuss the separation of the C_2–C_8 nitriles.

Appropriate derivatives were prepared by Gehrke *et al.* (1966) prior to the quantitative analysis of amino acids.

Other nitrogen-containing compounds subject to quantitative analysis by gas chromatography include amphetamines, alkaloids, carbamates, urea derivatives, hydrazines, cyanates and cyanides, nitroparaffins, pyridine bases, and vitamin mixtures.

e. SULFUR COMPOUNDS. Freedman (1968) used an open tubular column for the analysis of C_1–C_4 saturated alkyl thiols and sulfides. Pentafluorobenzyl derivatives were used by Kawahara (1971) to determine mercaptans in surface waters. Nagai *et al.* (1970) used gas chromatography to separate and determine alpha-olefin sulfonates. The samples were hydrogenated and converted to sulfonyl chlorides prior to analysis. Dagnall *et al.* (1969) studied a microwave-excited emissive detector for the determination of specific sulfur compounds. Using a microcoulometric detector, Ripperger (1968) was able to report a 5-ppm detection limit for sulfur compounds. Mizany (1970) discusses the use of a flame-photometric detector for the determination of sulfur compounds.

f. HALOGENATED COMPOUNDS. Faris and Lehman (1969) have reported on the use of porous polymer supports for the separation of C_1–C_2 halocarbons. Castello and D'Amato (1971) have studied the analysis of all of the isomers of C_1–C_6 alkyl iodides. Karasek (1970) used a coated bentone-34 column for the analysis of brominated benzene compounds in 5–10 min. Williams and Umstead (1968) determined the traces of chlorinated hydrocarbons in air.

An indium flame detector was employed by Gutsche and Herrmann (1970) for the specific detection of bromine compounds. A flame emission detector that is specific for halogenated organic compounds has been described by Overfield and Winefordner (1970).

A great deal of the interest in gas-chromatographic analyses has been applied to pesticide analysis, commercial aerosol formulations, thyroid hormones, chemical warfare agents, and the analysis of polychlorinated biphenyls.

g. ORGANOMETALLIC AND NONGASEOUS INORGANICS. Svob (1970) shows the separation and analysis of alkyl leads. Pommier and Guiochon (1970) describe the analytical conditions for the determination of metal carbonyls. Methyl mercury compounds in fish have been analyzed by Johansson *et al.* (1970). Organophosphorous compounds were analyzed on silicone oil columns by Ives and Giuffrida (1970).

Other inorganic compounds and elements that have been analyzed by gas chromatography include uranium hexafluoride, silicates, phospholipids, Mo, Te, Al, Cr, Cd, Zn, Cu, Fe, Sb, As, Ge, Ti, Sn, Sc, V, Be, Bo, C, Pb, Mn, Mg, Ru, and Se.

5 Auxiliary Techniques

The technique of gas chromatography is rapid, precise, and quantitative; and the instrumentation required is not overly complex. Its one serious drawback is that in the majority of its applications it does not provide unequivocal identification of the sample components. This negative aspect has been compensated for by combining gas chromatography with other analytical techniques. Thus two or more methods, including gas chromatography, may be selected so that they complement one another and provide analytical information that is superior to what either is capable of providing alone. The three areas to which the majority of effort has been applied are gas chromatography–mass spectrometry, gas chromatography–spectrophotometry, and gas chromatography–chemical methods.

5.1 Gas Chromatography–Mass Spectrometry

The gas chromatograph–mass spectrometer (GC/MS) is an almost ideal marriage of analytical techniques. A GC/MS system consists of three basic components: (1) a gas chromatograph to resolve the components of the sample; (2) an interface device to allow direct coupling of the high-pressure gas chromatograph to the low-pressure mass-spectrometer ion source; (3) the mass spectrometer that provides unequivocal identification of the pure sample components owing to the unique ion fragments formed as a result of its observation of the sample components.

The key factor in a GC/MS system is the interface between the two individual system components. This interface may be a splitting arrangement whereby most of the column effluent is led to the gas-chromatograph detector, the remainder being admitted to the ion source of the mass spectrometer. An alternative arrangement is to pass all of the column effluent through an interface device that removes the major portion of the carrier gas and allows only enriched sample components to enter the ion source of the mass spectrometer. In this technique the mass-spectrometer ion source is modified to measure total ion current, in addition to yielding fragments for subsequent spectrometric identification. This total ion-current measurement serves the normal function of the gas-chromatograph detector. The emerging sample component in the column effluent is scanned by the mass spectrometer in only a fraction of the time required for its complete elution from the column. A single scan over a wide range of sample masses can be accomplished in only a few seconds with modern instrumentation. Normal procedure is to obtain several scans for a single chromatograph peak to ensure that a single peak does not contain more than one sample component.

With the proper use of a GC/MS system, one can obtain in a single analysis a chromatogram of the sample showing the number of compounds present and the amounts of each, along with a mass spectrum of each component that provides absolute identification of that component. This qualitative and quantitative information would be difficult if not impossible to obtain using either instrument alone. The time required for such an analysis seldom exceeds 1 hr.

For detailed information and references on the use of GC/MS, the reader is directed to the comprehensive review by Junk (1972), who discusses the various combinations of gas chromatography–mass spectrometry and applications of the technique.

5.2 Gas Chromatography–Spectrophotometry

The reason for the union of gas chromatography with spectrophotometry is again to supplement the limited qualitative capabilities of the gas chromatograph. The spectrophotometer segment of the system may measure in either the infrared or ultraviolet region of the spectrum, depending upon the suspected nature of the sample. The use of an integrated system is faster and more convenient than trapping the individual components and subsequently identifying them by spectrophotometric techniques.

The direct coupling of a gas chromatograph with either an infrared or ultraviolet spectrophotometer is not easy. A splitting arrangement must be used on the column effluent, with a portion being directed to a conventional gas-chromatograph detector and the remainder being sent to the measuring cell of the spectrophotometer. Generally, the spectrophotometric techniques are not so sensitive as mass spectrometry, thus requiring larger samples or multiple spectrophotometric scans for easy identification of the sample components. Also, the spectrophotometer is limited as to the time in which it can scan a sample over a reasonable wavelength and still maintain adequate resolution to yield sample identification. Often this time is greater than the time required for complete emergence of the gas-chromatograph peak, thus requiring interruption of the gas flow to allow the sample component to remain in the spectrophotometer measuring chamber. One additional problem is that unless the sample components are well resolved on the gas-chromatograph column, two or more may be present in the spectrophotometer simultaneously, thus confusing the qualitative interpretation.

In spite of the above difficulties, various combinations of a gas chromatograph–spectrophotometer have been successfully employed. Littlewood (1968a,b) has discussed the direct coupling of a gas chromatograph with infrared and ultraviolet spectrophotometers. Freeman (1969) has reviewed the procedure for combining gas chromatography and infrared spectrophotometry. This review includes tandem combination of gas chromatography and infrared spectrophotometry, interrupted elution gas chromatography, and rapid-scan infrared spectrophotometry.

5.3 Gas Chromatography–Chemical Methods

The combination of gas chromatography with chemical methods is of a more limited value than the foregoing techniques. It generally consists of allowing the effluent from the column to enter a reaction chamber containing specific reagents. A positive reaction with the reagent should indicate the presence of a particular class of compound. Since most reagents are not absolutely specific, the qualitative information obtained is more indicative of sample type than of a specific compound present in the sample. However, coupled with other information regarding the sample, this may be sufficient to yield the desired qualitative information.

In cases where the chromatographic detector is nondestructive of the sample, the reaction chamber may be placed after the detector. When using a destructive gas-chromatograph detector, it is necessary to employ a splitting arrangement that directs a portion of the effluent from the gas-chromatograph column to the detector and the remainder to the chemical-reaction chamber.

A good review of chemical methods coupled with gas chromatography has been given by Littlewood (1968c).

5.4 Other Techniques in Combination with Gas Chromatography

Gas chromatography has been combined with other analytical techniques to solve specific analysis problems. However these combinations have received limited use to date. Examples include the combining of gas chromatography with atomic-absorption spectrophotometry, nuclear magnetic resonance, emission spectrometry, polarography, and subsequent pyrolysis or chemical modification of the sample to yield a second chromatogram giving elemental or chemical derivative information.

6 Special Literature on Gas Chromatography

There exists a special body of literature on gas chromatography that enables the analyst to rapidly search out solutions to problems involving systematic materials analysis. The proper utilization of this special literature allows the analyst efficiently and expeditiously to advance his knowledge on the subject so that he remains up-to-date on the state of the art and can thus solve his own unique analysis problems. Therefore in the systematic analysis of materials, one need not be an expert in gas chromatography to take advantage of its latest developments, applications, and techniques.

The available special literature pertaining to gas chromatography may be generally classified into one of five categories, exclusive of reference books: 1. technical journals, 2. reviews, 3. symposia proceedings, 4. bibliographies, and 5. abstracts. Each has a slightly different function and often provides the user with varying degrees of information in different formats.

6.1 Technical Journals

As in any field of science, the technical journal provides the means whereby the latest accomplishments are communicated to other workers with similar interests. Since gas chromatography is almost universal in its application, references to its uses may be found in nearly all technical journals. However, the bulk of the primary or original information regarding gas chromatography and its applications may be found in the following journals:

Analyst,
Analyticia Chemicia Acta,
Analytical Chemistry,
Journal of Chromatography,
Chromatographia,
Journal of Chromatographic Science (Formerly *Journal of Gas Chromatography*), and
Zeitschrift für Analytishe Chemie.

In seeking the solution to a specific analysis problem it is worthwhile to review the articles published in the these journals.

6.2 Reviews

Reviews attempt to provide references and brief commentaries on all work reported in the technical literature. They may be general in nature or deal with only one aspect or application of the technique. Each review covers only a finite time period, generally only that period of time since a prior review. Reviews are normally included in the technical journals or in a special journal issue. The discussion of each reference in a review is limited and often less organized and indexed than that provided by the abstract services. Nonetheless, such discussions are useful and should be considered when seeking solutions to a particular analysis problem. The most encompassing review specific to gas chromatography is that given in the journal, *Analytical Chemistry: Annual Reviews*. In alternate years the *Annual Reviews* are dedicated to specific analytical techniques. The section on gas chromatography is an excellent summary of works published since the previous review. In addition, the aforementioned technical journals periodically review some specific aspect of gas chromatography or its applications (see Section 6.1).

6.3 Symposia Proceedings

Symposia proceedings are publications of the complete texts of papers that have been presented. Procedure varies with the different symposia. For some the papers presented are later independently submitted to technical journals by the authors, while for others the symposia proceedings are published in a single technical journal, often in one volume. Other symposia sponsors publish the proceedings in a single bound volume or book. The abstracting services provide detailed information on the references to all gas-chromatography symposia. By nature, the symposia proceedings do not provide the broad coverage afforded by reviews or abstracts, but quite often present the detailed treatment of a given topic required by the analyst.

6.4 Bibliographies

A bibliography is more useful to the analyst intimately concerned with gas chromatography than to the occasional user. The degree of usefulness depends primarily on the requirements of the analyst and the organization and indexing of the bibliographic material by the compiler. The most comprehensive bibliographies on gas chromatography are those prepared by the Preston Technical Abstracting Services and published in the *Journal of Chromatographic Science*.

6.5 Abstracts

For the analyst seeking a solution to a specific materials analysis problem, the most complete and valuable presentation of information on gas chromatography is that given by the abstracting services. The abstracts that they provide allows one to review all of the work reported in any given area of gas chromatography with a minimum of effort. Items of specific interest may then be obtained for detailed consideration. The most prominent gas-chromatography abstracts are provided by the following:

1. *Gas Chromatography Abstracts* (C.E.H. Knapman, ed.), Elsevier, Amsterdam. These abstracts are sponsored by the Gas Chromatography

Discussion Group of the Institute of Petroleum, London, England. They have been published annually since 1958.

2. *Gas Chromatography Abstracting Service,* Preston Technical Abstracts Co., Evanston, Illinois. This service provided coded, punched cards containing the abstracts through 1969. Since 1970 these abstracts have been available only in bound booklet form. The abstracts are issued weekly and are available beginning with the literature published since 1952.

Because of the detailed cross indexing used by the abstracts, information on any aspect of gas chromatography, including specific materials analysis, is easily obtained by the analyst. Typical section headings are (a) General Articles; (b) Theory, Definitions, and Retention Data; (c) Apparatus and Techniques; (d) Carrier Gas and Column-Packing Material; (e) Sample Type; (f) Applications and Specialized Techniques; and (g) Related Methods and Techniques. Each section is then further subdivided to facilitate locating the information pertinent to a specific analysis problem.

References

Amy, J. W., and Baitenger, W. E. (1963). *In* "Lectures on Gas Chromatography, 1962" (H. A. Szymanski, ed.), pp. 19–31. Plenum, New York.

Applebury, T. E., and Schaer, W. J. (1970). *J. Air Pollut. Contr. Ass.* **20,** 83.

Ashley, J. W. (1967). *Dissert. Abstr.* **27B** (11), 3804B.

Aue, W. A., and Teli, P. M. (1971). *J. Chromatogr.* **62,** 15.

Ayers, C. W. (1969). *Talanta* **16,** 1085.

Back, R. A., Friswell, N. J., and Boden, J. C. (1969). *J. Chromatogr. Sci.* **7,** 708.

Barakat, M. F. (1967). *J. Sci. Instrum.* **44,** 1031.

Barker, P. G., and Purnell, J. H. (1970). *Trans. Faraday Soc.* **66,** 163.

Berezkin, V. G. (1968). "Analytical Reaction Gas Chromatography." Plenum, New York.

Berezkin, V. G., and Tatarinsky, V. S. (1971). *Zh. Anal. Khim.* **25,** 398.

Beroza, M., and Sarmiento, R. (1964). *Anal. Chem.* **36,** 1744.

Blackmore, D. J. (1970). *Analyst* **95,** 439.

Blandenet, G., and Robin, J. P. (1966). *J. Gas Chromatogr.* **4,** 288.

Bowman, M. C., and Beroza, M. (1968). *Anal. Chem.* **40,** 1448.

Brookman, D. J., and Sawyer, D. T. (1968a). *Anal. Chem.* **40,** 1368.

Brookman, D. J., and Sawyer, D. T. (1968b). *Anal. Chem.* **40,** 1847.

Brookman, D. J., and Sawyer, D. T. (1968c). *Anal. Chem.* **40,** 2013.

Brookman, D. J., and Sawyer, D. T. (1968d). *Anal. Chem.* **40,** 106.

Brown, I. (1962). *Conf. Chromatography Univ. New South Wales, Kennsington.*

Bruner, F. A., and Cartoni, G. P. (1964). *Anal. Chem.* **36,** 1522.

Bruner, F., and DiCorcia, A. (1969). *J. Chromatogr.* **45,** 304.

Brunnock, J. V., and Luke, L. A. (1968). *Anal. Chem.* **40,** 2158.

Brunnock, J. V., and Luke, L. A. (1969). *Anal. Chem.* **41,** 1126.

Bryzgalova, N. I., Gavrilova, T. B., Kiselev, A. V., and Khokholva, T. D. (1968). *Neftekhimiya* **8,** 915.

Byrnes, P. W. (1972). *Ind. Res.* **14,** 44.

Callery, I. M. (1970). *J. Chromatogr. Sci.* **8,** 408.
Castello, G., and D'Amato, G. (1971). *J. Chromatogr.* **58,** 127.
Castello, G., and Munari, S. (1970). *Anal. Abstr.* **19,** 1806.
Chizhkov, V. P. (1970). *Anal. Abstr.* **18,** 4443.
Chovin, P., and Lebbe, J. (1962). *In* "Separation Immediate et Chromatographie" (J. Tranchant, ed.), p. 90. GAMS, Paris.
Christofferson, D. J. (1967). *Dissert. Abstr.* **28B,** 517B.
Condon, R. D., and Ettre, L. S. (1968). *In* "Instrumentation in Gas Chromatography" (J. Krugers, ed.), pp. 87–109. Centrex, Eindhoven.
Coudert, M., and Vergnaud, J. M. (1971). *J. Chromatogr.* **54,** 1.
Coudert, M., Larrat, J., and Vergnaud, J. M. (1971). *J. Chromatogr.* **58,** 159.
Cramers, C.A.M.G. (1967). "Some Problems Encountered in High Resolution Gas Chromatography." Technische Hogeschool, Eindhoven.
Cramers, C.A.M.G., and Van Kessel, M. M. (1968). *J. Gas Chrom.* **6,** 577.
Creitz, E. C. (1969). *J. Chromatogr. Sci.* **7,** 137.
Cronin, D. A. (1970). *J. Chromatogr.* **48,** 406.
Crossley, J. (1970). *J. Chromatogr. Sci.* **8,** 426.
Dagnall, R. M., Pratt, S. J., West, T. S., and Deans, D. R. (1969). *Talanta* **16,** 797.
Darbre, A., and Islam, A. (1970). *J. Chromatogr.* **49,** 293.
Dardenne, G. A., Severin, M., and Marlier, M. (1970). *J. Chromatogr.* **47,** 182.
Dave, S. B. (1969). *J. Chromatogr. Sci.* **7,** 389.
Desty, D. H. (1967) *In* "The Gas Liquid Chromatography of Steroids" (J. K. Grant, ed.), p. 7. Cambridge Univ. Press, London and New York.
Devaux, P., and Guiochon, G. (1970). *J. Chromatogr. Sci.* **8,** 502.
DiCorcia, A., and Bruner, F. (1970). *J. Chromatogr.* **49,** 139.
Dietz, R. N. (1968). *Anal. Chem.* **40,** 1576.
Dodd, H. C. (1971). *Anal. Chem.* **43,** 1724.
Doelle, H. W. (1969). *J. Chromatogr.* **42,** 541.
Dubansky, A. (1968). *Chem. Abstr.* **69,** 12163.
Dubsky, H. (1970). *J. Chromatogr.* **47,** 313.
Dunlop, A. S., and Pollard, S. A. (1971). *Anal. Chem.* **43,** 1344.
Ettre, L. S. (1965). "Open Tubular Columns in Gas Chromatography." Plenum, New York.
Ettre, L. S. (1969). *Proc. Nat. Symp. Heterogenous Catal. Controll Air Pollution, 1st Cincinati, Ohio,* p. 349.
Ettre, L. S., and Averill, W. (1961). *Anal. Chem.* **33,** 680.
Faris, A., and Lehman, J. G. (1969). *Septn. Sci.* **4,** 225.
Fenimore, D. C., Zlatkis, A., and Wentworth, W. E. (1968). *Anal. Chem.* **40,** 1594.
Folmer, O. F. (1971). *Anal. Chem.* **43,** 1057.
Franc, J., and Kolowskova, V. J. (1965). *J. Chromatogr.* **17,** 221.
Freedman, R. W. (1968). *J. Gas Chromatogr.* **6,** 495.
Freeman, S. K. (1969). *In* "Ancillary Techniques of Gas Chromatography" (L. S. Ettre and W. H. McFadden, eds.), p. 227. Wiley(Interscience), New York.
Frolov, I. A. (1968). *Anal. Abstr.* **15,** 1940.
Garzo, G. (1967). *Magyar Kem. Lapja.* **22,** 463.
Gaugh, T. A., and Walker, E. A. (1970). *Analyst* **95,** 1.
Gehrke, C. W., Lamkin, W. M., Stalling, D. L., and Shahrokhi, F. (1966). *Biochem. Biophys. Res. Commun.* **19,** 328.
Gershbein, L. L., and Singh, E. J. (1969). *J. Amer. Oil Chem. Ass.* **46,** 554.
Giddings, J. C. (1964). *Anal. Chem.* **36,** 1170.
Giddings, J. C. (1966). Principles and theory. *In* "Dynamics of Chromatography," Pt. I. Dekker, New York.

Giddings, J. C., Seager, S. L., Stucki, L. R., and Stewart, G. H. (1960). *Anal. Chem.* **32,** 867.
Gill, J. M., and Perone, S. P. (1969). *J. Chromatogr. Sci.* **7,** 709.
Gladney, H. M., Dowden, B. F., and Swallen, J. D. (1969). *Anal. Chem.* **41,** 883.
Gluekauf, E. (1955). *In* "Ion Exchange and Its Applications," p. 34. Soc. Chem. Ind., London.
Golay, M. J. E. (1958). *In* "Gas Chromatography" (V. J. Coates, ed.), p. 1. Academic Press, New York.
Golay, M. J. E. (1960). *In* "Gas Chromatography, 1960" (R. P. W. Scott, ed.), p. 139. Butterworths, London and Washington, D.C.
Goltz, H. L., and Moffat, J. B. (1970). *J. Chromatogr. Sci.* **8,** 596.
Graham, R. J., and Stevenson, F. D. (1970). *J. Chromatogr.* **47,** 555.
Grob, R. L., and McGaugh, E. J. (1971). *J. Chromatogr.* **59,** 13.
Gronendyk, H., and Van Kemenade, A. W. C. (1969). *Chromatographia* **2,** 107.
Guichard, N., and Buzon, J. (1969). *In* "Practical Manual of Gas Chromatography" (J. Tranchant, ed.), p. 181. Elsevier, Amsterdam.
Guran, B. T., O'Brian, R. J., and Anderson, D. H. (1970). *Anal. Chem.* **42,** 115.
Gutsche, B., and Herrmann, R. (1970). *Z. Anal. Chem.* **249,** 168.
Halasz, I., and Horvath, C. (1963). *Anal. Chem.* **35,** 499.
Halasz, I., Bruderreck, H., and Schneider, W. (1970). U.S. Patent 3,513,636 (May 26, 1970).
Hamilton, C. H. (1968). *In* "Instrumentation in Gas Chromatography" (J. Krugers, ed.), pp. 33–52. Centrex, Eindhoven.
Handy, P. R., and MacDonald, A., Jr. (1970). *Anal. Chem.* **53,** 780.
Hargrove, G. L., and Sawyer, D. T. (1968). *Anal. Chem.* **40,** 409.
Harris, R. J. (1968). U.S. Patent 3,385,113 (May 28, 1968).
Harris, W. E., and Habgood, H. M. (1966). "Programmed Temperature Gas Chromatography." Wiley, New York.
Hartman, C. H. (1969). *J. Chromatogr. Sci.* **7,** 163.
Harvey, D., and Chalkey, D. E. (1955). *Fuel* **34,** 191.
Helzel, M. (1970). *Anal. Abstr.* **19,** 28.
Herout, V., Streibl, M., and Holasova, M. (1971). *Flavour Ind.* **1,** 673.
Hoffman, R. L., and Evans, C. D. (1966). *Anal. Chem.* **38,** 1309.
Isbell, A. F., and Sawyer, D. T. (1969). *Anal. Chem.* **41,** 1381.
Ives, N. F., and Giuffrida, L. (1970). *J. Ass. Off. Anal. Chem.* **53,** 973.
James, A. T., and Martin, A. J. P. (1952). *Biochem J.* **50,** 679.
Johansson, B., Ryhage, R., and Westoo, G. (1970). *Anal. Chem. Scand.* **24,** 2349.
Jones, W. L. (1961). *Anal. Chem.* **33,** 829.
Jones, C. N. (1967). *Anal. Chem.* **39,** 1858.
Junk, G. A. (1972). *Int. J. Mass Spectrom.* **8,** 1.
Juvet, R. S., and Cram, S. P. (1970). *Anal. Chem.* **42,** 1R.
Juvet, R. S., and Dal Nogare, S. (1968). *Anal. Chem.* **40,** 33R.
Juvet, R. S., Smith, J. L., and Li, K. P. (1971). *In* "Advances in Chromatography, 1971" (A. Zlatkis, ed.), p. 166. Univ. of Houston Publishers, Houston, Texas.
Kaiser, R. (1968). *Chromatographia* **1,** 34.
Kaiser, R. (1969). *Chromatographia* **2,** 383.
Kaiser, R. (1970). *Chromatographia* **3,** 38.
Karas, E. C. (1967). U.S. Patent 3,339,582 (September 5, 1967).
Karasek, F. W. (1970). *J. Chromatogr. Sci.* **8,** 282.
Kawahara, F. K. (1971). *Environm. Sci. Technol.* **5,** 235.
Kieselback, R. (1960). *Anal. Chem.* **32,** 380.
Kim, B. S., Fetterman, C. P., and Walker, J. Q. (1970). U.S. Patent 3,498,107 (March 3, 1970).

Kirkland, J. J., and De Stefano, J. J. (1970). *J. Chromatogr. Sci.* **8,** 309.
Kiselev, A. V. (1970). *J. Chromatogr.* **49,** 84.
Kiselev, A. V., and Yashin, Y. I. (1969). *In* "Gas Adsorption Chromatography," pp. 1–45. Plenum, New York.
Klinkenberg, A., and Sjenitzer, F. (1956). *Chem. Eng. Sci.* **5,** 258.
Kouzentsov, A. V., and Scherbakova, K. D. (1970). *J. Chromatogr.* **49,** 21.
Kovats, E. (1958). *Helv. Chem. Acta* **41,** 1915.
Kovats, E. (1966). *Advan. Chromatogr.* **I,** 23.
Krugers, J. (1968). *In* "Instrumentation in Gas Chromatography" (J. Krugers, ed.), pp. 53–70. Centrex Publ., Eindhoven.
Lavoue, G. (1968). *J. Gas Chromatogr.* **6,** 233.
Lawson, A., and McAdie, H. G. (1970). *J. Chromatogr. Sci.* **8,** 731.
Lawson, A. E., and Miller, J. M. (1966). *J. Gas Chromatogr.* **4,** 273.
Lebedeva, N. P., Frolov, I. I., and Yashin, Y. I. (1971). *J. Chromatogr.* **58,** 11.
Leveque, R. E. (1967). *Anal. Chem.* **39,** 1811.
Levins, R. J., and Ottenstein, D. M. (1967). *J. Gas Chromatogr.* **5,** 539.
Levy, R. C., Murray, P. A., Gesser, H. D., and Hougen, F. W. (1968). *Anal. Chem.* **40,** 459.
Levy, R. C., Fanter, D. L., and Wolf, C. J. (1971). *In* "Advances in Chromatography, 1971" (A. Zlatkis, ed.), p. 155. Univ. of Houston Publ., Houston, Texas.
Liberti, A., Nota, G., and Goretti, G. (1968). *J. Chromatogr.* **38,** 282.
List, G. R., Hoffman, R. L., and Evans, C. D. (1967). *Nature* (*London*) **213,** 380.
Littlewood, A. B. (1962). *In* "Gas Chromatography" p. 351. Academic Press, New York.
Littlewood, A. B. (1968a). *Chromatographia* **1,** 133.
Littlewood, A. B. (1968b). *Chromatographia* **1,** 223.
Littlewood, A. B. (1968c). *Chromatographia* **1,** 37.
Littlewood, A. B. (1970). "Gas Chromatography. Principles, Techniques, and Applications." Academic Press, New York.
Llewellyn, P. W. (1970). U.S. Patent 3,507,147 (April 21, 1970).
Louis, R. (1969). *Z. Anal. Chem.* **244,** 81.
Lovelock, J. E. (1961). *Anal. Chem.* **33,** 162.
Lovelock, J. E., Shoemake, G. R., and Zlatkis, A. (1964). *Anal. Chem.* **36,** 1410.
McKinney, C. B., and Sheppard, W. M. (1968). U.S. Patent 3,374,660 (March 26, 1968).
Mahadevan, V., and Zieve, L. (1969). *J. Lipid Res.* **10,** 338.
Mallik, K. L., and Khurana, M. L. (1971). *Z. Anal. Chem.* **253,** 125.
Mamaril, J. C., and Meloan, C. E. (1965). *J. Chromatogr.* **17,** 23.
Masuda, Y., and Hoffman, D. (1969). *J. Chromatogr. Sci.* **7,** 694.
Maume, B. (1965). *These*, Dijon.
Meyers, H. S., and Rosman, A. (1969). *J. Chromatogr. Sci.* **7,** 751.
Mindrup, R. F., and Taylor, J. H. (1970). *J. Chromatogr. Sci.* **8,** 723.
Mizany, A. I. (1970). *J. Chromatogr. Sci.* **8,** 151.
Mohnke, M., and Saffert, W. (1962). *In* "Gas Chromatographie 1961" (M. Van Swaay, ed.), p. 216. Butterworths, London and Washington, D.C.
Moreno, A. M., and Llama, M. (1970). *Anal. Abstr.* **18,** 4364.
Myakishev, G. Y. (1967). *Usp. Khim.* **36,** 1484.
Nagai, T., Shigeru, H., Yamane, I., and Mori, A. (1970). *J. Amer. Oil Chem. Ass.* **47,** 505.
Nebbia, L., and Bellotti, V. (1970). *Anal. Abstr.* **19,** 2211.
Nelson, K. H., and Lysyj, I. (1969). *Water Res.* **3,** 357.
Nichikova, P. R., Kotel'Nikov, B. P., Martynushkina, A. V., and Kiseleva, N. S. (1970). *Chem. Abstr.* **72,** 6080.
Nonaka, A. (1970). *Anal. Abstr.* **18,** 4096.
Novak, J., Ruzickova, J. G., and Wicar, S. (1971). *J. Chromatogr.* **60,** 127.

Novotny, M., Bartle, K. D., and Blomberg, C. (1969). *J. Chromatogr.* **46,** 469.
Nowak, A. V. (1968). *Dissert. Abstr.* **29B,** 510B.
Oberholtzer, J. E. (1969). *J. Chromatogr. Sci.* **7,** 720.
Oberholtzer, J. E., and Rogers, L. B. (1969). *Anal. Chem.* **41,** 1234.
Olson, R. J., Oro, J., and Zlatkis, A. (1967). *Geochim. Cosmochim. Acta.* **31,** 1935.
Onuska, F., and Janak, J. (1968). *Chem. Zvesti* **22,** 929.
Overfield, C. V., and Winefordner, J. D. (1970). *J. Chromatogr. Sci.* **8,** 233.
Palframan, J. F., and Walker, E. A. (1967). *Analyst* **92,** 71.
Parkinson, R. T., and Wilson, R. E. (1968). *J. Chromatogr.* **36,** 553.
Pattee, H. E., Wiser, E. H., and Singleton, J. A. (1970). *J. Chromatogr. Sci.* **8,** 668.
Perrett, R. H. (1965). *Anal. Chem.* **37,** 1346.
Petitjean, D. L., and Leftault, C. J. (1963). *J. Gas Chromatogr.* **1,** 18.
Pitt, P. (1969). *Chromatographia* **2,** 304.
Pommier, C., and Guiochon, G. (1970). *J. Chromatogr. Sci.* **8,** 486.
Powell, R. A. (1969). *Lab. Equip. Digest.* **7,** 73.
Prescott, B. O., and Wise, H. L. (1966). *J. Gas Chromatogr.* **4,** 80.
Prescott, B. O., and Wise, H. L. (1969). U.S. Patent 3,451,780 (June 24, 1969).
Preston, S. T. (1970). *J. Chromatogr. Sci.* **8,** 18A.
Prevot, A. F. (1969). *In* "Practical Manual of Gas Chromatography" (J. Tranchant, ed.). Elsevier, Amsterdam.
Purer, A., Kaplan, R. L., and Smith, D. R. (1969). *J. Chromatogr. Sci.* **7,** 504.
Ripperger, W. (1968). *Chem. Abstr.* **69,** 4994.
Roberts, D. R. (1968). *J. Gas Chromatogr.* **6,** 27.
Rohrschneider, L. (1959). *Z. Anal. Chem.* **170,** 256.
Rohrschneider, L. (1965). *Z. Anal. Chem.* **211,** 18.
Ross, W. D., and Jefferson, R. T. (1970). *J. Chromatogr. Sci.* **8,** 386.
Samarkina, V. I., and Vereschagina, M. A. (1970). *Chem. Abstr.* **74,** 121.
Sanchez, L. G. (1968). *Anal. Abstr.* **15,** 2366.
Schmitt, D. L., and Jonassen, H. B. (1970). *Anal. Chim. Acta* **49,** 580.
Schussler, P. W. H. (1969). *J. Chromatogr. Sci.* **7,** 763.
Schwartz, R. D. (1967). *In* "The Practice of Gas Chromatography," pp. 51–69. Wiley (Interscience), New York.
Schwartz, R. D., Brasseaux, D. J., and Shoemake, G. R. (1963). *Anal. Chem.* **35,** 496.
Science Guide to Scientific Instruments (1971–1972).
Seide, H., Serfas, O., and Assman, K. (1970). *Anal. Abstr.* **19,** 1871.
Shoor, S. K., Walker, R. D., and Gubbins, K. E. (1969). *J. Phys. Chem.* **73,** 312.
Simmonds, P. G., Fenimore, D. C., Pettit, B. C., Lovelock, J. E., and Zlatkis, A. (1967). *Anal. Chem.* **39,** 1428.
Singliar, M., and Dykyj, J. (1970). *Anal. Abstr.* **19,** 312.
Smith, J. R. L., and Waddington, D. J. (1968). *Anal. Chem.* **40,** 522.
Smith, J. R. L., and Waddington, D. J. (1969). *J. Chromatogr.* **42,** 183.
Smith, E. D., Johnson, J. L., and Oathout, J. M. (1964). *Anal. Chem.* **36,** 1750.
Snyder, P. S., and Thomas, J. F. (1968). *J. Chem. Eng. Data* **13,** 527.
Stuckey, C. L. (1969). *J. Chromatogr. Sci.* **7,** 177.
Supina, W. R. (1963). *In* "Lectures on Gas Chromatography, 1962" (H. A. Szymanski, ed.), p. 33. Plenum, New York.
Supina, W. R., and Rose, L. P. (1969). *J. Chromatogr. Sci.* **7,** 192.
Svob, V. (1970). *Chem. Abstr.* **72,** 81038.
Swinnerton, J. W., Linnenbrom, V. J., and Cheek, C. H. (1968). *Chem. Abstr.* **69,** 5113.
Szonntagh, E. L. (1968). *Dechema. Monogr.* **62,** 205.
Takacs, J., and Kralik, D. (1970). *J. Chromatogr.* **50,** 379.

Taylor, J. H., and Shoemake, G. R. (1972). *J. Chromatogr. Sci.* **10,** 48.
Teranishi, R. (1967). *Perfum. Essent. Oil Rec.* **58,** 172.
Terrada, O. (1971). *Chem. Abstr.* **74,** 88997.
Thompson, C. J., Coleman, H. J., Hopkins, R. L., and Roll, H. T. (1964). *J. Chem. Eng. Data.* **9,** 293.
Tochner, M., Magnuson, J. A., and Soderman, L. Z. (1969). *J. Chromatogr. Sci.* **7,** 740.
Todd, J. W., and Courneya, C. G. (1968). U.S. Patent 3,393,551 (July 23, 1968).
Tranchant, J. (1969). *In* "Practical Manual of Gas Chromatography," pp. 1–28. Elsevier, Amsterdam.
Tsuda, T., Tokoro, N., and Ishii, D. (1969). *J. Chromatogr.* **46,** 241.
Tucknott, O. G., and Williams, A. A. (1969). *Anal. Chem.* **41,** 2086.
Van Cleemput, O. (1969). *J. Chromatogr.* **45,** 317.
Van Deemter, J. J., Zuiderweg, F. J., and Klinkenberg, A. (1956). *Chem. Eng. Sci.* **5,** 271.
Van Ling, G. (1969). *J. Chromatogr.* **44,** 175.
Viska, J., Kiss, F., Pollak, M., and Pospichal, O. (1970). *J. Chromatogr.* **51,** 103.
Walker, J. Q. (1968). *Anal. Chem.* **40,** 226.
Walker, J. Q., and Wolf, C. J. (1970a). *Anal. Chem.* **42,** 1652.
Walker, J. Q., and Wolf, C. J. (1970b). *J. Chromatogr. Sci.* **8,** 513.
Waters Associates (1972). Catalog of Chromatographic Supplies.
Watkin, B. H. L., Evans, N., and Arnold, J. F. K. (1969). Ger. Offen. 1,909,053.
Williams, F. W., and Umstead, M. E. (1968). *Anal. Chem.* **40,** 2232.
Willis, D. E. (1968). *Anal. Chem.* **40,** 1597.
Wolf, C. J., Levy, R. L., and Walker, J. Q. (1971). *Ind. Res.* **13,** 40.
Yannone, M. E. (1968). *J. Gas Chromatogr.* **6,** 465.
Yasuda, K., and Nakashima, K. (1970). *Anal. Abstr.* **19,** 341.
Yu, J., Hedlin, C. P., and Green, G. H. (1970). *J. Chromatogr. Sci.* **8,** 840.
Zielenski, W. L., Fishbein, L., and Thomas, R. O. (1967). *Anal. Chem.* **39,** 1674.
Zlatkis, A., and Walker, J. Q. (1963). *Anal. Chem.* **35,** 1359.
Zlatkis, A., Oro, J., and Kimball, A. P. (1960). *Anal. Chem.* **32,** 162.

Selected Additional Reading

Ambrose, D., and Ambrose, B. (1962). "Gas Chromatography." Wiley, New York.
Berezkin, V. G. (1968). "Analytical Reaction Gas Chromatography." Plenum, New York.
Burchfield, H. P., and Storrs, E. E. (1962). "Biochemical Applications of Gas Chromatography." Academic Press, New York.
Dal Nogare, S., and Juvet, R. S. (1962). "Gas-Liquid Chromatography. Theory and Practice." Wiley (Interscience), New York.
Domsky, I., and Perry, J. (eds.) (1971). "Recent Advances in Gas Chromatography." Dekker, New York.
Eik-Nes, K. B., and Horning, E. C. (eds.) (1968). "Gas Phase Chromatography of Steroids." Springer-Verlag, Berlin and New York.
Ettre, L. S. (1965). "Open Tubular Columns in Gas Chromatography." Plenum, New York.
Ettre, L. S., and McFadden, W. H. (1969). "Ancillary Techniques of Gas Chromatography." Wiley (Interscience), New York.
Ettre, L. S., and Zlatkis, A. (eds.) (1967). "The Practice of Gas Chromatography." Wiley (Interscience), New York.
Gehrke, C. W., Roach, D., Zumwalt, R. W., Stalling, D. L., and Wall, L. L. (1968). Quantitative Gas-Liquid Chromatography of Amino Acids in Proteins and Biological Substances. Macro, Semi-Micro and Micro Methods. Anal. Biochem. Lab., Columbia, Missouri.

Giddings, J. C., and Keller, R. A. (eds.) (1966–1970). *Advan. Chromatogr.* **1–9**.
Grant, J. K. (ed.) (1967). "The Gas Liquid Chromatography of Steroids." Cambridge Univ. Press, London and New York. Cambridge, England.
Gudzinowicz, B. J. (1967). "Gas Chromatographic Analysis of Drugs and Pesticides." Dekker, New York.
Harris, W. E., and Habgood, H. W. (1966). "Programmed Temperature Gas Chromatography." Wiley, New York.
Jeffrey, P. G., and Kipping, P. J. (1964). "Gas Analysis by Gas Chromatography." Pergamon, Oxford.
Jones, R. A. (1970). "An Introduction to Gas-Liquid Chromatography." Academic Press, New York.
Kaiser, R. (1964). "Gas Phase Chromatography." Butterworths, London and Washington, D.C.
Keulemans, A. I. M. (1959. "Gas Chromatography." Van Nostrand Reinhold, Princeton, New Jersey.
Kiselev, A. V., and Yashin, Ya. I. (1969). "Gas-Adsorption Chromatography." Plenum, New York.
Knox, J. H. (1962). "Gas Chromatography." Wiley, New York.
Kroman, H. S., and Bender, S. R. (1968). "Theory and Application of Gas Chromatography in Industry and Medicine." Grune and Stratton, New York.
Krugers, J. (ed.) (1968). "Instrumentation in Gas Chromatography." Centrex Publ., Eindhoven.
Littlewood, A. B. (1970). "Gas Chromatography. Principles, Techniques, and Applications." Academic Press, New York.
McNair, H. M., and Bonelli, E. J. (1969). Basic Gas Chromatography. Varian Aerograph Co., Walnut Creek, California.
Moshier, R. W., and Sievers, R. E. (1965). "Gas Chromatography of Metal Chelates." Pergamon, Oxford.
Phillips, C. (1956). "Gas Chromatography." Academic Press, New York.
Purnell, J. H. (1962). "Gas Chromatography." Wiley, New York.
Purnell, J. H. (ed.) (1968). "Progress in Gas Chromatography." Wiley (Interscience), New York.
Scholler, R., and Joyle, M. F. (eds.) (1967). "Gas Chromatography of Hormonal Steroids." Gordon and Breach, New York.
Schupp, O. E. (1968). "Gas Chromatography." Wiley (Interscience), New York.
Stevens, M. P. (1969). "Characterization and Analysis of Polymers by Gas Chromatography." Dekker, New York.
Szepesy, L. (1970). "Gas Chromatography." CRC Press, Cleveland, Ohio.
Szymanski, H. A. (ed.) (1968). "Biomedical Applications of Gas Chromatography," Vol. II. Plenum, New York.
Tranchant, J. (ed.) (1969). "Practical Manual of Gas Chromatography." Elsevier, Amsterdam.
Wotiz, H. H., and Clark, S. J. (1966). "Gas Chromatography and the Analysis of Steroid Hormones." Plenum, New York.
Zlatkis, A., and Pretorius, V. (1971). "Preparative Gas Chromatography." Wiley, New York.

CHAPTER 25

Ion-Scattering Spectrometry for Surface Analysis

Robert S. Carbonara

Battelle Memorial Institute
Columbus Laboratories
Columbus, Ohio

Introduction

Ion-scattering spectrometry (ISS) is a new and unique technique for analyzing the elements present on the outermost atomic monolayer of a real surface. This technique simultaneously analyzes and sputters atoms from the surface. Thus by recording successive scattering spectra, a depth profile of the elemental composition can be obtained. ISS has been used on a wide variety of materials, including both insulators and conductors.

1 Theory

In ion-scattering spectrometry (ISS), a surface is bombarded with a beam of noble or inert gas ions ($^{3}He^{+}$, $^{4}He^{+}$, $^{20}Ne^{+}$ or $^{40}A^{+}$). A fraction of these ions ($\sim$1 in 10^{5}) experience a single binary elastic collision with a surface atom. This interaction, which takes place in roughly 10^{-16} sec, changes the energy and momentum of these primary ions and scatters them. It is these scattered ions which are analyzed and contain the information revealing the elemental composition of the outermost atomic monolayer of the surface. The energy change (energy loss) is related to the mass of the scattering center (atom) on the surface by

$$\frac{E_1}{E_0} = \frac{M_0^2}{(M_0 + M_s)^2}\left[\cos\theta + \left(\frac{M_s^2}{M_0^2} - \sin^2\theta\right)^{1/2}\right]^2 \qquad (1)$$

where

M_0 = mass of the analyzing gas ion;
M_s = mass of atom on the sample surface;
E_0 = kinetic energy of M_0 before collision;
E_1 = kinetic energy of M_0 after collision; and
θ = scattering angle for M_0.

This equation is derived without imposing any conditions of quantum or relativistic mechanics, or Coulombic interactions; only the basic concepts of conservation of energy and momentum, and binary scattering are necessary. For a scattering angle of $\theta = 90°$, which is employed in the commercial instrument manufactured by the 3M Company, Eq. (1) becomes

$$E_1/E_0 = (M_s - M_0)/(M_s + M_0) \qquad (2)$$

where $M_s \geq M_0$ is a prevailing condition for $\theta = 90°$.

Equation (2) can be rearranged to solve for M_s as follows:

$$M_s = M_0\left(1 + \frac{E_1}{E_0}\Big/ 1 - \frac{E_1}{E_0}\right) \qquad (3)$$

As can be seen from Eq. (3), M_s is a single-valued function of E_1/E_0, which means that for each M_s there is a unique value of E_1/E_0. Subsequently, the energy spectrum of the scattered ions has a single peak for each element

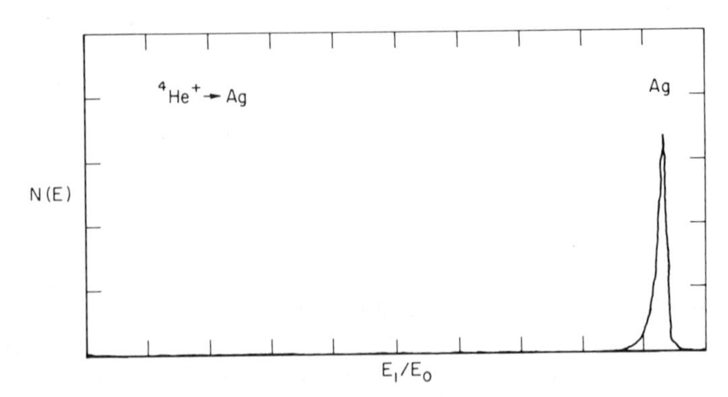

Fig. 1. Typical energy spectrum for a single element, in this case pure silver.

(atomic mass) on the surface (see Fig. 1). Some other aspects of the kinematics of collisions of ions with solid surfaces have been considered by Smith (1967a).

The theory developed above is based on the *single* binary elastic collision model. If the primary ion penetrates the surface beyond the first atomic layer, it has a very low probability of escaping from the surface without undergoing multiple collisions (elastic or inelastic). If it collides more than once, it is lost to the single binary elastic collision model. Therefore, when surface atoms are detected by the ISS technique, they must reside almost exclusively in the first average atomic layer. This inherent sensitivity to the first, or outermost, atomic monolayer is unique to ISS and offers an entirely new dimension for the examination of surfaces, since the precise location of the detected element is known.

2 Experimental Technique and Apparatus

Smith (1971), Goff and Smith (1970), and Carbonara (1973) have given the basic details of the experimental technique and apparatus that are at present commercially available. Several authors, (Smith, 1967b; Strehlow and Smith, 1968; Mashkova *et al.*, 1970; Ball *et al.*, 1972a,b; Brongersma and Mul, 1972; Armour and Carter, 1972), have discussed methods for applying ion scattering to surface analysis, as well as various other aspects of ion scattering. Figure 2 is a schematic representation of the ISS apparatus.

The ion gun, perhaps the most important component of the ISS instrument, utilizes Bayard–Alpert guage geometry for the production of inert gas ions. These primary ions are extracted axially from one end of the gun at any desired kinetic energy between 300 and 3000 eV by a set of electrostatic pinhole apertures and focused onto the sample. Nominal beam diameters from 10 to 0.5 mm are obtained by additional electrostatic focusing adjustments. This focusing gives a variation in the beam current intensity from 1 to 50 $\mu A/cm^2$. The energy spread of the primary ion beam is less than 0.5% at full width half maximum (FWHM) of the primary peak. Beam currents are typically around 100 to 200 nA (10^{-9} A). Uniaxial manipulation of the beam is possible by employment of electrostatic deflection electrodes placed just in front of the sample.

The inert gas ions from the ion gun strike the sample surface and are reflected or scattered. A 2-in.-radius, 127° parallel-plate electrostatic energy analyzer with a 5-mil acceptance aperture is placed at 90° from the incident beam to monitor the scattered inert gas ion leaving the sample surface. Ions that pass through the electrostatic analyzer have undergone a single binary elastic collision with an atom in the outermost atomic mono-

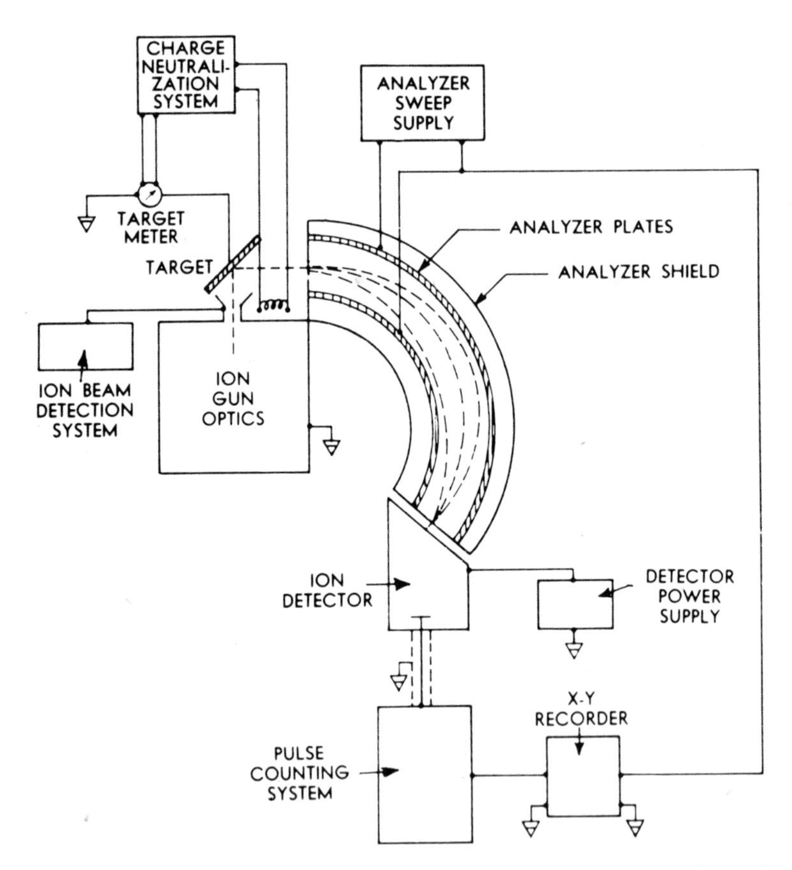

Fig. 2. Schematic of ISS apparatus exclusive of the vacuum system and electronics.

layer of the surface and thereby have their energy altered according to Eq. (2). Detection of the scattered ions that have passed through the energy analyzer is by a Bendix Model 4028 channel electron multiplier with a special cathode and a gain factor of $\sim 10^8$. This type of detector is very stable and withstands vacuum-to-atmosphere cycling without degrading.

The compact design of the ion optics of the ISS system enable it to be mounted on an 8-in. vacuum flange, which is mounted in an all-metal vacuum system capable of reaching 10^{-9} Torr. This system is rough-pumped from atmospheric pressure down to $\sim 10^{-4}$ Torr by a zeolite cryosorption pump. Higher vacuum is achieved with a 20-liter/sec differential ion pump and a titanium sublimation pump surrounded by an annular cryopanel. This clean vacuum pumping technique is necessary to reduce the chance of sample surface contamination that frequently occurs in more conventional diffusion-pump systems.

Normally the vacuum chamber is pumped down to $\sim 5 \times 10^{-8}$ Torr. The vac-ion pump is then valved off, and the titanium sublimation is turned off. Only the titanium-coated cryopanel remains activated in order to remove any active gases that may remain or are generated by the ion bombardment of the sample.

The ISS data display is a X–Y recorder plot of intensity versus atomic mass (as in Fig. 1) and shows the presence or absence of all the elements that have an atomic mass greater than that of the bombarding ion (M_0). Using helium, all elements except hydrogen and helium are analyzed. At present, there are no surface techniques that are capable of detecting hydrogen or helium. When neon or argon are used as the analyzing gas, the elements from hydrogen to fluorine and hydrogen to chlorine, respectively, are not detectable.

Utilization of neon or argon is necessary to obtain higher mass resolution for the heavy elements, as can be seen from the following:

$$\frac{d(E_1/E_0)}{dM_s} = \frac{2M_0}{(M_s - M_0)^2} \tag{4}$$

The left-hand side of Eq. (4) is related to the mass resolution and is proportional to M_0. Thus, increasing M_0 increases the resolution, and subsequently the motivation to use either neon or argon. Figure 3 shows this effect.

The primary ions that strike the surface can be involved in several types of interactions with the atoms in the surface. Binary elastic collisions are of

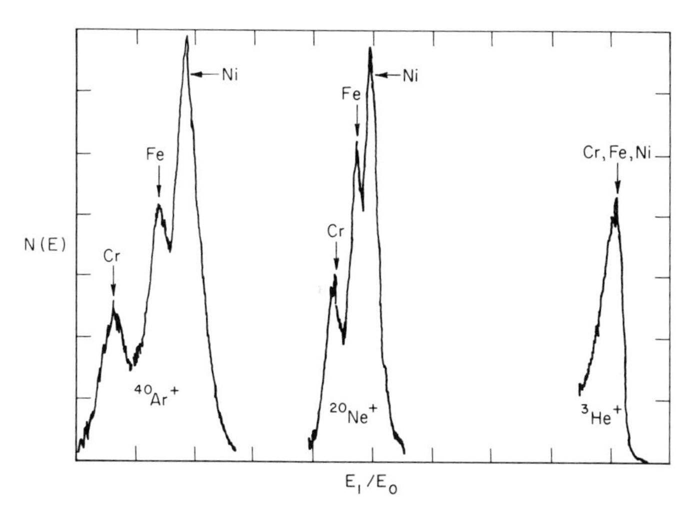

Fig. 3. ISS spectrum of chromium–iron–nickel alloy using 1500 eV $^{40}Ar^+$, $^{20}Ne^+$, and $^3He^+$ showing the effect of increasing mass resolution with increasing mass of the primary ion.

prime importance because they provide information concerning the elemental composition of the surface; however of almost equal importance is the simultaneous sputtering interaction. Sputtering is important because it reveals the underlying atomic layers and makes analysis of successively deeper layers possible. This simultaneous action of analysis and sputtering gives a depth profile of the elemental composition, as shown in Fig. 4.

Depth profile analysis, with possible monolayer resolution, is extremely useful in studies involving surface composition gradients, studies relating surface to bulk properties, studies of film thickness and composition as a function of thickness, and studies of contamination and penetration of contamination into the sample. This unique ability to examine a surface, layer by layer, is particularly important on most technological surfaces where the greatest compositional changes occur in approximately the first 20 monolayers.

Sputtering rates vary with the analyzing gas and depend on several parameters, including the nature of the sample, some of which are not yet well understood. For helium ions the sputtering rate generally ranges from 3 to 50 monolayers/hr. With the relatively slow removal rate (3 monolayers/hr) it is possible to examine carefully a single monolayer. When higher sputtering rates are desired, more massive noble gas ions, either neon or argon, are used. Removal rates are increased by approximate factors of 5 and 10, respectively.

Qualitative information on the conductivity of a surface can also be de-

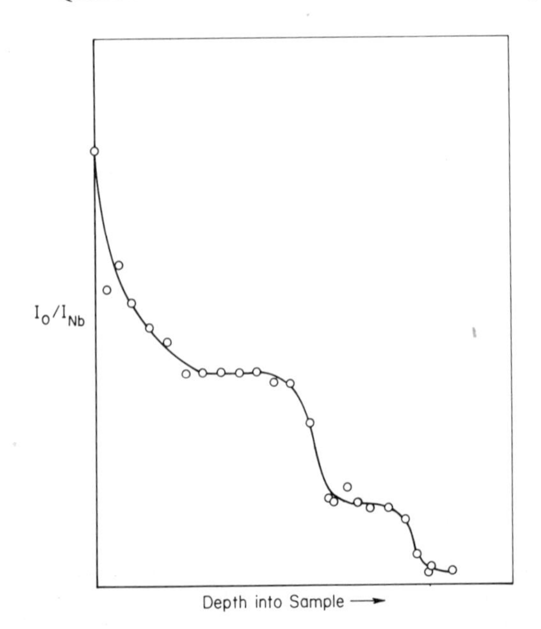

Fig. 4. Depth profile of Nb_2O_5 on NbO on Nb.

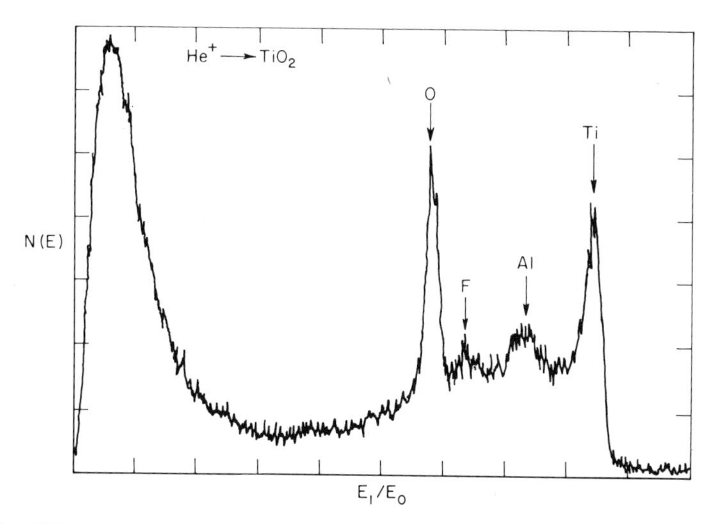

Fig. 5. ISS energy spectrum of the nonconductive surface of TiO_2 showing the displacement of the low-energy sputter threshold at left-hand end of the spectrum.

termined by the ISS technique. Figure 5 is the spectrum of a nonconducting material that shows a shift in the spectrum of the low-energy sputter threshold. This shift is not present in conductive material.

All types of surfaces, from insulator to metallic; crystalline, amorphous, and even liquids, can be analyzed with the ISS technique. The only stipulation is that the material be vacuum worthy (i.e., have a sufficiently low vapor pressure).

Nonconductive surfaces are routinely examined by ISS with the aid of the charge neutralization system, which eliminates the surface charge buildup by bathing the sample with thermally emitted electrons while the sample remains at ground potential. Insulating surfaces are examined directly, without the need for a thin shadowing. The charge neutralization system operates via feedback stabilization between the electron source and the total measured target current, thereby assuring a high degree of stability in the total ion beam current.

Because ISS analyzes the outermost surface of a sample, it is imperative to maintain that surface in its technologically applied condition. Placing any foreign object (fingers, tweezers, adhesives, etc.) on the surface can make subsequent analysis ambiguous and unreliable. Great care should be taken when preparing and handling the samples, and during placement in the sample holder. Preferably, only the edges should be touched. The standard sample holder can accommodate six samples with areal dimensions of roughly $\frac{5}{16} \times \frac{1}{2}$ in.; thicknesses of up to $\frac{1}{2}$ in. can be accommodated.

The area of the sample analyzed is determined by the acceptance angle of the receiving slits (5 mils) and the beam diameter. For a 1-mm beam diameter, the analyzed area is 1×0.25 mm. This entire area is continuously irradiated by the beam, as opposed to a raster scan by a much smaller diameter beam.

3 Applications

In itself ion scattering can be considered and utilized as a research tool. However, up to this point its greatest use and value have been its ability to solve real-world problems by elucidating the chemistry of surfaces. Carbonara (1973) has listed numerous applications of ISS. General categories such as adhesion, failure analysis, corrosion, catalysis, friction and wear, coatings, semiconductors, medicine, biochemistry and biophysics, effects of metallographic preparations and etchants, chemical treatment and cleaning monitoring, electrochemical effects, and surface contamination are a few of the areas to which ISS has been applied.

Since the intensity of a given scattering peak is proportional to the number of scattered ions of a given energy, and since each element gives rise to a single peak of scattered ions of a given energy (single-energy ions), the peak intensity is directly related to the amount of the given element on the surface. Evidence that ISS can be used in a quantitative, as well as qualitative, manner has been shown by Smith (1967b). To perform absolute quantitative analysis by ISS requires a knowledge of the two parameters that affect peak intensity, namely, the cross section for scattering of the primary ion by the atom in question and the probability for charge neutralization or charge exchange during the binary elastic collision. As of now there are only limited data on these parameters. Matrix effects on these parameters may also be of considerable importance, since the manner in which the surface atom is bonded to the matrix can affect the parameters. As yet, however, there is a dearth of information in this area.

Sensitivity, of course, is inherently related to cross section and neutralization. A better understanding of these parameters will define sensitivity limits more precisely. It may be that sensitivity is matrix dependent. Presently ISS can detect surface concentrations of 1.0 to 0.1% of a monolayer. Using standards to define sensitivity is not very realistic for any surface analytical technique, since the surface character of the standards is not normally the same as that of the sample, and the surface character can markedly affect sensitivity.

At this time it is very difficult to set absolute accuracy and precision limits for ISS. However, it is possible to obtain relative concentrations of surface constituents on almost all samples. Thus, it is reasonable to consider ISS as semiquantitative. This semiquantitative ability is extremely useful where relative amounts are of technological importance.

In addition to peak intensity and concentrations, quantitative information on sputtering rates is very important and useful. In order to determine accurate depth profiles of concentration, the removal rate of atomic layers must be known. Of course, this problem exists for all surface analytical techniques that use ion bombardment for sputtering. A limited amount of data on sputtering yields is available (Carter and Colligon, 1968; McDonald, 1970; Wehner, 1965; Kamansky, 1965). Most of the data were generated under specific conditions which are not necessarily those used in all experiments. Extrapolation of the available data to other experimental conditions is possible, but not always reliable or advisable. In cases where compositional gradients exist, especially where oxides are involved, the choice of sputtering yields from the data is usually very difficult. With the use of available data, some success has been realized in recent experiments on 1000-A nominally thick aluminum films which become oxidized. ISS data indicate the total film thickness to be 1200 Å. Other experiments indicate oxide film thicknesses on aluminum to be between 100 and 200 Å. Data should be obtained on each instrument by bombarding a sample under specific experimental conditions and using another technique, e.g., ellipsometry, scanning electron microscopy, interference microscopy, or profilometry to check the ISS results.

Even though sputtering of the sample surface takes place during the analysis, the ISS technique can be considered essentially nondestructive. However, there are samples where the removal of even 25 Å of the surface would be destructive. Also some samples must be cut in order to fit the sample holder thereby making the technique destructive. ISS is applicable as a quality control monitor where sputtering of the surface can be tolerated or where sample coupons are representative of entire lots of material. The designation of ISS as destructive or nondestructive depends on each individual case.

The time required for analysis depends largely on the sample depth for which data are required and the sputtering rate. A single scan of the entire spectrum usually requires 5 to 15 min. With helium ions it is possible to obtain data to a depth of $\sim$100 Å in approximately 1 hr.

Examples of ISS applications for solving real-world problems are numerous. Some examples from the author's own experience include the ISS analysis of (1) a red stain on pressed and sintered zinc oxide powder, which was found to be iron oxide contamination picked up during the pressing operation; (2) passivation films on superalloys, which were found to be compositionally complex; (3) fracture surfaces of a T-111 alloy, which were found to have larger amounts of zirconium, sodium, magnesium, aluminum, silicon, calcium, and iron than the bulk; (4) coating composition and thickness, which were determined for laser mirrors; (5) glass surfaces cleaned by various chemicals which in some cases showed residual amounts of those chemicals and in others leaching of some of the glass constituents; (6) glass bottles, which having undergone a pasturization and/or sterilization process had their surface chemistry altered in such a fashion as to affect label adherence; (7) lubricated bearing surfaces, which showed evidence of a chemical reaction between the lubricant and the bearing; (8) lunar material from Apollo 16, which revealed thin layers of silicates; (9) fusion-bonded parts, which were found to have significant amounts of contaminants in failure interfaces; (10) paint, whose failure to adhere to galvanized sheet metal was traced to compositional variations in the galvanizing process; (11) several oxidized metal surfaces, in which the oxide:metal ratio was measured as a function of depth.

These examples of ISS applications are only a few of the many that have been undertaken in the relatively short time the technique has been available.

4 Conclusion

ISS is just one of several surface analytical techniques, among which are Auger electron spectroscopy (AES), electron spectroscopy for chemical analysis (ESCA), the ion microprobe mass analyzer (IMMA), and low-energy electron diffraction (LEED). All of these techniques are valuable tools for surface analysis, and the choice of a technique for problem solving is usually determined by the problem itself. In some cases one method may be unsuccessful, whereas another will readily provide the answer. Awareness of surface problems cannot be overemphasized. Some general references on ion scattering and ISS (in addition to those previously cited) include review articles (Mott and Massey, 1965; Snoek and Kistemaker, 1965; Cawthorn, 1969; Palmer, *et al.*, 1970), articles on instrumentation and experiments (Smith, 1968; Smith and Goff, 1969; McCracken and Freeman, 1969; Van Der Weg and Bierman, 1969; Cawthorn *et al.*, 1969; Dahl and Sandager, 1969; Suurmeÿer, *et al.*, 1970; Suurmeÿer and Boers 1971; Goff, 1971; Eckstein and Verbeek, 1972; Ball *et al.*, 1972a,b), cross sections (Bingham, 1966, 1967; Rice and Bingham, 1972; Abrahamson, 1969; Robinson, 1970), and applications (Heiland and Taglauer, 1972; Malm and Vasile, 1972; Harrington and Hong, 1972; Begemann and Boers, 1972).

References

Abrahamson, A. A. (1969). *Phys. Rev.* **178,** 76.
Armour, D. G., and Carter, G. (1972). *J. Phys. E: Sci. Inst.* **5,** 2.
Ball, D. J., Buck, T. M., Caldwell, C. G., McNair, D., and Wheatley, G. H. (1972a). *J. Vac. Sci. Technol.* **9,** 611.
Ball, D. J., Buck, T. M., McNair, D., and Wheatley, G. H. (1972b). *Surface Sci.* **30,** 69.
Begemann, S. H. A., and Boers, A. L. (1972). *Surface Sci.* **30,** 134.
Bingham, F. W. (1966). Tabulation of Atomic Scattering Parameters Calculated Classically from a Screened Coulomb Potential. Sandia Res. Rep. SC-RR-66-506, TID-4500 Physics.
Bingham, F. W. (1967). *J. Chem. Phys.* **46,** 2003.
Brongersma, H. H., and Mul, P. M. (1972). *Chem. Phys. Lett.* **14,** 380.
Carbonara, R. S. (1973). *In* "Microstructural Analysis: Tools and Techniques" (J. L. McCall and W. M. Mueller, eds.), p. 315. Plenum, New York.
Carter, G., and Colligon, J. S. (1968). "Ion Bombardment of Solids." Amer. Elsevier, New York.
Cawthorn, E. R. (1969). Thesis, Australian Nat. Univ., Canberra, A.C.T., Australia.
Cawthorn, E. R., Cotterell, D. L., and Oliphant, Sir M. (1969). *FRS Proc. Roy. Soc. London* **A314,** 39.
Dahl, P., and Sandager, N. (1969). *Surface Sci.* **14,** 305.
Eckstein, W., and Verbeek, M. (1972). *J. Vac. Sci. Technol* **9,** 612.
Goff, R. F. (1971). *Proc. Nat. Conf. Electron Probe Anal., 6th, Pittsburgh.*
Goff, R. F., and Smith, D. P. (1970). *J. Vac. Sci. Technol.* **7,** 72.
Harrington, W. L., and Honig, R. E. (1972). *Proc. ASMS Ann. Conf. Mass. Spectrometry, 20th.*
Heiland, W., and Taglauer, E. (1972). *J. Vac. Sci. Technol.* **9,** 620.
Kamansky, M. (1965). "Atomic and Ionic Impact Phenomenon on Metal Surfaces." Academic Press, New York.
Malm, D. L., and Vasile, M. J. (1972). *Proc. ASMS Ann. Conf. Mass Spectrometry, 20th.*
McCracken, G. M., and Freeman, N. J. (1969). *J. Phys. B* **2,** 661.
McDonald R. J. (1970). *Adman Phys.* **19,** 457.
Mashkova, E. S., Molchanov, V. A., and Skripa, Yu. G. (1970). *Phys. Lett.* **33A,** 373.
Mott, N. F., and Massey, H. S. W. (1965). "The Theory of Atomic Collisions." Oxford Univ. Press, (Clarendon), London and New York.
Palmer, D. W., Thompson, M. W., and Townsend, P. D. (eds.) (1970). "Atomic Collision Phenomenon in Solids." North-Holland Publ., Amsterdam.
Rice, J. K., and Bingham, F. W. (1972). *Phys. Rev. A* **5,** 749.
Robinson, M. T. (1970). Tables of Classical Scattering Integrals. Oak Ridge Nat. Lab. Rep. No. ORNL-4556, UC-34-Physics.
Smith, D. P. (1967a). *J. Appl. Phys.* **38,** 340.
Smith, D. P. (1967b). *Ann. Conf. Mass Spectrometry Allied Topics, 15th Denver, 1967.*
Smith, D. P. (1968). *Proc. Ann. Phys. Electron. Conf., 28th, Minneapolis; Bull. Amer. Phys. Soc.* **13,** 947.
Smith, D. P. (1971). *Surface Sci.* **25,** 171.
Smith, D. P., and Goff, R. F. (1969). *Proc. Ann. Phys. Electron. Conf., 29th, Yale; Bull. Amer. Phys. Soc.* **14,** 788.
Snoek, C., and Kistemaker, J. (1965). *Advan. Electron. Electron Phys.* **21,** 67.
Strehlow, W. H., and Smith, D. P. (1968). *Appl. Phys. Lett.* **13,** 34.

Suurmeijer, E. P. T. M., and Boers, A. L. (1971). *J. Phys. E.* **4,** 663.
Suurmeijer, E. P. T. M., Boers, A. L., and Begemann, S. H. A. (1970). *Surface Sci.* **20,** 424.
Van Der Weg, W. F., and Bierman, D. J. (1969). *Physica* **44,** 177.
Wehner, G. K. (1965). *Advan. Electron. Electron Phys.* **7,** 239.

CHAPTER 26

Mössbauer Spectrometry

P. A. Pella

Institute for Materials Research
Analytical Chemistry Division
National Bureau of Standards
Washington, D. C.

Introduction

Mössbauer spectrometry is a nuclear resonance technique which uses gamma photons of an appropriate frequency from a radioactive source to populate low-lying nuclear energy levels in the sample. Nuclear resonance absorption (or scattering) is exhibited in nuclides with excited-state energies typically less than 150 keV with lifetimes between 10^{-10} to 10^{-7} sec. The energy width of the gamma photons is small enough so that no nuclei other than those of the Mössbauer nuclide can be resonantly excited in the sample. For example, a commonly studied Mössbauer isotope is ^{57}Fe. If one uses a radioactive source which populates the ^{57}Fe-14.4-keV excited state, e.g., ^{57}Co in palladium metal, then regardless of what other elements are present in the sample only the ^{57}Fe can be resonantly excited. Hence, in this

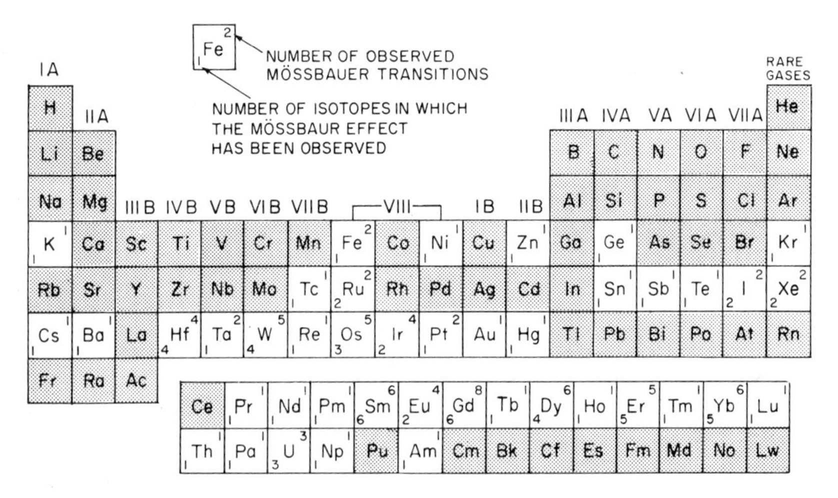

Fig. 1. Periodic table of the Mössbauer isotopes, showing the number of isotopes (lower left corner) and nuclear transitions (upper right corner) for each element exhibiting the Mössbauer effect (nonshaded). [From J. G. Stevens and V. E. Stevens, 1972. "Mössbauer Effect Data Index Covering the 1970 Literature." Plenum Press, New York.

respect, this method is completely specific. Because of the conditions nature has imposed, this method is limited to the study of the Mössbauer isotope(s) in solids (crystals or glasses) and in frozen solutions.

Perusal of the literature over the past ten years clearly shows the interdisciplinary nature of this technique for the structural characterization of materials. The applications cover a diversity of fields such as nuclear physics, metallurgy, biology, mineralogy, solid-state physics, and chemistry. This nuclear resonance phenomenon has been demonstrated for many elements, as shown in Fig. 1. However, about 70% of the applications have focused on ^{57}Fe and ^{119}Sn. This is because these isotopes have the most favorable nuclear characteristics. Also, the availability of commercial Mössbauer sources and the relatively simple experimental methodology have contributed to their popularity. It should be mentioned that for many of the elements shown in the chart, low temperatures using liquid helium are required.

1 Theory

In this spectrometry one observes hyperfine interactions such as isomer shift, quadrupole splitting, and magnetic splitting. These spectral parameters can be considered to originate from the interaction of nuclear moments with extranuclear moments arising from the ground-state electronic configuration and the surrounding lattice.

The isomer shift arises from the Coulombic interaction between the nuclear charge and the effective electron density at the nucleus. Its magnitude is proportional to the difference in the nuclear charge radii of the ground the excited states and the total electron density at the nucleus. An expression for the isomer shift (IS) can be written as

$$\text{IS} = F(Z)\,\Delta R/R[|\Psi_s(0)|^2 - |\Psi_a(0)|^2] \tag{1}$$

where $F(Z)$ represents a composite of nuclear terms, $\Delta R/R$ is the relative change in the average radius of the nuclear charge distribution from the ground to the excited state, and $|\Psi_s(0)|^2$ and $|\Psi_a(0)|^2$ are the s-electron densities at the nucleus for the source and the absorber, respectively. This hyperfine interaction is primarily a measure of the s-electron density, since only these electrons overlap the nuclear volume appreciably. But the presence of p, d, and f electrons will indirectly influence the effective s-electron density at the nucleus, thereby affecting the isomer shift. Because the nuclear term is constant for a particular nuclide, the isomer shift can be indicative of the formal oxidation state and/or degree of covalent bonding in molecules.

The quadrupole splitting, ΔE_q, is derived from the electrostatic interaction of an aspherical nucleus (i.e., the nuclear quadrupole moment) with an asymmetric charge distribution. Mathematically, the charge distribution is described by an electric-field gradient (EFG) tensor whose principal components are V_{xx}, V_{yy}, and V_{zz}. For ^{57}Fe the quadrupole splitting is

$$\Delta E_q = [(eQ)(eq)/2][1 + \eta^2/3]^{1/2} \tag{2}$$

where eQ is the nuclear quadrupole moment, $eq = V_{zz} = \delta^2 V/\delta Z^2$, and η is the asymmetry parameter given by $\eta = (V_{xx} - V_{yy})/V_{zz}$ where $|V_{zz}| \geq |V_{yy}| \geq |V_{xx}|$. The value of η ranges from 0 to 1. From measurements of ΔE_q information about chemical structure can be obtained.

The magnetic hyperfine interaction or nuclear Zeeman effect arises when the magnetic moment of the nucleus interacts with the effective magnetic field produced by the extranuclear electrons. For a nucleus with $I > 0$, the magnetic interaction splits the nuclear levels into $2I + 1$ components. From studies of the internal magnetic field, some knowledge can be gained about electronic spin states, atomic spin distributions (ordering), and chemical bonding. In Fig. 2 the energy level diagram is shown for each of the hyperfine interactions for ^{57}Fe.

In addition to these hyperfine interactions, the intensity of the resonance absorption can yield some information about the dynamics of the Mössbauer

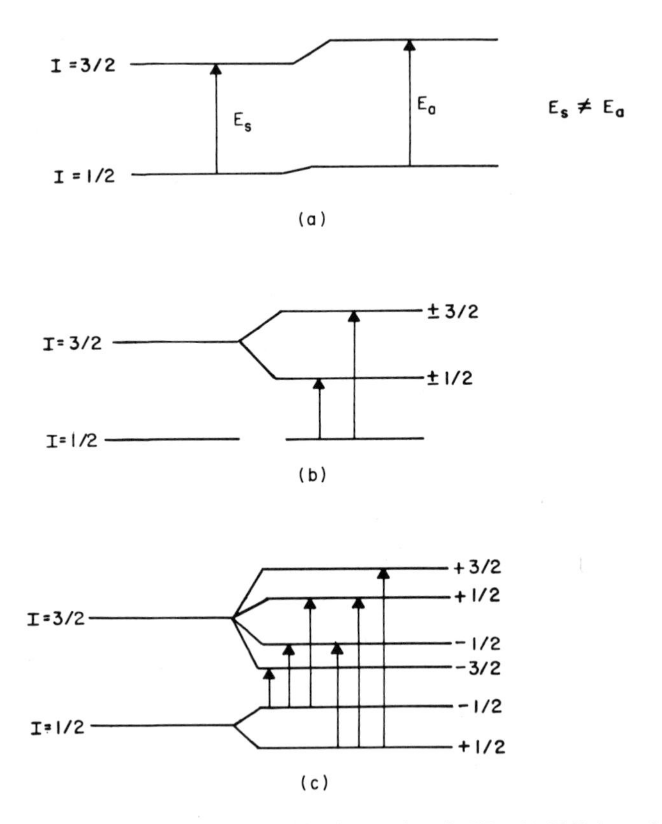

Fig. 2. Energy level diagrams for hyperfine interactions in ^{57}Fe. (a) Shift in nuclear energy levels due to electric monopole interaction (isomer shift); (b) splitting of nuclear excited state due to quadrupole interaction; (c) splitting of nuclear states due to magnetic hyperfine interaction.

nuclide in the solid state. The magnitude of resonant absorption ϵ, for a single energy transition, can be written as

$$\epsilon = [I_{(\infty)} - I_{(0)}]/I_{(\infty)} = f_s[1 - e^{-T_A|2}J_0(iT_A|2)] \tag{3}$$

where $I_{(\infty)}$ is the count rate at which no resonance absorption occurs; $I_{(0)}$ is the count rate at the resonance maximum; and f_s, f_a are the recoil-free fractions for the source and absorber (sample), respectively. That is, they represent the fraction of Mössbauer photons that are resonantly emitted (f_s) or absorbed (f_a). T_A is the effective thickness of the absorber, where $T_A = f_a\sigma_0 na$, σ_0 being the resonance cross section in square centimeters, n the number of atoms per unit volume, and a the isotopic abundance of the Mössbauer nuclide. The zero-order Bessel function with an imaginary argument is represented by J_0. The recoil-free fraction permits the measurement of the mean-square displacement $\langle x^2 \rangle$ of the Mössbauer atom from its equilibrium position and can yield information about chemical bonding in solids.

A number of reviews and books have been published over the years which describe the theory and practice of Mössbauer spectrometry. The review articles of DeVoe and Spijkerman (1966, 1968) cite a comprehensive bibliography of applications, in addition to the data indexes of Muir *et al.* (1958–1966), Stevens and Stevens (1972), and Stevens *et al.* (1972). Recent texts include an introduction to the technique (May, 1971) and chemical applications (Gol'danskii and Herber, 1968). Also conference proceedings are included in the *Mössbauer Effect Methodology* series (Gruverman, 1965–1970), and in a text by Dézsi (1971).

2 Instrumentation

A Mössbauer spectrometer basically consists of a velocity drive unit capable of producing a Doppler motion with sufficient accuracy and precision, and a gamma-ray detection system. The Doppler shift provides the necessary change in gamma-ray frequency (about 10^{-9} eV for ^{57}Fe) so that resonance absorption can occur. A recent trend in instrumentation appears to favor electromechanical drive units coupled with multichannel analyzers operating in the constant-acceleration mode. This kind of operation enables one to obtain an entire energy spectrum. Constant-velocity motions are also used, especially when limited regions of a spectrum are of interest. The radiation detectors available include the proportional counter, the NaI(Tl) scintillation detector, and the solid-state detector. The proportional counter is generally employed in the energy region from 1 to 20 keV, and is used extensively in ^{57}Fe spectrometry. For higher energies such as the 23.8-keV gamma-ray of $^{119}Sn^{m}$, the scintillation detector is preferred. The solid-state detectors are not as popular in this field as one might expect, although capable of greater resolution. Both the cost and relatively low counting efficiencies as compared to the more conventional detectors have prevented a more widespread use.

Because most Mössbauer sources are not monochromatic but emit a number of discrete gamma and x rays, some means of discriminating between the Mössbauer gamma rays and the unwanted radiation must be devised. This is usually done by a single-channel analyzer (SCA). Presented in Fig. 3 is a schematic diagram illustrating the basic components of a Mössbauer spectrometer for both transmission and backscattering experiments. In materials studies additional information can be obtained by

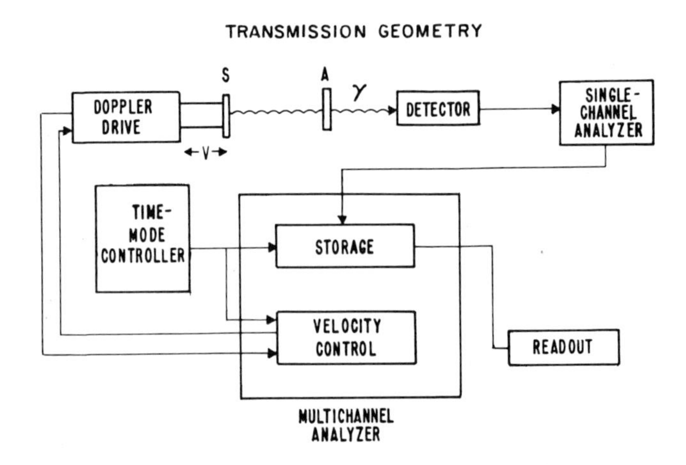

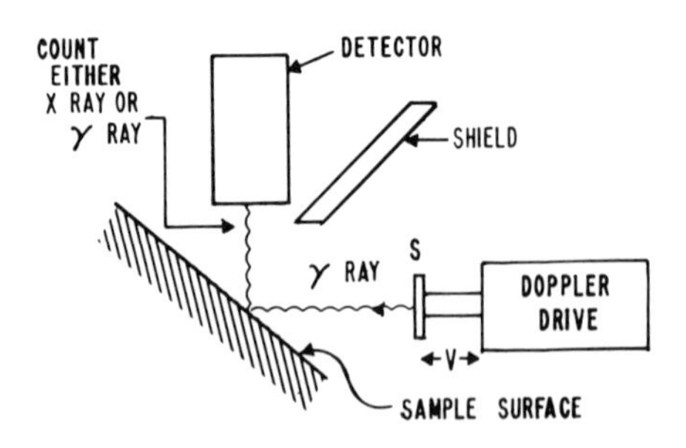

Fig. 3. Block diagram of Mössbauer instrumentation.

observing the temperature dependence of the isomer shift, quadrupole splitting, and magnetic hyperfine splitting. Therefore, cryogenic equipment is often required to cool samples to 77°K (liquid nitrogen) or 4°K (liquid helium). Also, the use of applied magnetic fields is desirable for obtaining the sign of the EFG for structural studies, for magnetic polarization studies, and for other applications. Weak magnetic fields, on the order of 1 kG, are relatively easy to apply using a permanent magnet. However, large fields, on the order of 50 to 100 kG, are often needed and require the use of superconducting magnets.

3 Sample Preparation

For resonance absorption measurements using transmission geometry, the sample can be in the form of a single crystal or powder finely ground to eliminate any preferred orientation effects. May and Snediker (1967) have described a technique for mounting powdered materials. Since the Mössbauer measurement is nondestructive, the sample can be easily recovered for further studies. For example, samples of iron-containing mineral phases were measured at the NBS by sprinkling the powdered specimen between layers of cellophane tape. The sample was later recovered by dissolving the tape adhesive with acetone.

Often the materials scientist wishes to know what quantity of sample is required to obtain a "good" Mössbauer spectrum in the shortest amount of time. The answer to this question depends basically on the concentration of Mössbauer nuclide present, how the Mössbauer nuclide is chemically bound in the solid material, and what matrix elements are present. If enough sample is available for measurement, then the effective absorber thickness, i.e., T_A in Eq. (3), can be optimized for a given sample. Some compounds, e.g., organometallic compounds, have such low recoil-free fractions, that adequate spectra can only be obtained at temperatures of 80°K or lower. If the sample is prepared to give too "thick" an absorber, then line broadening will result in decreased spectral resolution. Line broadening can be estimated by the approximation

$$\Gamma_{\mathrm{exp}} = (2 + 0.27\Gamma)\Gamma_{\mathrm{nat}} \tag{4}$$

where Γ_{exp} and Γ_{nat} are the experimental and natural line widths, respectively. Margulies and Ehrman (1961) have prepared a table for predicting the line broadening as a function of effective absorber thickness. In some cases, the sample must be diluted to obtain an adequate absorber thickness. These procedures usually involve mixing a portion of the sample with a low Z element or compound such as boron nitride or aluminum oxide. A convenient method used in some laboratories consists of casting a portion of sample in a plastic material which can easily be pressed into disks of a desired thickness. Another factor to be considered in sample preparation is the attenuation of the transmitted gamma photons because of atomic absorption by the sample. The attenuation is given by

$$I | I_0 = \exp\left[-(\mu | \rho)(\rho x)\right] \tag{5}$$

where $\mu | \rho$ is the mass absorption coefficient in square centimeters per gram and ρx is the sample thickness in grams per cubic centimeter. In addition, if elements which produce large Compton scattering (e.g., Br, I) are present in the sample matrix, then the sample thickness must again be adjusted to minimize this interference. Usually if enough sample is available, thickness optimization can frequently be accomplished with just a few preliminary Mössbauer measurements.

From the considerations just described, the time required to obtain a Mössbauer spectrum can vary from as little as 15 min to hours, days, or weeks depending on the patience of the investigator. In general, the signal-to-background ratio improves as the square root of time. For example, it requires an additional 3 days to double the quality of a day-old spectrum. Generally, times on the order of hours and even a few days are not uncommon, particularly when samples with low values of T_A are measured or when the counting error must be minimized for certain applications.

For studies of liquid samples (i.e., frozen solutions), Herber (1971) has summarized some of the important criteria which the solvents must satisfy as follows: (1) They must be chemically inert, with good solvating power for the solutes in question. (2) They should be glass formers at low temperature so that the solvent accommodates the structure of the solute. (3) The solvent should exhibit a high lattice temperature in order to obtain the largest possible resonance effect. This same author has found EPA (16:42:42% ethanol–1-propanol–diethyl ether) and MTA (methyl tetrahydrofuran) suitable solvents for studies of organoiron carbonyl compounds in the frozen state.

4 Applications

The scope of the applications presented in this chapter will be limited basically to examples employing ^{57}Fe and ^{119}Sn nuclides. For materials characterization studies employing other specific Mössbauer nuclides, the reader should refer to the literature. The objective of this limited presentation is to provide sufficient knowledge so that one can intelligently decide whether the technique will be applicable in his particular problem area. It should be emphasized that whenever possible, information obtained by Mössbauer spectrometry can and should be correlated with data from other techniques such as NMR, NQR, x-ray diffraction, infrared and Raman spectroscopy, specific-heat and magnetic susceptibility measurements, and more recently, ESCA. Correlations of this kind are important in many instances, particularly in chemical structure and bonding studies. This is because the Mössbauer spectral parameters yield mainly indirect information about structure and bonding.

4.1 Mineralogy

The Mössbauer method has been used to obtain information about the distribution of atoms and their site occupancies in crystal structures when the structure of the mineral phase(s) is known. Many applications are concerned with the iron distribution in complex silicate structures. The unequal distribution of iron over nonequivalent crystallographic sites can provide data for estimating thermodynamic quantities such as distribution isotherms and the free energy of cation order–disorder. For example, the

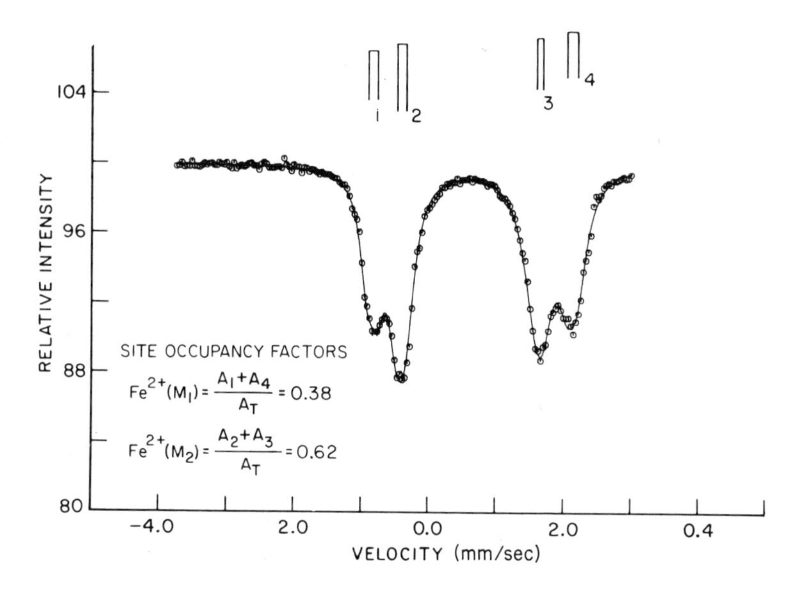

Fig. 4. Mössbauer spectrum of orthopyroxene.

orthopyroxenes are rock-forming minerals with compositions similar to the $MgSiO_3$–$FeSiO_3$ system. The structure consists of single silicate chains (SiO_3) held together by Mg^{2+} and Fe^{2+} in octahedral coordination. There are two nonequivalent octahedral positions, i.e., M_1 M_2, which can be occupied by Mg^{2+} and Fe^{2+} in all proportions (Evans *et al.*, 1967). A Mössbauer spectrum of an orthopyroxene sample is shown in Fig. 4. Each pair of quadrupole split lines is assigned to a nonequivalent Fe^{2+} octahedral site in the crystal structure. Assuming that the recoil-free fractions are the same for Fe^{2+} in each of the crystallographic sites, the site distribution can be calculated.

Mössbauer studies of the thermal decomposition of such minerals as amosite and crocidolite have permitted assignment of Fe^{2+} and Fe^{3+} ions to the appropriate sites in the crystal structure on the basis of the isomer shift and quadrupole splitting values (Whitfield and Freeman, 1967). Spectral parameters of many other important minerals have been measured by Bancroft *et al.* (1967) and Herzenberg and Toms (1966). It has also been possible to distinguish between Sn(II) and Sn(IV) in some tin-containing minerals (Smith and Zuckerman, 1967).

4.2 Metallurgy

In this area of materials science, there arise many important questions, such as phase identification, the characterization of precipitates from solid solution, and the detection and study of particular phases undergoing transformation. Phases identifiable by their Mössbauer patterns include ferrite (α-Fe), martensite, austenite, cementite, and ϵ-carbide. The quantitative determination of austenite in steel is already becoming a popular application (Veits *et al.*, 1970).

Metallic iron (α-Fe or ferrite) at room temperature has a body-centered cubic crystal structure (bcc) and is ferromagnetic. The Mössbauer spectrum consists of six Zeeman lines, as shown in Fig. 5, with an internal magnetic field of 330 kOe. γ-Fe is a face-centered cubic (fcc) form of iron which is stable in the neat metal (i.e., $< 0.1\%$ C) only between 910 and 1400°C. The addition of carbon to γ-Fe from above zero to 2% results in an interstitial primary solid solution called austenite, which is not ferromagnetic at room temperature. When carbon is retained in solid solution by fast quenching from the austenite phase, the Mössbauer spectrum should consist of a single line centered about zero velocity. In practice, however, those iron atoms which are neighboring a carbon atom have the cubic symmetry of their environment destroyed, resulting in a quadrupole splitting Δ. In addition, the presence of carbon atoms causes a redistribution of the s-electron density, producing an isomer shift. Genin and Flinn (1968a) have analyzed their austenite pattern, shown in Fig. 6, in terms of three unresolved lines. The central peak is located at $+0.10$ mm/sec versus α-Fe and

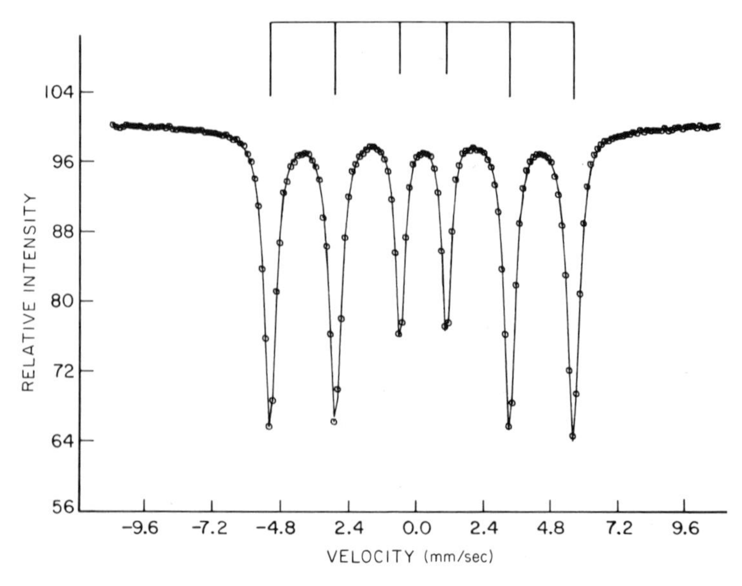

Fig. 5. Mössbauer spectrum of α-iron.

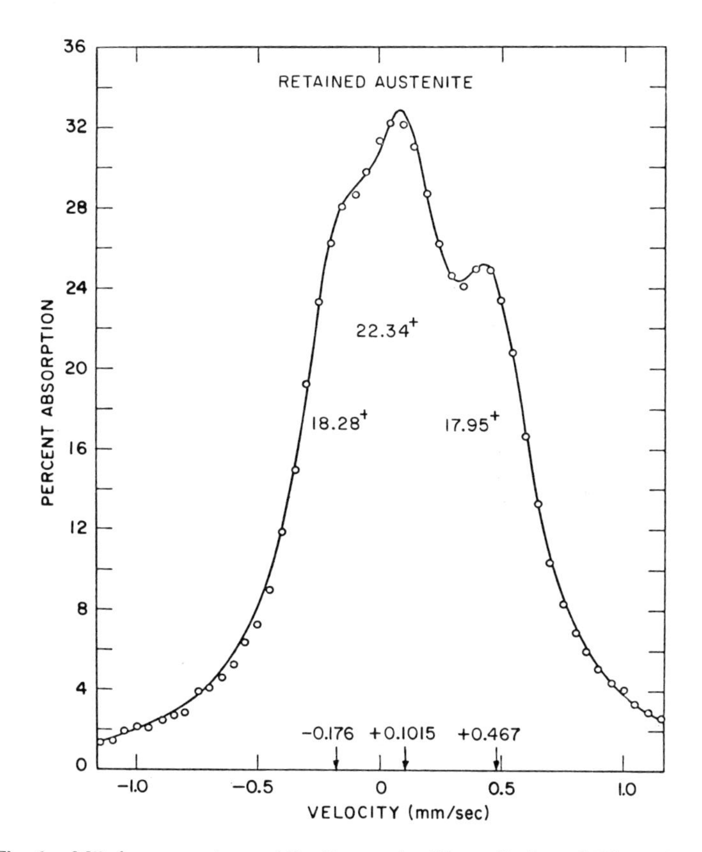

Fig. 6. Mössbauer spectrum of Fe–C austenite. [From Genin and Flinn, 1968a.]

the quadrupole split pair (iron atoms with one carbon neighbor) at +0.145 mm/sec with $\Delta = 0.643 \pm 0.005$ mm/sec.

When metallic iron containing from 0.3 to 1 wt% carbon is heat treated, the carbon precipitates from solid solution as cementite, Fe_3C. A Mössbauer pattern of a heat-treated steel is shown in Fig. 7. The cementite pattern has an isomer shift of about +0.15 mm/sec versus α-Fe with an internal magnetic field of about 220 kOe.

Moriya *et al.* (1968) obtained spectra of a 4.2 at.% carbon alloy containing both austenite and martensite phases. The spectra were resolved into three magnetic patterns characteristic of martensite with first, second, third, and fourth nearest-neighbor carbon atom configurations and a paramagnetic singlet of austenite in the center. Hence, in addition to phase identification, the Mössbauer patterns may be used to identify different atomic environments within a phase, with the limitation of spectral reso-

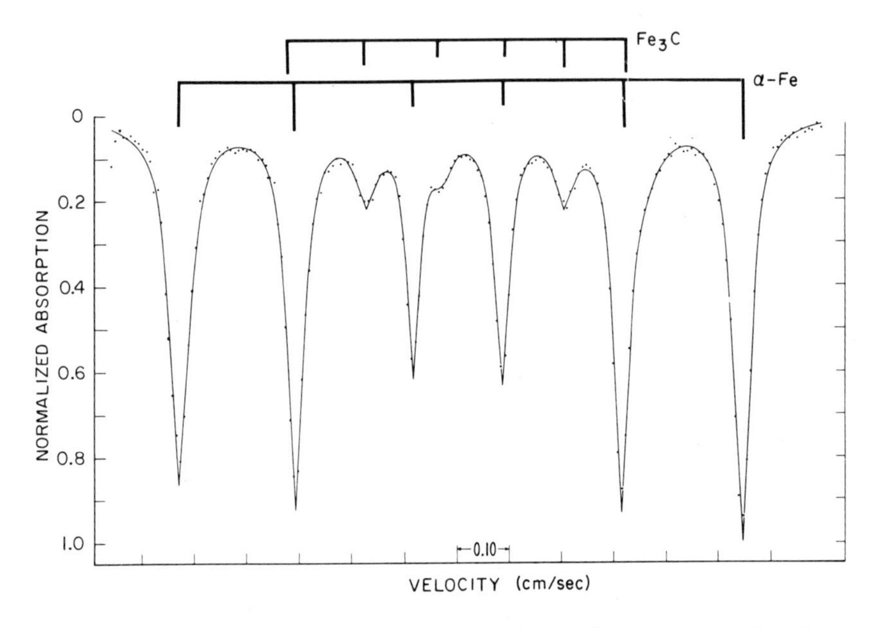

Fig. 7. Mössbauer pattern of 10105 steel heat treated to precipitate cementite. Sample was quenched from 893°C, then held 96 hr at 140°C and 3 hr at 300°C. [From B. W. Christ and P. M. Giles, 1965–1970. *In* "Mössbauer Effect Methodology" (I. J. Gruverman, ed.), Vol. 3, p. 48. Plenum, New York.]

lution in mind. Ino *et al.* (1967, 1968) studied spectral changes in iron–carbon martensitic steels during the process of tempering. The martensitic structure of high-carbon steels decomposes into cementite and an alpha-solid solution through a series of transient metastable carbides. Yamamoto (1964) observed the aging of Fe–Cr alloys at 500°C where the chromium content was varied from 20 to 46.5% by weight. Samples were heated for one hour at 1100°C, and then quenched in water. Comparison of quenched with aged specimens clearly showed a large increase in the value of the internal magnetic field for the aged samples.

4.3 Backscattering Studies in Metallurgy

For ^{57}Fe Mössbauer analyses of bulk materials, such as steel plates and gauge blocks, scattering geometry is particularly useful. An important application is in the study of the compounds formed on metal surfaces as a result of corrosion. Studies have shown that various iron oxides and oxyhydroxides as little as 10 Å thick are formed on iron surfaces. Terrell and Spijkerman (1965) have tabulated the Mössbauer parameters for the α, β, γ, and δ forms of FeOOH at room temperature and at 77°K as shown in Table 1. The isomer shifts for all forms have been shown to fall within the range characteristic of Fe^{3+}. The quadrupole splitting was marked in the cases of β- and γ-FeOOH. At 77°K, β-FeOOH splits into a six-line magnetic pattern with an internal magnetic field of 475 ± 5 kOe.

The absorption of the 14.4-keV gamma photons by the sample is followed by a reemission of only 10% of these photons; the remaining 90% are internally converted and emitted as 6.4-keV x rays and 8-keV conversion electrons. Detection systems have been designed for use in the backscattering geometry to count either the 14.4-keV gamma photons (Chow *et al.*, 1969) or the conversion x rays or electrons. Swanson and Spijkerman (1970) have constructed a proportional counter optimized to count either conversion x rays or electrons, depending on the flow gas used. For detecting the x rays a 10% CH_4–90% Ar mixture at atmospheric pressure is used. Conversion electrons are detected by placing the metal surface to be studied on one side of the counter as shown in Fig. 8. The flow gas mixture is 6% CH_4–94% He. A backscattering spectrum of a tool steel surface using conversion x-ray counting is shown in Fig. 9. The

TABLE 1

MÖSSBAUER PARAMETERS OF IRON OXIDES AND IRON OXYHYDROXIDES AT 300°K[a]

Compound	Chemical shift[b] (mm/sec)	Quadrupole split (mm/sec)	Internal magnetic field (kOe)
FeO	1.37	0.6	0
α-Fe_2O_3	0.61	0.4	517 ± 5
γ-Fe_2O_3			
Td site	0.53[d] ± 0.04	0	488 ± 5
Oh site	0.67 ± 0.04	0	499 ± 5
α-FeOOH	0.70[d]	0 ± 0.1	364 ± 37
β-FeOOH	0.640 ± 0.006	0.700 ± 0.008	0
	—	—	475 ± 5[c]
γ-FeOOH	0.648 ± 0.006	0.594 ± 0.006	0
δ-FeOOH	0.76 ± 0.2	0. ± 0.1	0
Td site[c]	—	—	525 ± 5
Oh site[c]	—	—	505 ± 5

[a] From Terrell and Spijkerman (1965).
[b] Chemical shift relative to sodium nitroprusside standard.
[c] Values are for 77°K.
[d] Corrected, by adding 0.17 mm/sec, to sodium nitroprusside standard.

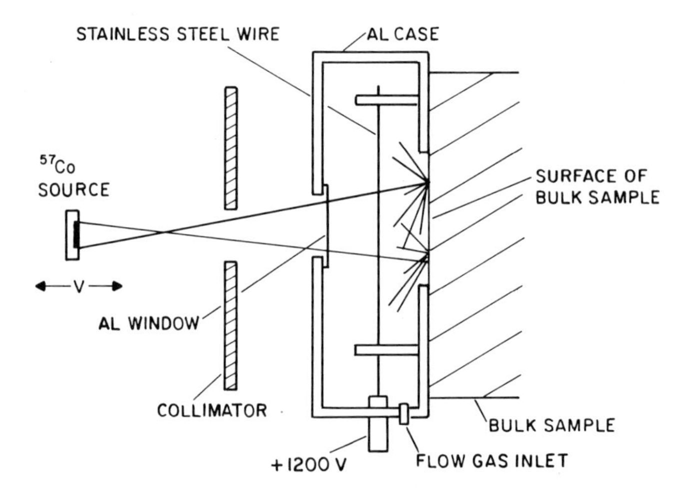

Fig. 8. Detector for obtaining conversion electron spectra from surfaces. [From Swanson and Spijkerman, 1970.]

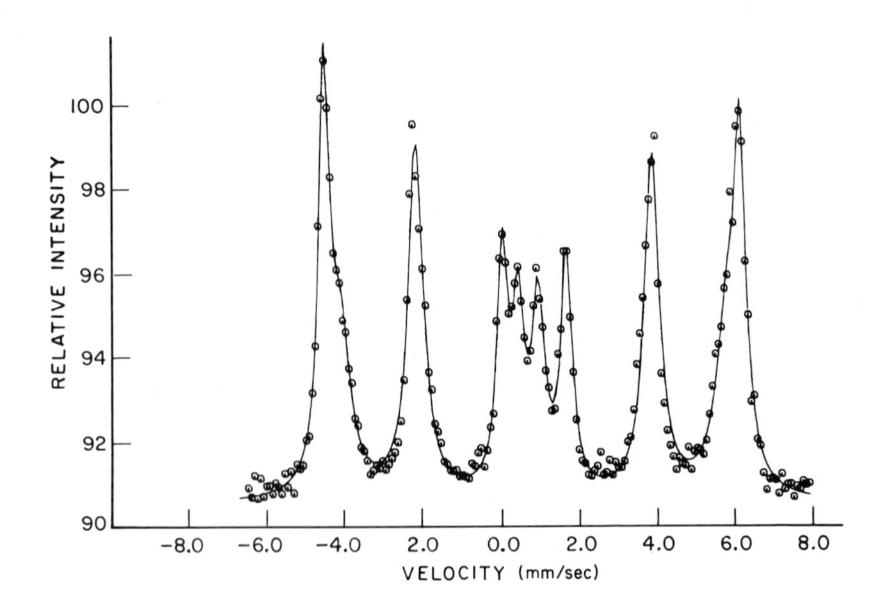

Fig. 9. Mössbauer spectrum of tool steel surface using conversion x-ray detection. [From Spijkerman *et al.*, 1971.]

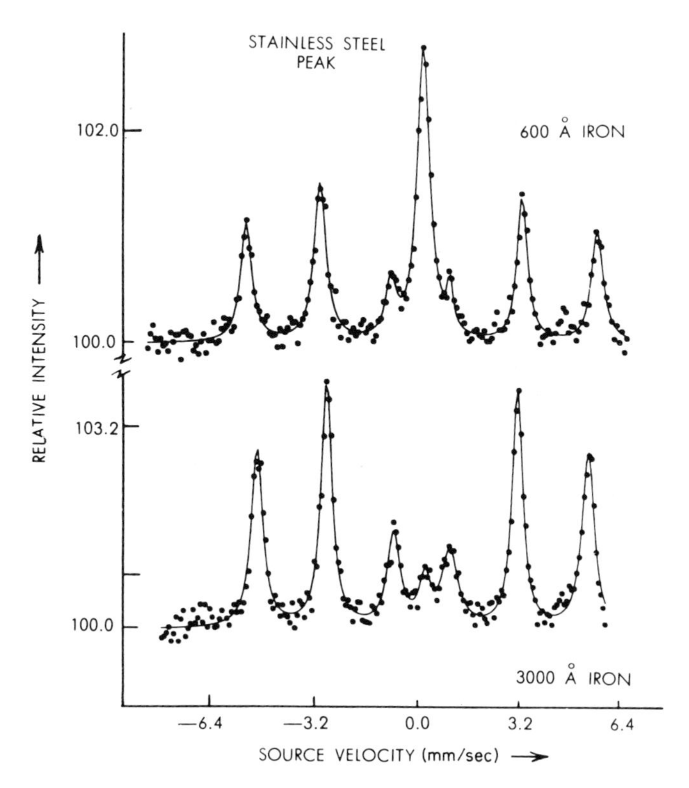

Fig. 10. Conversion electron Mössbauer spectra of iron films on stainless steel. [From DeVoe, 1967.]

spectrum consists of the α-Fe magnetic hyperfine pattern and pattern(s) probably characteristic of alloy carbides. Because conversion electrons have a finite penetration depth, only the outermost surface layers (i.e., about 3000 Å thick) can be characterized by this technique. This mode of detection offers a distinct advantage in cases where such thin layers are present on surfaces and results in a considerable enhancement of the signal-to-background ratio. For example, oxide layers containing as little as 10 $\mu g/cm^2$ of iron have been measured at the NBS. Figure 10 shows spectra of vacuum-deposited iron on stainless steel foil. The stainless steel singlet in the center practically disappears with an iron overlay of 3000 Å and can be used as a measure of the penetration depth of the conversion electrons.

4.4 Magnetic Properties

In addition to the study of phase identification and related phenomena, as discussed above, a systematic study of the hyperfine magnetic field(s) in materials can reveal the presence of several types of magnetic ordering. For this reason, this technique has been used to study magnetic properties and to detect transitions of metals, alloys, and compounds, including such phenomena as ferromagnetism (where the electron spins are aligned in the same direction), antiferromagnetism (where the spins are equal in magnitude but aligned in opposite directions), ferrimagnetism, and paramagnetism.

The effective magnetic field at the iron nucleus H_n, consists of the sum of basically three contributions—H_s, H_l, and H_d. H_s is the spin density field or core polarization field. It originates by an electron-spin exchange ("Fermi contact") interaction between polarized 3d electrons and the inner core 1s, 2s, and 3s electrons. An exchange interaction can produce a magnetic asymmetry between the inner core s electrons, depending on the relative electron-spin orientation in the respective shells. This gives rise to a net spin density at the nucleus. H_s is also the primary contribution to the ^{57}Fe effective magnetic field for ferro-, ferri-, and paramagnetic materials. H_1 is the field produced by the orbital angular momentum of the electron, and H_d is the dipole field produced by the valence electron spin.

In iron-containing alloys, the temperature above which magnetic ordering disappears in ferromagnetic materials, i.e., the Curie point, or in antiferromagnetic materials, i.e., the Neél point, can in many cases be easily determined from Mössbauer spectra. At or above the Curie or Neél points, the Zeeman pattern will collapse into either a singlet or unresolved quadrupole split doublet(s) (Preston *et al.*, 1962; Craig *et al.*, 1965; Petitt and Forester, 1971).

The magnetic properties of hematite, or α-Fe_2O_3, have been studied extensively (Ono and Ito, 1962; Gilad *et al.*, 1963; Cinader *et al.*, 1967; Van der Woude, 1966; Simkin and Bernheim, 1967). The interest in this compound was inspired by the fact that it exhibits a magnetic transition from an antiferromagnetic to a weakly ferromagnetic state. This transition became commonly known as a "spin-flop." Below a characteristic temperature called the Morin temperature ($T_M = 260°K$), α-Fe_2O_3 is antiferromagnetic, and the magnetic moments lie parallel to the [111] crystallographic plane. Above 260°K the moments of the paired atoms lie in the basal [111] plane and are not precisely antiparallel but canted slightly toward one another in the basal plane. This gives rise to a weak ferromagnetism. In addition, the "spin-flop" can be brought about by the application of an external magnetic field between 60 to 70 kOe along the [111] direction below the Morin temperature. The observation of the "spin-flop" is based on the angular dependence of the relative spectral line intensities of a single crystal of α-Fe_2O_3. The two lines in the spectral pattern corresponding to $\Delta m = 0$ transitions have an angular dependence of

$\sin^2 \theta$, where θ is the angle between the spin orientation axis and the gamma-ray direction. When the crystal is oriented so that the gamma rays propagate along the [111] axis, the two lines corresponding to $\Delta m = 0$ will be absent below T_M but will reappear on warming through T_M. Figure 11 demonstrates the changes in the Mössbauer spectrum upon hematite's going through the Morin temperature.

Magnetite, Fe_3O_4, has some interesting magnetic properties that were investigated by Verwey (Verwey and de Boer, 1936; Verwey and Haaijman, 1941; Verwey and Heilmann, 1947; Verwey *et al.*, 1947.) Marked changes in the electrical conductivity, magnetization, and specific heat behavior were observed between 100 and 120°K. Magnetite has an inverse spinel structure and can be represented by the formula $Fe^{3+}{}_{tetra}$ $[Fe^{2+}, Fe^{3+}]_{oct}O_4$.

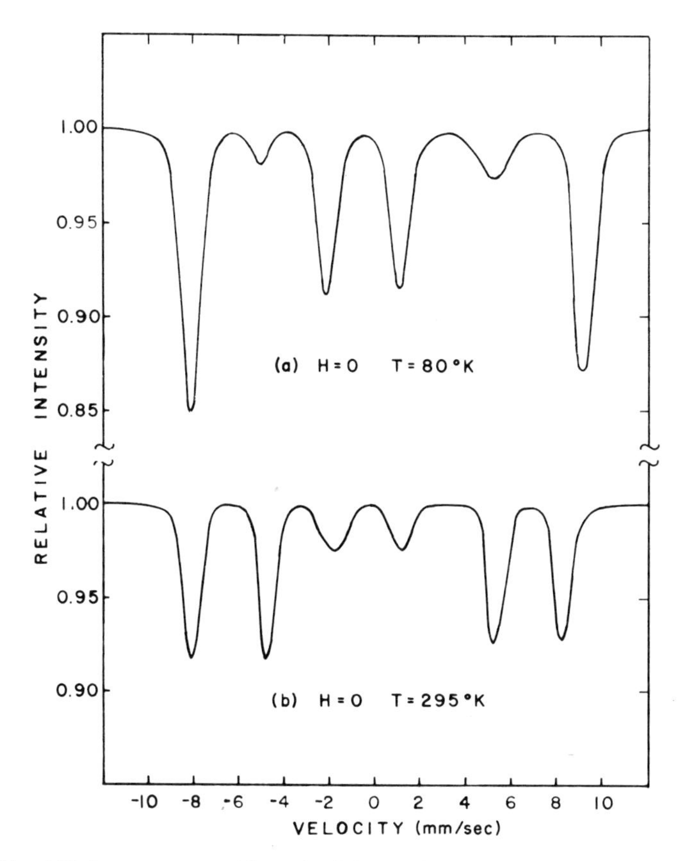

Fig. 11. Mössbauer spectra of hematite below and above the Morin transition. [From N. Blum, A. J. Freeman, and J. W. Shaner, 1965. *J. Appl. Phys.* **36**, 1169.]

The unit cell consists of eight ferric and eight ferrous ions in octahedral sites (B sites), each with six oxygen neighbors, and eight ferric ions at tetrahedral sites (A sites), each with four oxygen nearest neighbors. Verwey proposed that above the transition region there is a dynamic order caused by a fast electron exchange between the ferric and ferrous ions in the octahedral sites, which leads to an increase in electrical conductivity. Below the transition region, the ferric and ferrous ions in the octahedral sites are separately ordered. Bauminger *et al.* (1961) measured the internal magnetic field acting on the iron nuclei. Two distinct magnetic fields were measured. At 300°K, H_{eff} of 500 kOe was attributed to the ferric ions in tetrahedral sites. A comparison of the line intensities of the two superimposed Zeeman patterns indicated that only half of the ferric ions in Fe_3O_4 contributed to the 500-kOe field. A value of 450 kOe was assigned to both the ferrous and the remaining ferric ions. These observations from Mössbauer spectra were clearly consistent with Verwey's hypothesis that a rapid $Fe^{2+} \longleftrightarrow Fe^{3+}$ electron exchange did occur in octahedral sites. A spectrum of stoichiometric magnetite is shown in Fig. 12. Kündig and Hargrove (1969) measured the relaxation time τ for the "electron hopping" by measuring the broadening of the lines corresponding to the B site. A value of $\tau = 1.1 \pm 0.2 \times 10^{-9}$ sec was calculated at room temperature. There have been reported some interesting studies in which the Mössbauer technique has been used to measure the sublattice magnetizations of Fe_3O_4 when ions

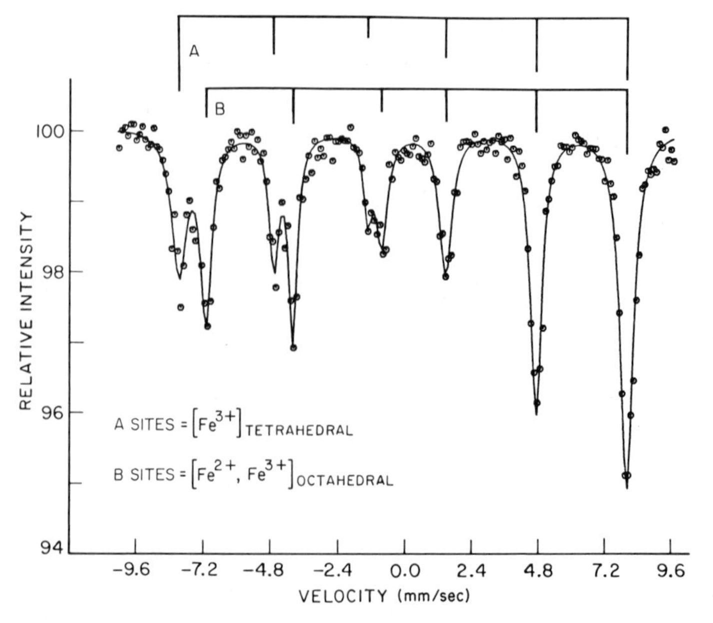

Fig. 12. Mössbauer spectrum of stoichiometric magnetite.

such as Zn^{+2} and Ni^{+2} have been substituted for Fe^{2+} in octahedral sites (Ziemniak, 1970; Gol'danskii *et al.*, 1965).

When the particle size of magnetically ordered powdered materials decreases to the region of 50 Å, the magnetization disappears because a relaxation of magnetic moments occurs. This property is known as superparamagnetism and has been studied principally in ultrafine antiferromagnetic particles of α-Fe_2O_3 and FeOOH (Syzdalev, 1970). A number of magnetic hyperfine studies have also been reported for paramagnetic compounds and for the rare earths (Nowik, 1966; Clifford, 1967).

4.5 Chemical Structure and Bonding

4.5.1 Quadrupole Splitting

Essentially two factors contribute to producing an electric field gradient (EFG) at an ^{57}Fe nucleus. These are contributions due to the surrounding electronic configuration and to distortions within a crystal lattice. Two most common coordination compounds of iron contain either four (tetrahedral coordination) or six (octahedral coordination) ligands surrounding the central iron ion. It is possible to predict *a priori* a quadrupole interaction for different d orbital electron populations shown in Fig. 13 at the limit of zero temperature using the data from Table 2. These data include values of

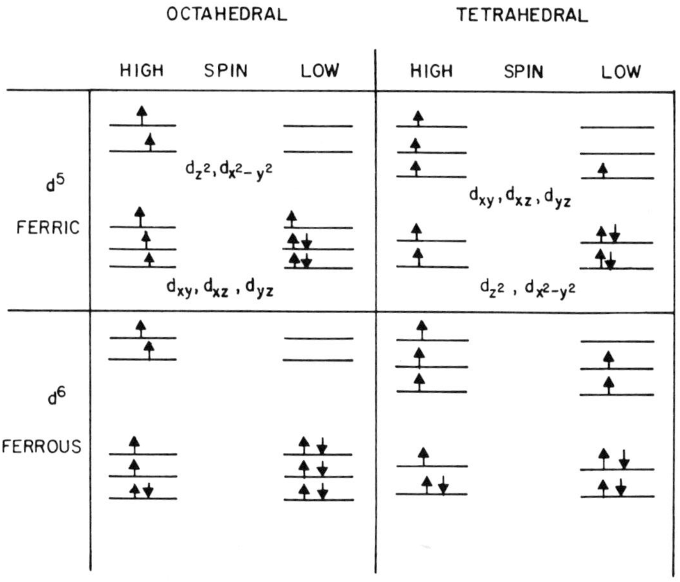

Fig. 13. Electron population of d orbitals for various electronic configurations of iron in octahedral and tetrahedral ligand fields. [From "Critical Reviews in Analytical Chemistry," J. J. Spijkerman and P. A. Pella, Vol. 1, p. 12, 1970, © CRC Press, Inc., 1970. Used by permission of CRC Press, Inc.]

TABLE 2

3d ORBITAL CONTRIBUTION TO THE EFG ELEMENTS[a,b]

Orbital	$V_{xx}/(qr^{-3})$d	$V_{yy}/(qr^{-3})$d	$V_{zz}/(qr^{-3})$d
d_{z^2}	$-2/7$	$-2/7$	$+4/7$
$d_{x^2-y^2}$	$+2/7$	$+2/7$	$-4/7$
d_{xy}	$+2/7$	$+2/7$	$-4/7$
d_{xz}	$+2/7$	$-4/7$	$+2/7$
d_{yz}	$-4/7$	$+2/7$	$+2/7$

[a] From "Critical Reviews in Analytical Chemistry," J. J. Spijkerman and P. A. Pella, Vol. 1, p. 12, 1970. © CRC Press, Inc. used by permission of CRC Press, Inc.

[b] The charge q is equal to $-e$ for an electron and $+e$ for an "electron hole."

the principal components of the EFG tensor for the five d orbitals. For example, octahedral and tetrahedral high-spin ferric complexes each have one electron in each orbital corresponding to a half-filled shell. The sum of each column in the table for one electron in each orbital is zero. Therefore, no quadrupole splitting is expected from these valence electrons. The same holds true for octahedral low-spin ferrous compounds with two electrons in each d_ϵ orbital, and none in d_γ. In octahedral low-spin ferric compounds, one of the d_ϵ orbitals would contain only one electron, whereas the other two orbitals would each be filled. Hence, these compounds as well as tetrahedral low-spin ferric and octahedral high-spin ferrous have EFG contributions arising from the uncompensated 3d electrons. These uncompensated electrons cause relatively large quadrupole splittings on the order of 3 mm/sec in high-spin ferrous compounds. A direct application of the EFG tensor to structural studies would indeed be more straightforward if no valence electron contribution existed. In addition to the valence electron contribution to the EFG, a distortion of the crystal lattice, for example, from pure octahedral symmetry, can give rise to quadrupole splitting as a consequence of dissimilar ligands, e.g., in MA_4B_2 complexes, or from a Jahn–Teller effect, or a packing distortion produced by ions of different sizes.

Ingalls (1964) has treated the problem of both the valence electron and the ligand charge distribution in a study of the EFG in high-spin ferrous compounds. Ingalls describes the total charge distribution q by the equation

$$q = V_{zz}/e = (1 - R)q_v + (1 - \gamma_\infty)q_{lat} \quad (6)$$

where eq is the principal component of the EFG tensor, q_v and q_{lat} are the EFG values produced by the valence electron and ligand charge distributions, respectively, and R and γ_∞ are shielding parameters proposed and discussed at length by Sternheimer (1950). Webb (1969) has compared crystal field distortion parameters calculated from Mössbauer data with those found by other methods for various compounds.

Edwards *et al.* (1967) studied a series of high-spin Fe (II) tetrahedral compounds of the type $R_2\,FeX_4$ when X is a halide or pseudohalide, and R is a quaternary ammonium or phosphonium cation. The degree of distortion from cubic symmetry was dependent on the ligand X and cation R. For large R the quadrupole splitting was studied as a function of temperature. These results indicated that the distortion of the crystal removed the degeneracy of the d_{z^2} and $d_{x^2-y^2}$ orbitals. These orbitals were separated by an energy $\Delta = 470\ cm^{-1}$. Bearden *et al.* (1965) reported on a study of the Fe^{2+} porphyrins (hemes) and the Fe^{3+} porphyrins (hemins) which are important constituents in hemoproteins such as hemoglobin. These compounds containing the ethyl, vinyl, and acetyl groups as substituents in the 2 and 4 positions were measured over a temperature range from 2.5 to 360°K. A large quadrupole splitting from 0.6 to 1.0 mm/sec was due principally to a tetragonal distortion from cubic symmetry because of the nonequivalence between in-plane porphyrin nitrogens and two out-of-plane pyridine nitrogens.

The sign of V_{zz} (an independent parameter of the EFG tensor) can also give some information about chemical structure, at least in iron coordination compounds. The sign can give the ground-state orbital directly in compounds where axial distortion is present when $V_{xx} = V_{yy}$ (i.e., $\eta = 0$). The sign of V_{zz} can be determined by measurements on single crystals using various crystal orientations relative to the gamma-ray propagation direction. Since in most cases single crystals are not available, measurements on polycrystalline materials involve the application of an external magnetic field on the order of 30 kOe. It can be seen from Fig. 14 that the measurement of the sign of V_{zz} determines whether the $\pm 3/2$ or the $\pm 1/2$ energy level is highest. The application of the magnetic field to ^{57}Fe results in a splitting of one of the quadrupole split lines into a doublet, and the other into a triplet, if $\eta \approx 0$. The doublet corresponds to the $\pm 3/2 \rightarrow \pm 1/2$ energy transition and the triplet to the $\pm 1/2 \rightarrow \pm 1/2$ transition. A doublet at positive velocities means that the $\pm 3/2$ energy level is highest and $V_{zz} > 0$. When $V_{zz} < 0$, the doublet occurs at negative velocities. Dale *et al.* (1968) determined the distortion in low-spin Fe(II) compounds such as $PcFe(Py)_2$, $Fe(niox)_2(Im)_2$, and $Fe(niox)_2(NH_3)_2$ at 4.2°K.* The results indicated that these compounds were elongated along the axial z axis where $V_{zz} > 0$.

* Pc: phthalocyanine; niox: 1,2-cyclohexanedioxime mono-anion; Im: imidazole; Py: pyridine.

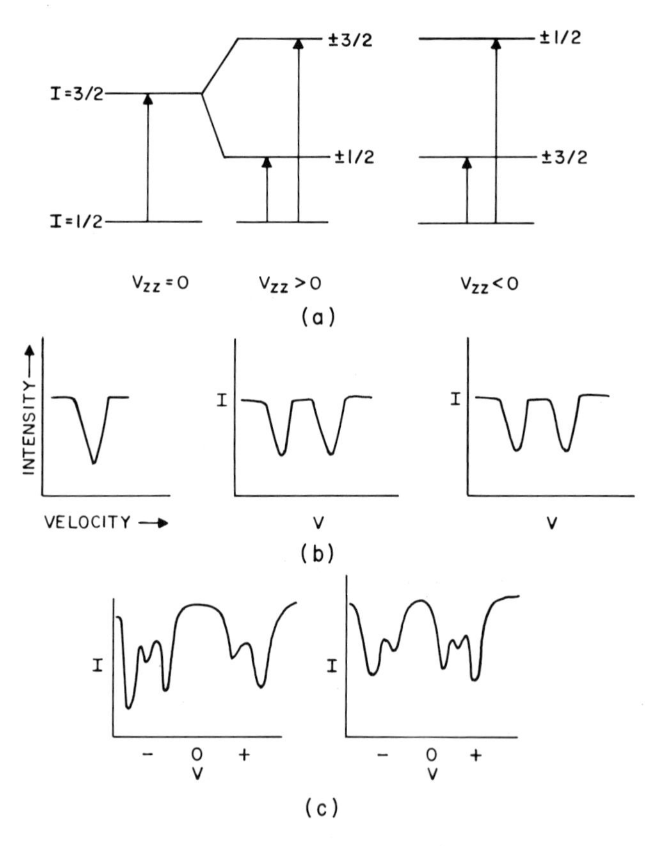

Fig. 14. Energy levels of [57]Fe. (a) Nuclear energy states for $V_{zz} = 0$, $V_{zz} > 0$, and $V_{zz} < 0$. (b) Mössbauer spectra with no applied field and $\eta = 0$. (c) Mössbauer spectra with 30-kG applied field; $\eta = 0$.

4.5.2 Isomer Shift

For compounds of iron in the high-spin configuration, the isomer shift is usually indicative of the formal oxidation state. The same, however, is not true for iron compounds in the low-spin state. For high-spin compounds, the s-electron density at the nucleus depends strongly upon the shielding effect of the 3d electrons. As the population of the 3d orbitals increases, so does the effect of shielding. Because the sign of $\Delta R/R$ is negative for iron, a decrease in s-electron density means an isomer shift toward positive velocities. In low-spin compounds the 3d electrons of the iron ion are delocalized with ligand orbitals of the appropriate symmetry. This covalent bonding results in a reduced shielding efficiency of the 3d electrons. Consequently, the observed isomer shift is much smaller than for the high-spin compounds. Danon (1963) has classified the isomer shift for low- and high-spin compounds in the order

$$Fe^{2+}(\text{ionic}) > Fe^{3+}(\text{ionic}) > Fe^{2+}(\text{cov}) > Fe^{3+}(\text{cov})$$

For the complexes $Fe(CN)_6^{-3}$, $Fe(CO)_5$, and $[Fe(CN)_5NO]^{-2}$, containing isoelectronic ligands, the order of s-electron density is given as $CN < CO < NO^+$. This trend also follows the order of increasing ligand electronegativity. Collins and Pettit (1963) have observed large isomer shifts in compounds of the type $L \rightarrow Fe(CO)_4$. The variation in isomer shift is explained by σ-bonding between ligand and metal due to the Lewis base (electron-donating) strength of the ligands. A large change in isomer shift was observed when the sulfur ligand was replaced by phosphorus in the cyclopentadienyl (Cp) derivatives $Cp(OC)Fe(SMe_2)_2Fe(CO)Cp$. It was found that sulfur is a less effective σ-electron donor than phosphorus in these compounds.

Cordey-Hayes (1964) has classified 20 or more inorganic tin compounds into two distinct groups. Positive isomer shifts correspond to Sn(II), while negative isomer shifts correspond to Sn(IV) compounds measured with respect to β-Sn metal. The compound Cs_2SnF_6 came closest to being completely ionic Sn(IV), i.e., $4d^{10}$, while $SnCl_2$ was nearest the covalent $4d^{10}5s^2$ configuration. The interpretation of the isomer shift for tin compounds should take into account important screening effects produced by (a) shielding of 5s by 5p electrons, (b) mutual shielding of paired 5s electrons characteristic of the divalent state, and (c) changes in the shielding of the inner electrons by outer valence electrons.

4.6 Miscellaneous Solid-State Studies

In the Debye approximation the recoil-free fraction is related to the mean-square displacement of the emitting or absorbing atom from its equilibrium position, i.e., $\langle x^2 \rangle$, by the equation

$$f = \exp(-\langle x^2 \rangle / \lambda\!\!\!^{-2}) \tag{7}$$

where $\lambda\!\!\!^{-} = \lambda/2\pi$, with λ the gamma radiation wavelength. For lattice dynamical studies Mössbauer nuclides can be incorporated as minor constituents into a host matrix (Abragam, 1964). Experimentally, relative measurements of f are made by comparing "thin" absorbers of the same material. The so-called black absorber technique has been used to make absolute f measurements (Gol'danskii and Markarov, 1968). Because of the difficulties in making absolute measurements, uncertainties in f no better than $\pm 15\%$ are usually reported. The major contributions to the uncertainty arise from the background and peak area measurements. Housley (1965) has described the principal errors encountered in making both relative and absolute measurements.

Bukshpan and Herber (1967) and Hazony (1968) have studied the temperature dependence of the recoil-free fraction to get an estimate of the normal-mode vibrations in the tetrahedral molecular crystal, SnI_4. Measurements were made on both the Mössbauer nuclei ^{119}Sn and ^{129}I over a temperature range 85–220°K. The tin atom was considered as being in the center of mass where its motion could be described by two characteristic frequencies, represented by two effective temperatures. One temperature θ_1 characterized the intermolecular translational lattice vibrations. From $\ln f$ versus T data, θ_1 was estimated to be 44°K and in agreement with the value of 48°K reported by infrared spectroscopy.

Stöckler *et al.* (1967) in a study of polymeric tin compounds have obtained data which indicate that the recoil-free fraction at room temperature for compounds of tin with intermolecular chemical bonds is much larger than for similar nonpolymeric compounds. For polymeric compounds of the type $(\phi_2SnO)_n$, Gol'danskii *et al.* (1963) found a sharp increase in the recoil-free fraction when a halide substituent was substituted in the *para* position on the phenyl ring.

An interesting study of ^{57}Fe imbedded in single crystals of sodium chloride showed that the recoil-free fraction of the iron ions associated with the vacancies in the Neél lattice was approximately four times greater than that for the isolated substitutional iron atoms in the lattice. This suggested that the iron ions associated with the vacancies were Fe^{2+} and Fe^{3+}. These ions strongly attract the neighboring chloride ions leading to an increase in the binding of the iron atom and effectively reducing its mean-square displacement (Mullen 1963). In crystals where the mean-square displacement is anisotropic, i.e., $\langle x^2 \rangle$, $\langle y^2 \rangle$, and $\langle z^2 \rangle$ are unequal, the recoil-free fraction becomes anisotropic also. This gives rise to the so-called Gol'danskii–Karyagin effect. Spectra of powdered samples exhibiting this phenomenon usually consist of spectral lines whose intensities are asymmetric. Measurements of the recoil-free fraction along the three principal crystallographic axes in single crystals will yield information about the mean-square displacements in the respective directions. Some precise measurements of the anisotropy in single crystals of sodium nitroprusside have been made by Danon and Iannerella (1967).

5 Data Processing

Since in most cases a multichannel analyzer is used to store Mössbauer spectral data, the output consists of intensity data in digital form which can easily be punched on paper tape. This is a convenient data format for processing by a computer. Although the spectral parameters can be obtained from the raw data by curve fitting by hand when the resolution is adequate, most laboratories involved in research in this area have made ex-

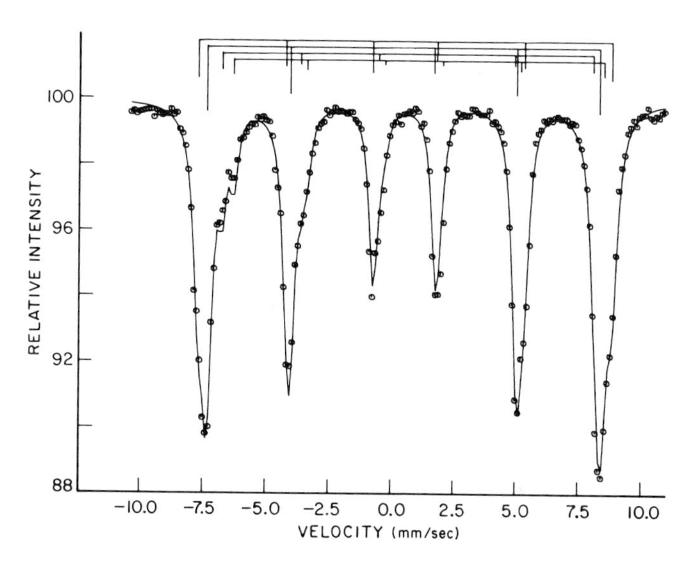

Fig. 15. Magnetic sites in nonstoichiometric nickel ferrite.

tensive use of computers. In this spectrometry the spectral line width makes up a significant fraction of the entire band/width over which resonance absorption (or scattering) is observed. Therefore, in many cases, reliable estimates of the spectral parameters can only be made with the aid of machine digital computation. Computer programs are available and consist basically of three types. The first consists of the computation of Mössbauer spectra from theoretical considerations assuming a model(s) (Collins and Travis, 1967; Gabriel and Ruby, 1965; Kündig, 1967). The second is a curve-fitting routine consisting of a linear conbination of Lorentzian line profiles. This type of program requires an *a priori* input of the estimates of the line width and peak positions and can handle as many as 12 peaks per spectrum (DeVoe, 1967). The third type consists of fitting entire spectral patterns using theoretically justified constraints (Spijkerman *et al.*, 1971). This constrained-type program is probably the only means of resolving several magnetic interactions (sites) in the same material. For example, a constrained program was used to resolve four of the several magnetic sublattice interactions in a nonstoichiometric nickel ferrite compound shown in Fig. 15. Most of these programs are based upon a least squares procedure employing a linearization of the basic Lorentzian profile $L(x)$, where

$$L(x) = A/\{1 + [(p - x)/\Gamma]^2\} \tag{8}$$

and A is the amplitude, Γ is the full width at half height, and p is the line position.

References

Abragam, A. (1964). "L'Effect Mössbauer." Gordon and Breach, New York.
Bancroft, G. M., Maddock, A. G., and Burns, R. G. (1967). *Geochim. Cosmochim. Acta* **31,** 2219.
Bauminger, R., Cohen, S. G., Morinov, A., Ofer, S., and Segal, E. (1961). *Phys. Rev.* **122,** 1447.
Bearden, A. J., Moss, T. H., Caughey, W. S., and Beaudreau, C. A. (1965). *Proc. Nat. Acad. Sci. U.S.* **53,** 1246.
Blum, N., Freeman, A. J., and Shaner, J. W. (1965). *J. Appl. Phys.* **36,** 1169.
Bukshpan, S., and Herber, R. H. (1967). *J. Chem. Phys.* **46,** 3375.
Chow, H. K., Weise, R. F., and Flinn, P. A. (1969). USAEC Rep. NSEC-4023-1.
Cinader, G., Flanders, P. J., and Shtrikman, S. (1967). *Phys. Rev.* **162,** 419.
Clifford, A. F. (1967). *In* "The Mössbauer Effect and Its Application in Chemistry" (R. Gould, ed.), Ch. 8, p. 113. Amer. Chem. Soc., Washington, D.C.
Collins, R. L., and Pettit, R. J. (1963). *J. Chem. Phys.* **39,** 3433.
Collins, R. L., and Travis, J. C. (1967). *In* "Mössbauer Effect Methodology" (I. J. Gruverman, ed.), Vol. 2, p. 23. Plenum Press, New York.
Cordey-Hayes, M. (1964). *J. Inorg. Nucl. Chem.* **26,** 915.
Craig, P. P., Perisho, R. C., Segnan, R., and Steyert, W. A. (1965). *Phys. Rev.* **138,** A1460.
Dale, B. W., Williams, R. J. P., Edwards, P. R., and Johnson, C. E. (1968). *Trans. Faraday Soc.* **64,** 3011.
Danon, J. (1963). *J. Chem. Phys.* **39,** 236.
Danon, J., and Iannerella, L. (1967). *J. Chem. Phys.* **47,** 382.
DeVoe, J. R. (ed.) (1967). NBS Tech. Note 421. U.S. Gov. Printing Office, Washington, D.C.
DeVoe, J. R., and Spijkerman, J. J. (1966). *Anal. Chem.* **38,** 382R.
DeVoe, J. R., and Spijkerman, J. J. (1968). *Anal. Chem.* **40,** 472R.
Dézsi, I. (ed.) (1971). *Proc. Conf. Application of the Mö*ssbauer Effect. Akad. Kiado, Budapest.
Edwards, P. R., Johnson, C. E., and Williams, R. J. P. (1967). *J. Chem. Phys.***47,** 2074.
Evans, B. J., Ghose, S., and Hafner, S. (1967). *J. Giol.* **75,** 306.
Gabriel, J. R., and Ruby, S. L. (1965). *Nucl. Instrum. Methods* **36,** 23.
Genin, J. R., and Flinn, P. A. (1968a). *Trans. Met. Soc. AIME* **242,** 1419.
Genin, J. R., and Flinn, P. A. (1968b). *Trans. Met. Soc. AIME* **242,** 1421.
Gilad, P., Greenshpan, M., Hillman, P., and Shechter, H. (1963). *Phys. Lett.* **4,** 239.
Gol'danskii, V. I., and Herber, R. H. (eds.) (1968). "Chemical Applications of Mössbauer Spectroscopy." Academic Press, New York.
Gol'danskii, V. I., and Makarov, E. F. (1968). *In* "Chemical Applications of Mössbauer Spectroscopy" (V. I. Gol'danskii and R. H. Herber, eds.), Ch. 1, p. 24. Academic Press, New York.
Gol'danskii, V. I., Makarov, E. F., Stukan, R. A., Trukhtanov, V. A., and Khrapov, V. V. (1963). *Dokl. Akad. Nauk SSSR* **151,** 357.
Gol'danskii, V. I., Belov, V. F., Devisheva, M. N., and Trukhtanov, V. A. (1965). *J. Exp. Theoret. Phys.* (*U.S.S.R.*) **49,** 1681.
Gruverman, I. J. (ed.) (1965–1970). "Mössbauer Effect Methodology." Plenum Press, New York.
Hazony, Y. (1968). *J. Chem. Phys.* **49,** 159.
Herber, R. H. (1971). *In* "An Introduction to Mössbauer Spectroscopy" (L. May, ed.), pp. 148–150. Plenum Press, New York.
Herzenberg, C. L., and Toms, D. (1966). *J. Geophys. Res.* **71,** 2661.

Housley, R. M. (1965). *Nucl. Instrum. Methods* **35,** 77.
Ingalls, R. (1964). *Phys. Rev.* **133,** A787.
Ino, H., Moriya, T., Fujita, F. E., and Maeda, Y. (1967). *J. Phys. Soc. Japan* **22,** 346.
Ino, H., Moriya, T., Fujita, F. E., Maeda, Y., Ono, Y., and Inokuti, Y. (1968). *J. Phys. Soc. Japan* **25,** 88.
Kündig, W. (1967). *Nucl. Instrum. Methods* **48,** 219.
Kündig, W., and Hargrove, R. S. (1969). *Solid State Commun.* **7,** 223.
Margulies, S., and Ehrman, J. R. (1961). *Nucl. Instrum. Methods* **12,** 131.
May, L. (ed.) (1971). "An Introduction to Mössbauer Spectroscopy." Plenum Press, New York.
May, L., and Snediker, D. K. (1967). *Nucl. Instrum. Methods* **55,** 183.
Moriya, T., Ino, H., Fujita, F. E., and Maeda, Y. (1968). *J. Phys. Soc. Japan* **24,** 64.
Muir, A. H. Jr., Ando, K. J., and Coogan, H. M. (1958–1966). "Mössbauer Effect Data Index." Wiley (Interscience), New York.
Mullen, J. G. (1963). *Phys. Rev.* **131,** 1415.
Nowik, I. (1966). *In* "Mössbauer Effect Methodology" (I. Gruverman, ed.), p. 147. Plenum Press, New York.
Ono, K., and Ito, A. (1962). *J. Phys. Soc. Japan* **17,** 1012.
Petitt, G. A., and Forester, D. W. (1971). *Phys. Rev. B* **4,** 3912.
Preston, R. S., Hanna, S. S., and Heberle, J. (1962). *Phys. Rev.* **128,** 2207.
Simkin, D. J., and Bernheim, R. A. (1967). *Phys. Rev.* **153,** 621.
Smith, D. L., and Zuckerman, J. J. (1967). *J. Inorg. Nucl. Chem.* **29,** 1203.
Spijkerman, J. J., and Pella, P. A. (1970). *In* "Critical Reviews in Analytical Chemistry," Vol. 1, p. 12. Chem. Rubber Co., Cleveland, Ohio.
Spijkerman, J. J., Travis, J. C., Pella, P. A., and DeVoe, J. R. (1971). NBS Tech. Note 541. U.S. Gov. Printing Office, Washington, D.C.
Sternheimer, R. M. (1950). *Phys. Rev.* **80,** 102.
Stevens, J. G., and Stevens, V. E. (1972). "Mössbauer Effect Data Index, Covering the 1970 Literature." Plenum Press, New York.
Stevens, J. G., Travis, J. C., and DeVoe, J. R. (1972). *Anal. Chem.* **44,** 384R.
Stöckler, H. A., Sano, H., and Herber, R. H. (1967). *J. Chem. Phys.* **47,** 1567.
Swanson, K. R., and Spijkerman, J. J. (1970) NBS Tech. Note 501 (J. R. DeVoe, ed.). U.S. Gov. Printing Office, Washington, D.C.
Syzdalev, I. P. (1970). *Sov. Phys. Solid State* **12,** No. 4, 775.
Terrell, J. H., and Spijkerman, J. J. (1965). *Appl. Phys. Lett.* **13,** 12.
Van der Woude, F. (1966). *Phys. Status Solidi.* **17,** 417.
Veits, B. N., Grigalis, V. Ya, Lisin, Yu D., and Taksar, I. M. (1970). *Ind. Lab. (USSR)* **36,** 700.
Verwey, E. J. W., and de Boer, J. H. (1936). *Rec. Trav. Chim.* **55,** 531.
Verwey, E. J. W., and Haaijman, P. W. (1941). *Physica.* **8,** 979.
Verwey, E. J. W., and Heilmann, E. L. (1947). *J. Chem. Phys.* **15,** 174,
Verwey, E. J. W., Haaijman, P. W., and Romeijn, F. C. (1947). *J. Chem. Phys.* **15,** 181.
Webb, G. A. (1969). *Coord. Chem. Rev.* **4,** 107.
Whitfield, H. J., and Freeman, A. G. (1967). *J. Inorg. Nucl. Chem.* **29,** 903.
Yamamoto, H. J. (1964). *J. Appl. Phys. (Japan)* **3,** 745.
Ziemniak, S. E. (1970). KAPL-M-6261 (NCT-69) Rep. Part I.

CHAPTER 27

Optical Microscopy

J. H. Richardson

The Aerospace Corporation
El Segundo, California

Introduction

The optical microscope was probably the first instrument used in scientific analysis. It is therefore not surprising that it has reached a very high degree of sophistication in each of the areas in which it can be applied.

This chapter includes a brief description of the various types of optical microscopes and a discussion of their use in the analysis of materials. The optical microscope can be used for three different types of materials analysis, i.e., chemical analysis, phase identification, and textural measurement. Each analysis requires a somewhat different type of instrumentation, depending on the application.

1 Optical Microscopes

1.1 General

The optical microscope has been popularized as a means of magnifying objects; this, however, is not its main function. The microscope is designed to resolve objects, i.e., to permit two closely spaced objects to be viewed as separate and distinct objects.

To achieve the resolution of objects, these objects must be distinguishable from the field of view surrounding them; this requires contrast. Thus, the object or sample may be observed as (1) a bright object in a dark field; (2) a dark object in a bright field; (3) a light or dark object in an identical field but with dark or light outlines, respectively, and (4) a multicolored object in a multicolored field. Contrast may be inherent in the sample, or it may be generated either by the use of sample preparation techniques or by the use of special accessories affixed to the microscope.

Effective use of the optical microscope requires sufficient contrast, which can be produced by either sample preparation techniques or the use of accessories, so that the eye or film can produce images of objects that are distinctly resolved.

1.2 Transmitted-Light Bright-Field Microscope

The operation of the optical microscope can be most easily explained by the use of the transmitted-light bright-field microscope (Fig. 1). The light from a source, usually an incandescent lamp, is focused on the sample by means of a condenser lens systems. Note that an inadequate condenser system or one that is incorrectly adjusted can be a major cause of inadequate images produced by the optical microscope. To achieve the highest resolution and even illumination, the Koehler method of illumination is used; see Richardson (1971, p. 26) for more detailed information. Samples used in transmitted-light bright-field observations must be at least partially translucent to permit passage of the light from the condenser system. Furthermore, the contrast, either intensity or color, must be inherent in the sample so that the structure of the sample can be resolved. Generally, the

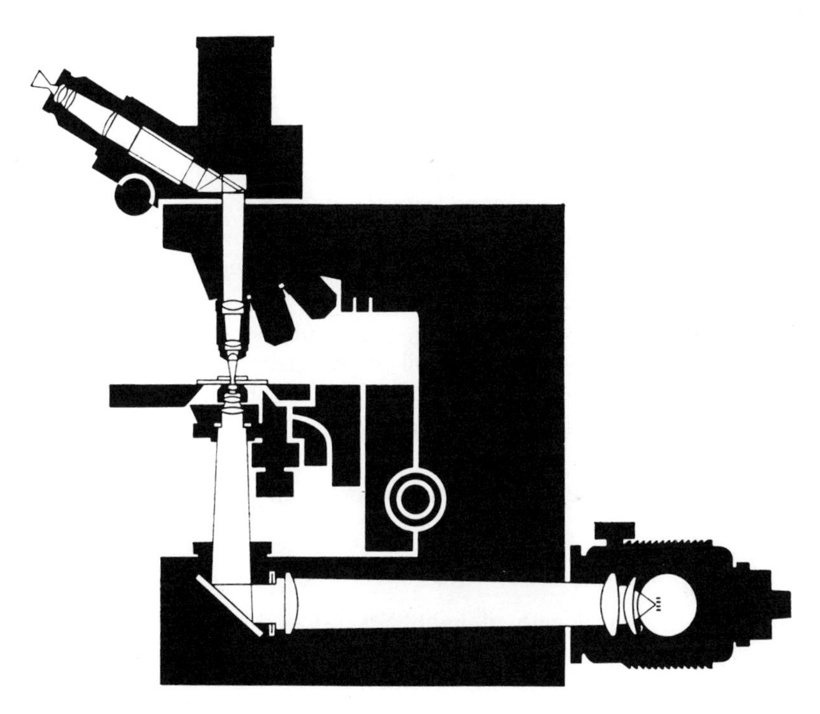

Fig. 1. Schematic cross section of transmitted-light bright-field microscope with standard configuration stand. This instrument is used in the examination of thinned transparent materials or particles. [Courtesy of E. Leitz.]

bright-field sample is 30 μm or less in thickness and is mounted on a glass microscope slide. A glass cover slip is affixed to solid samples for protection. The light transmitted by the sample is accepted by the objective lens.

The objective lens is the most important part of the optical system. It resolves the structure of the specimen and magnifies it.

Various properties of the objective lens must be considered with regard to the application of the optical microscope in materials analysis. Among these properties are numerical aperture, magnification, optical correction, flatness of field, and parfocality.

1.2.1 Numerical Aperture

Numerical aperture is a measure of the light-gathering capability of a lens. This is very important because the ability of a microscope to resolve a structure depends on the amount of light collected from each area of the specimen. Numerical aperture (NA) is defined as

$$\mathrm{NA} = \eta \sin \alpha \tag{1}$$

where η is the minimum refractive index encountered by the light in its path from the condenser to the objective lens, and α is the half-angle of the most oblique ray to enter the objective lens. It can be seen from Eq. (1) that the maximum value of NA for a standard or dry objective is NA = 1 with the η of air = 1.0 and sin α = 90°. This theoretical value is, however, not obtained in practice.

The NA, and hence the resolving power of the objective lens, can be improved somewhat by the use of oil- or water-immersion lenses; i.e., the air in the path of the light is replaced by a fluid that has a significantly higher refractive index. NAs of 1.32 to 1.40 are common when oil-immersion objectives are used.

The resolving power of the microscope can be calculated from the expression

$$\text{resolving power} = 2\mathrm{NA}/\lambda \tag{2}$$

where λ is the wavelength of light. Several of the properties of microscope objective lenses, including the NA and resolving power, are given in Table 1.

1.2.2 Magnification

Resolving a structure as an enlarged or magnified image of the object is accomplished by the objective lens. The magnification along with the NA of the objective is engraved on the lens mount. This magnified image is

TABLE 1

NUMERICAL APERTURE, RESOLVING POWER, AND SUITABLE EYEPIECES FOR SELECTED OBJECTIVE LENSES

Objective magnification	NA	Resolving power (μm^{-1})	Optimum range of magnification for eyepieces (×)
2.5	0.08	0.3	10–25
10	0.25	1.0	8–25
25	0.50	1.9	6–20
50	0.85	3.3	6–15
95	1.32	5.1	6–10

further magnified by the eyepiece before it is viewed by the eye or film. The total magnification is

$$M_T = M_O \cdot M_I \cdot M_E \tag{3}$$

where M_O is the magnification of the objective lens, M_I is the magnification of any intermediate optics, e.g., a zoom system, and M_E is the magnification of the eyepiece.

Upper and lower limits for M_T can be given for each objective lens. The lower limit is imposed as a result of the finite resolving power of the eye. This limit, expressed in terms of the NA of the objective lens, is the minimum M_T for the microscope or $\sim$300NA.

The upper limit for the M_T of the microscope is $\sim$1000NA. Beyond this value the resolution is not increased. This condition is often referred to as empty magnification. These upper and lower limits are the criteria for eyepiece selection as given in Table 1.

1.2.3 Optical Correction

To minimize the effects of the two main lens defects, i.e., spherical and chromatic aberration, two or more types of optical materials are used in the fabrication of lens elements. The three degrees of optical correction encountered in the objective lens are achromatic, semiapochromatic, and apochromatic; the characteristics and applications of objective lenses with these corrections are given in Table 2.

1.2.4 Flatness of Field

When most objective lenses are critically focused on an object in the center of the field of view, the focus of other objects degrades progressively as their positions approach the edge of the field. These outer objects may only be seen clearly by readjusting the fine focus. In this way, the objects on concentric rings with increasing diameters become successively more sharply focused as the fine focus is gradually changed. This effect, although somewhat annoying, does not seriously detract from visual study since the microscopist constantly readjusts the focus to the area of interest. Successful photography can be accomplished in a number of ways (Richardson, 1971), the best of which involves the use of specially corrected objectives, referred to as plano- or flat-field objective, in combination with matching eyepieces.

1.2.5 Parfocality

If a microscope has several objectives mounted on a rotating nosepiece, the objectives should be roughly in focus so that the coarse focusing knob does not have to be affected when changing from one objective to another.

TABLE 2

OBJECTIVE LENS CORRECTIONS

Correction	Characteristics	Applications
Achromatic	Spherically corrected for green light. Chromatically corrected for red and blue. Strain can be minimized for polarized light investigations. New lens designs can yield flat fields (planachromat). Low cost	Achromats are generally the lowest cost objectives lenses, suitable for most visual applications. Photography is usually accomplished with black and white film and green filter. Most suitable for polarized light application if strain free. Planachromats yield large flat fields of view
Semiapochromatic	Spherically corrected for two colors. Chromatically corrected for two colors. Medium cost	Higher spherical and chromatic corrections result in improved applications to photomicrography, especially for use with color film. Usually not suitable for polarizing investigation
Apochromatic	Spherically corrected for two colors. Chromatically corrected for three colors. High cost	Yield the best correction for color photomicrographs. Usually not suitable for polarizing investigation

This speeds up the investigation and minimizes the possibility of damage to either the objective lens or the sample.

1.3 Normal-Incident-Light Bright-field Microscope

The normal-incident-light microscope is designed to be used in the study of sample surfaces; these may be either translucent or opaque.

To best utilize the full NA of the objective and still maintain the focus across the complete field, the illumination for the sample must come from the direction of the objective lens. This is accomplished by means of a vertical illumination system (Fig. 2). This system is a transmitted-light microscope in which the light path has been "folded" at the sample surface such that the objective lens also serves as the condenser lens.

The optical elements for normal-incident illumination function in the same way as their counterparts used in transmitted light. However, the optical correction of the objective lenses differs. In transmitted illumination, the light must pass through the glass cover slip; therefore, the spherical aber-

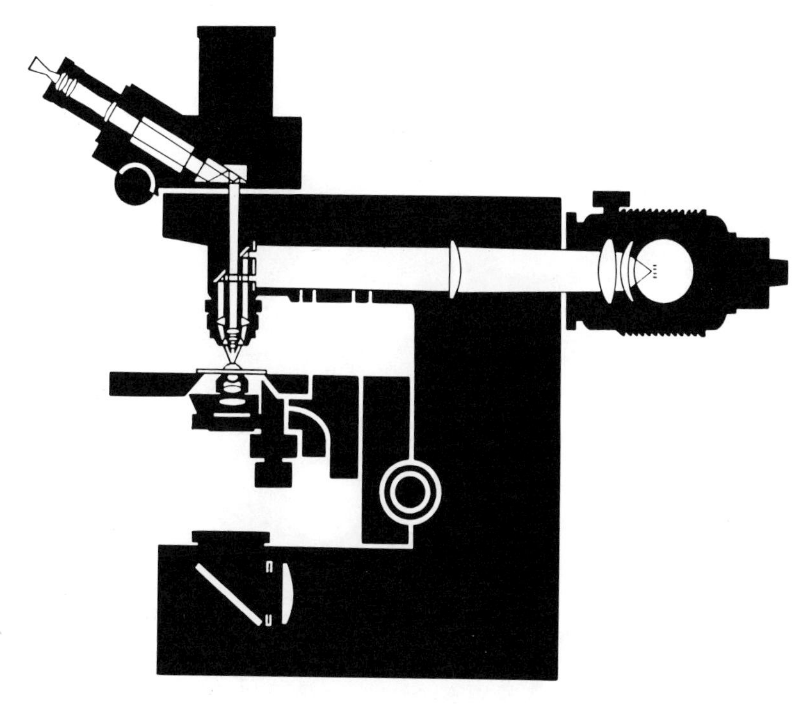

Fig. 2. Standard configuration microscope equipped for normal-incident bright-field illumination of polished and etched specimens. [Courtesy of E. Leitz.]

ration introduced by this additional element must be corrected in the objective lens. No cover glass is used in normal-incident illumination; therefore, this specific correction is not needed. These two types of objective lenses generally are not interchangeable.

The normal-incident-light microscope is available in two general configurations: (1) standard and (2) inverted or Le Chaltier. The latter is shown in Fig. 3. The standard configuration is often used with a microscope stand, which, incidently, can be used alternately for transmitted or incident illumination. The major disadvantage in the use of the standard normal-incident-illumination microscope results from the inevitable wedge angle introduced between the sample surface and the base surface of the mount during the grinding and polishing operations. This condition requires constant refocusing of the sample surface as the specimen is moved under the microscope. Leveling can be accomplished, but this requires additional time and care. The inverted configuration eliminates this problem

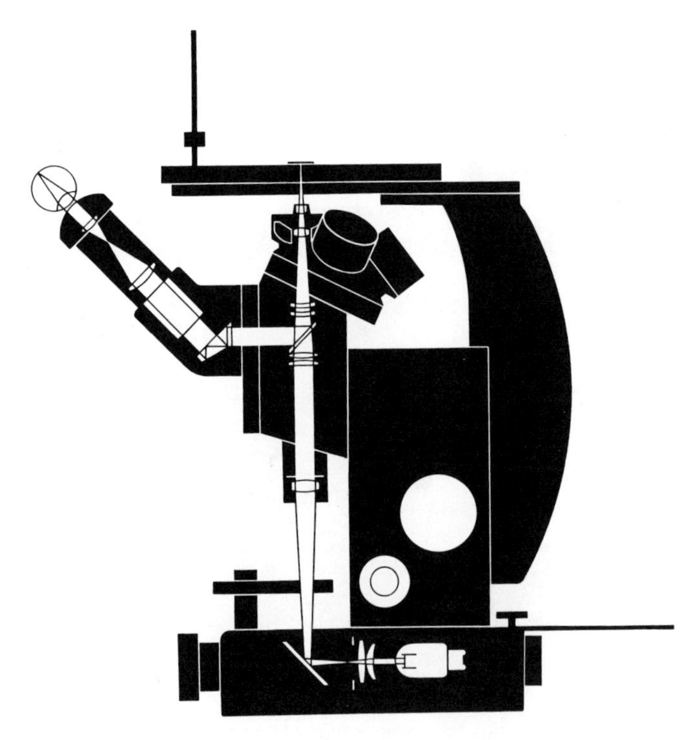

Fig. 3. Inverted or Le Chatelier stand for normal-incident bright-field microscopy. [Courtesy of Wild Heerbrugg, Ltd.]

because the sample surface rests on the microscope stage and is therefore parallel to the stage.

1.4 Phase-Sensitive Microscope

The previously described microscopes rely on brightness contrast for structure observation. Note, at this point, however, that the sample affects the light in yet another way; i.e., the phase of the illumination might be changed in one area of the sample with respect to another. The unaided eye, however, is not able to observe phase differences. Thus, these effects cannot be observed with a bright-field microscope.

Phase effects in transmitted light result from slight differences in refractive indices from one area to another. These effects can be independent of direction, as in the case of glassy or cubic materials, or dependent on direction, as in the case of noncubic crystalline materials.

Phase effects in normal-incident-illumination examination of materials result from differences in surface elevation and directional differences in reflectivity.

There are several types of specially equipped microscopes that can be conveniently grouped as phase-sensitive microscopes. These include the dark-field microscope, the phase-contrast microscope, the interference microscope, and the polarizing microscope. The theory and operation of each of these phase-sensitive microscopes are beyond the scope of this chapter. For detailed information, see Richardson (1971, pp. 107–173). Applications for the various phase-sensitive techniques for specimens in transmitted and incident illumination are given in Tables 3 and 4. The same type of stand used for the bright-field microscope is used for these microscopes. The so-called "universal"-type microscopes are bright-field instruments with appropriate accessories for achieving phase-sensitive operation.

1.5 Fluorescence Microscope

Fluorescence microscopy is another technique that can be used to produce brightness or color contrast in the examination of materials. In this type of microscopy, the images are formed almost entirely from the

TABLE 3

Effects Observed with Transmitted-Light Phase-Sensitive Microscopes

Type of illumination	Effect observed	Comments
Oblique	Direction independent. Refractive index changes	Illumination not uniform around periphery of grains
Dark-field	Direction independent. Refractive index changes	Very sensitive, but light levels are low for photomicrography
Phase-contrast	Direction independent. Refractive index changes	Very sensitive, but grains are outlined, making size measurements difficult
Interference-contrast (external reference)	Direction independent. Refractive index changes	Very sensitive to changes in refractive index and to mechanical vibration. Simple to interpret
Interference-contrast (internal reference)	Direction independent. Refractive index changes	Very sensitive. Not affected by mechanical vibration. Might be more difficult to interpret than with external reference
Polarizing	Directionally dependent. Refractive index changes in noncubic crystals	Very sensitive. Light levels usually low for photomicrography

TABLE 4

EFFECTS OBSERVED WITH INCIDENT-LIGHT PHASE-SENSITIVE MICROSCOPES

Type of illumination	Effect observed	Comments
Oblique	Differences in elevation on sample surface	Illumination not uniform around periphery of grains
Dark-field	Differences in elevation on sample surface	Very sensitive, but light levels are low for photomicrography
Phase-contrast	Differences in elevation on sample surface	Very sensitive, but grains are outlined making size measurements difficult
Interference-contrast (external reference)	Differences in elevation on sample surface	Very sensitive to changes in elevation and to mechanical vibration. Simple to interpret
Interference-contrast (internal reference)	Differences in elevation on sample surface	Very sensitive. Not affected by mechanical vibration. Might be more difficult to interpret than with external reference
Multiple-beam-interference	Differences in elevation on sample surface	Most sensitive technique. Measurements are quantitative. Can resolve differences in elevation as small as 0.5 nm (5Å)
Polarizing	Directionally dependent changes in reflectivity in noncubic crystals	Moderately sensitive. Light levels usually low for photomicrography

light (fluorescent) emitted from the specimen as a result of excitation of the material by light of shorter wavelength.

The fluorescence is observed in two distinct types of specimens. One type of specimen is composed of materials that have an intrinsic fluorescence or fluorescence due to traces of impurities. Another type of specimen exhibits secondary fluorescence, which is the result of treating the specimen with a solution of a fluorescing material (fluorochrome).

The fluorescence microscope in its simplest form is a bright-field microscope with two filters added: an excitation filter and a barrier filter. The excitation filter is placed between the lamp and specimen and blocks out all light except that which produces fluorescence of the specimen. The barrier filter is placed in the microscope at some point between the sample and eye. This filter allows the fluorescent light through, but blocks out the exciting

TABLE 5

REQUIREMENTS FOR MICROSCOPES SUITABLE FOR MATERIALS ANALYSIS

Requirement	Chemical analysis	Crystal structural analysis	Textural analysis
Objective lenses	Achromatic and strain free, usually low power	Achromatic, strain-free, low-to-intermediate common powers. Special objective required for dispersion staining (see Section 2.2)	Planachromatic. May require special objectives for phase or interference contrast. Must be strain free for polarized light
Illumination	Transmitted	Phase contrast optics may also be used for dispersion staining. Transmitted	Transmitted and normal incident
Substage condenser system	Simple with polarizer	Aplanatic–achromatic with polarizer	Aplanatic–achromatic. Polarizing phase contrast or interference dark-field accessories can be added
Stand	Standard stand equipped with rotating, calibrated stage. Insertable polarizing analyzer necessary	Standard stand equipped with rotating calibrated stage. Insertable polarizing filter and slot for inserting compensator plates necessary. Bertrand lens and Bertrand diaphragm are highly desirable	Inverted stand most desirable for incident light studies. Standard stand easiest to use for transmitted light studies. Rotating and translating stage necessary. TV attachment desirable for multiple or complicated textural analyses

radiation. The wavelength of the excitation radiation ranges from ~ 250 to ~ 480 nm. The fluorescent radiation may range throughout the entire visible spectrum.

1.6 Selection of an Optical Microscope

The optical microscope can be used for any of the three general classes of materials analysis, i.e., chemical identification, phase identification, and textural analysis. Each class of analysis has specific requirements; therefore the type of microscope and the accessories vary with the type of application. Some of the requirements for an optical microscope for the three types of materials analysis are given in Table 5. Universal instruments such as the Leitz Ortholux, the Zeiss Ultraphot, and the Reichert Zetapan are generally adequate to perform analyses in all of these classes. However, subjecting such expensive and complex instruments to possible corrosion and damage in the chemical analytical procedures is open to criticism.

2 Use of Optical Microscope for Identification of Crystalline Phases

The polarizing optical microscope provides a means of phase identification for solid materials. The trained operator can perform tests in a relatively short time and with a minimal sample. The branches of microscopy concerned with phase identification are optical crystallography, dispersion staining, petrographic microscopy, and ore microscopy.

2.1 Optical Crystallography

Optical crystallography is a powerful method of identifying nonopaque phases in the field of mineralogy. For optical crystallography, solids are grouped into five classes (Table 6). A sample can be as small as a single grain 1 to 3 μm in diameter; however, for large amounts of material, a crushed powder (200–325 mesh, 74–44 μm) is used. The grain (or several grains if available) is placed on a microscope slide and covered with a standard cover slip 1 cm^2 or less. A small amount of immersion oil of a known refractive index is placed at the edge of the cover slip and allowed to flow under to cover slip and around the unknown grains.

The prepared slide is then studied by means of a transmitted, polarized-light microscope, which is often referred to as a petrographic microscope. A suitable microscope for this purpose is described in Table 5.

By means of the Becke method (Richardson, 1971, p. 266) or the Schroder–van der Koch method (Hartshorne and Stuart, 1970), the refractive index of the grain is found to be greater or less than that of the immersion oil used. The oil is then carefully removed from a single-grain

TABLE 6

CLASSIFICATION OF NONOPAQUE SOLID MATERIALS FOR OPTICAL CRYSTALLOGRAPHY

Type of material	Class	Observations possible with microscope
Crystals in cubic system, glasses, and amorphous materials	Isometric	Color, cleavage, refractive index, dispersion, and optical activity
Crystals in tetragonal or hexagonal systems	Uniaxial +	Color, cleavage, optical sign, refractive two indices, dichrorism, two dispersions, sign of elongation, and optical activity
	Uniaxial −	Same as for uniaxial +
Crystals in orthorhombic, monoclinic, and triclinic systems	Biaxial +	Color, cleavage, optical sign, 2 V, 2 E, optical orientation, three refractive indices, pleochroism, three dispersions, sign of elongation, and optical activity
	Biaxial −	Same as biaxial +

mount by means of a small piece of blotter or filter paper. If considerable material is available for analysis, the oil and sample can be discarded after each examination. A second mount is made with an oil of higher or lower refractive index as dictated by the results of the first mount. The objective of this procedure is to match the refractive index of the sample and the oil. If a match is achieved, the refractive index of the particle is that of the oil. If a match is not achieved with the second mount, the process is repeated with oils of appropriate refractive index. Concurrent with the index-matching process, features such as color, morphology (shape and cleavage-dominated shapes), and optical characteristics that serve to classify the material are noted.

On the basis of the results of these tests, the material can be placed in one of five classes shown in Table 6. The various optical observations possible for each are also shown to indicate the amount of information obtainable for a single material. It is beyond the scope of this chapter to describe each of these observations; detailed discussions are given by Winchell (1937), Wahlstrom (1955), Shubnikov (1960), Heinrich (1965), El Hinnawi (1966), and Hartshorne and Stuart (1969).

The accumulated data should then be compared with the values for known materials. Winchell and Winchell (1951), Winchell (1957, 1964), and Larsen and Berman (1934) give data for a large number of minerals. Winchell and Winchell (1964) give data for artificial inorganic materials, and Winchell (1964) gives the values for a large number of organic solids.

Optical crystallography has been described as a qualitative technique; however, there are a number of quantitative applications of this method. A change in the refractive index can be used to measure the relative amounts of each of the compounds forming a complete solid solution series. Winchell and Winchell (1964, p. 324) give examples of such systems. Other optical properties such as 2V, extinction angles, or dispersion can also be used for the analysis of solid solutions (Poldervaart, 1950; Hess, 1949).

Optical activity, i.e., beams of plane-polarized light rotated during passage through the crystal, is a useful ancillary technique in the determination of the crystal structure by x-ray diffraction (Chapter 28). The x-ray diffraction process adds a center of symmetry to the other forms of symmetry present for a given structure; hence, a noncentrosymmetric structure and its centrosymmetric equivalent cannot be differentiated. Noncentrosymmetry results when crystals are formed from molecules devoid of a center of symmetry and/or structural elements are arranged in a spiral configuration. Optical activity is common to both types of crystals.

Optical activity may be observed in directions of a crystal in which the birefringence is zero. In this case the crystal will not undergo complete extinction in cross-polarized light. Extinction may be achieved by rotating one of the polarizing elements in the microscope.

The presence of optical activity as described above confirms the noncentrosymmetric nature of the crystal. However, a negative test for optical activity does not preclude noncentrosymmetry. The crystal might be too thin (some optically active crystals must be on the order of 1 mm thick to show an effect). The crystal could be a racemate, i.e., a nonoptically active solid solution of the two opposite-handed isomers; or the crystal could be of one of the noncentrosymmetric classes that do not exhibit optical activity. This is discussed in detail by Hartshorne and Stuart (1970) and by Kurtz and Dougherty (Chapter 38, Volume IV).

2.2 Dispersion Staining

Dispersion staining is a relatively new method for identification of crystalline phases and has its basis in the work of Crossman (1948, 1949) and Dodge (1948). Their radically simplified technique permits technical personnel untrained in optical crystallography to identify crystalline substances.

A particle as small as 1 to 10 μm (or several particles) is placed on a microscope slide, immersed in a liquid, and covered with a standard cover slip. The liquid is selected so that at some visible wavelength λ_0, the

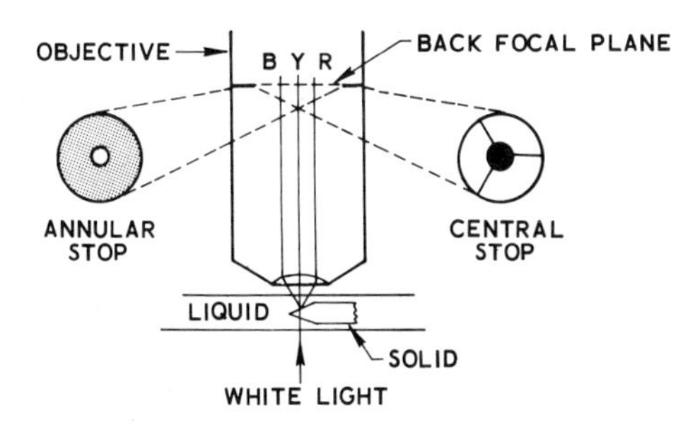

Fig. 4. Schematic cross section of dispersion-staining objective lens. The specimen shown in the immersion oil may be observed normally or with either of the two stops shown as insets. With the annular stop in place only the axial light is seen, and the specimen appears to be yellow, in this case, in a bright-field. With the central stop in place, the specimen is seen in the nonaxial light and appears to be magenta, i.e., red plus blue, in a dark field. [Courtesy of Walter C. McCrone and Associates, Inc.]

refractive indices of the particle and the immersion liquid are identical. The liquids are chosen for their high dispersion of refractive index; therefore, at all other wavelengths in the visible spectrum, the refractive indices of the particle and liquid do not match. The more extreme the difference in refractive index at wavelengths other than λ_0, the more vivid the effect. Thus at wavelengths other than λ_0, the edges of the grain are considered to be prisms that produce a spectrum.

The resulting effect on the image of the particle in the microscope can be viewed by means of dark-field (Dodge, 1948), phase-contrast (Crossman, 1949), and interference-contrast (Dodgson, 1963) techniques; however, a dispersion-staining objective lens described by Brown *et al.* (1963) permits viewing of the particle with either λ_0 (axial) illumination or non-λ_0 (deviated) illumination. This latter method is shown diagrammatically in Fig. 4.

If the annular stop is in place, the light from the particle edge that is undeviated by the prismatic effect remains; this corresponds to λ_0. If the central stop is in place, the remaining light, which is the light deviated by the prismatic effect, colors the edge of the particle. The color of this latter light is complimentary to λ_0.

A series of commercially available refractive index oils* were found to be satisfactory mediums for dispersion staining by Brown *et al.* (1963). On the basis of this series of oils, Brown *et al.* (1963), McCrone *et al.* (1967), Forlini (1969), Julian and McCrone (1970), Goodman (1970), and Forlini and McCrone (1971) have each developed extensive tables to be used for the qualitative identification of materials.

* Available from the R. P. Cargille Laboratories, Inc., Cedar Grove, New Jersey.

2.3 Petrographic Microscopy

Petrographic microscopy has been used primarily by the mineralogist in the quantitative analysis of rock types. In this type of analysis, it is often necessary to identify the various nonopaque minerals comprising the rock. The qualitative aspects of this technique are discussed here; the quantitative aspects of petrographic measurements are given in Section 5.

The theory and instrumentation (Table 5) of optical crystallography are used with a few changes in procedure. The specimen is cut into a slab approximately 1.5 × 1.5 × 0.1 cm. This slab is cemented to a glass microscope slide, then ground until it is from 10 to 30 μm thick; a cover slip is cemented in place to complete the specimen. This specimen is referred to as a thin section. Detailed descriptions of thin-section preparation are given by Richardson (1971, p. 266).

The refractive index of the cement must be known, since this is the "immersion oil" for the thin section. In this case the absolute refractive indices of a grain are not measured, but, rather, they are referenced against the refractive index of the cement or against the refractive index of adjacent grains. If the index of the grain is greater than that of the cement, the grain has positive (+) relief; if it is less than that of the cement, the grain has negative (−) relief. The degree of relief is related to the distinctness of the boundary. It can be absent, low, moderate, high, very high, or extreme. By means of the sign, the degree of relief, and additional parameters measured in the same manner as in optical crystallography, the identity of grains in the thin section can be ascertained as shown by Durrell (1949), Wahlstrom (1955), Moorhouse (1959), or Johannsen (1968).

2.4 Ore Microscopy

The role of the microscope in the study of ore and opaque minerals is twofold. The transmitted-illumination polarizing microscope is used in microchemical analysis of the individual minerals in the ore, and the normal-incident polarizing microscope is used to examine the optical characteristics of the minerals. The latter application is described here; microchemistry is discussed in Section 4.

The specimen for ore microscopy is prepared with a flat polished face; procedures are given by Short (1940), and Richardson (1971, p. 26). For examination with the normal-incident polarizing microscope, the materials are divided into two groups: (1) the isotropic group, which includes cubic and amorphous minerals, and (2) the anisotropic group, which includes the minerals in the tetragonal, hexagonal, orthorhombic, monoclinic, and triclinic systems. Short combines the results of this test with reflectivity measurements and chemical solubility tests to provide identification of the mineral in question. The identity of ore minerals in the polished section can be obtained from Short (1940), Cameron (1961), Schouten (1962), or Uytenbogaardt and Burke (1971).

3 Use of Optical Microscope for Detection of Crystal Imperfections

The optical microscope was the first instrument used for the "direct" observation of lineal crystal defects, i.e., dislocations. Direct methods permit visual localization of the dislocation. Indirect methods such as x-ray line broadening differ because they provide information as to number of dislocations, but not their location. The term "direct" is relative, since only the phenomena resulting from the presence of the dislocation is seen in the optical microscope.

The first method for the observation of dislocation could only be used for smooth growth surfaces on crystals. The intersection of favorably oriented dislocations* with this surface produces growth spirals. These spirals may be studied by means of normal-incident-light microscopy with some form of phase illumination, e.g., phase contrast (Amelinckx and Votava, 1953) or multiple-beam interferometry (Tolansky, 1970).

A second method for the examination of the intersection of dislocations with the surface of a sample involves etching with selected reagents to produce etch pits. These etch pits reflect the symmetry of the surface on which the examination is made. Any asymmetry in the shape of the etch pit may be used to deduce the direction of the dislocation.

Amelinckx (1967) gives reagents that produce etch pits on a number of materials. Although resolution of the smallest etch pits requires the use of the electron microscope, the optical microscope can be used effectively for larger pits.

Still another method used for observation of lineal defects involves the use of the "decoration" of the dislocation in select transparent crystals. Particles of a suitable impurity phase are precipitated out on the dislocation. These decorated dislocations can then be observed in a transmitted-light bright-field or dark-field microscope. The method is limited by the resolution of the microscope and the number of materials suitable for observation. A survey of systems suitable for decoration study is given by Amelinckx (1967).

Finally a method for the study of dislocations in translucent crystals involves the observation of the stress resulting from the dislocation by means

* Dislocations that cause growth spirals are those for which $b \cdot n \neq 0$, where n is the unit normal on the crystal face under examination and b is the Burgers vector of the dislocation.

of a transmitted-polarized-light microscope. Excellent correlation of this technique with decoration and etch pit methods was demonstrated by Indenbohm (1961, 1962).

4 Use of Optical Microscope in Chemical Analysis

4.1 General

When compared to the newer and more sophisticated methods of analysis, chemical microscopy might seem neither practical nor advantageous. However, by the use of any one of a number of procedures, tests can be performed in a minimum of time using a minimum amount of material. As a result, chemical microscopy has been and currently is very important to such diversified fields as mineralogy and drug identification. Chamot and Mason (1960) give three applications: (1) the determination of the chemical constitution of a material; (2) the determination of the presence or absence of an element or radical; and (3) the determination of the presence or nature of an impurity. Galopin and Henry (1972) were particularly concerned with the typically qualitative nature of chemical studies emphasizing microscopy and preferred to use the term "microchemical test" to describe the process. They reserve the term "microchemical analysis' for a quantitative analysis carried out on a microscale. Through this section "microchemical test" will be used in reference to chemical analysis performed under the optical microscope.

Results of a microchemical test may be (1) a color change in a solution; (2) the formation of a fine (unresolvable) precipitate, or (3) the formation of a coarser resolvable precipitate with crystals of characteristic form. The first two are often referred to as color tests. The crystals in the resolvable precipitate are amenable to analysis by means of the crystallographic techniques discussed in Section 2.1, thereby providing confirmatory information. This latter method is referred to as a crystal test. Fulton (1969) states that color and crystal tests should be used to complement each other and that any "controversy" over which is the "better" is absurd.

4.2 Inorganic Chemical Tests

Microchemical analyses of inorganic compounds have generally progressed in two more or less independent areas—the fields of general materials identification and ore mineral identification. As early as 1859, a number of microchemical tests for materials were performed by means of the polarizing microscope. Work in ore microscopy was pioneered by Murdoch (1916). Ore microscopy differs from general materials analysis because the elements tested are primarily those of economic importance. Also, ore microscopy is limited in range and configuration of samples; i.e., only polished specimens of ore minerals are examined.

Qualitative tests have been developed for most elements, the exceptions being oxygen, carbon, the rare earths, and the transuranium elements. Many systematic schemes are given in Benedetti-Pichler (1964) and in the above references for both anions and cations.

Most of the reactions in microchemical tests involve a precipitation; thus a positive reaction is obtained whenever the solubility product for the precipitate is exceeded. The sensitivity of the microchemical test is no greater than that involving the same reaction in a more macroscopic test. The apparent sensitivity of the microchemical test seems high, since the actual amount of material that gives a positive test is small. Typical quantities for a few elements are given in Table 7. Since the tests vary widely from one element to another, the sensitivities are significantly different. Be-

TABLE 7

APPROXIMATE SENSITIVITIES OF MICROCHEMICAL TESTS FOR SELECTED ELEMENTS

Element	Test reagent	Minimum detectable weight in 2-mm-diameter drop[a] (μg)
Antimony	Cesium iodide	11
Bismuth	Cesium iodide	14
Cadmium	Potassium mercuric thiocyanate	6.9
Cobalt	Potassium mercuric thiocyanate	4.2
Copper	Potassium mercuric thiocyanate	5.5
Gold	Pyridine double bromide	3.4
Iron	Potassium mercuric thiocyanate	5.9
Lead	Potassium iodide	5.7
Manganese	Sodium bismuthate	11
Mercury	Cobalt thiocyanate	7.8
Nickel	Dimethylglyoxime	8.6
Selenium	Potassium iodide	45
Silver	Chloride	27
Sulfur (as sulfate)	Calcium acetate	110
Tellurium	Cesium chloride	21
Tin	Cesium iodide	13
Zinc	Potassium mercuric thiocyanate	5.7

[a] Based on data from Short (1940).

cause of the numerous interferences, the sensitivities can be degraded by the presence of more than one ion in solution.

In *Identification of Materials* Benedetti-Pichler (1964) has given a concise, but complicated, notation for sensitivity. If a solid sample is employed for the test, a number in parentheses is listed with each specific test. This number is the negative logarithm of the limit of identification* pL.I. (in grams).

If a substance is in solution, the specific test is followed by a parenthetical notation containing the negative logarithm of the limiting concentration† minus the negative logarithm of the limit of identification: pL.C. – pL.I. (in mils). The subtraction of the two logarithms gives the logarithm of the volume of the test solution needed.

Furthermore, any interfering substances are listed after the test method and are followed by a number that is the logarithm of their limiting mass per unit mass of the substance to be tested.

Chamot and Mason (1960) caution, however, that in measurements of sensitivity, a negative test should never be interpreted as indicating that the actual concentration of the element sought is less than the reported sensitivity.

4.3 Organic Chemical Tests

The use of the microscope in the identification of organic material is becoming increasingly important in many fields. The microchemical test is used in chemical engineering to study the effects of process variables on the physical properties of a product. In air pollution studies the technique is used in the detection of finely dispersed solid and liquid airborne particulates. In forensic science the organic microchemical test provides for the identification of small particles that typically constitute evidence from the scene of a crime. In the clinical field microchemical tests have been necessary to the identification of crystalline material from living tissue. The technique has also been used to identify chemical compounds and to trace impurities in the pharmaceutical field.

Organic microchemical testing is extremely complex. There is an infinite number of possible organic compounds, many with unknown structures. The field of organic chemicals cannot be simply categorized by functional groups, since more than one functional group may be present in a given compound. Thus, as stated by Clarke (1969), "the microcrystal test‡ is unsuitable as a primary method of identification of an unknown compound

* He defines the limit of identification L.I. as the smallest absolute quantity (mass or volume) that will always give a positive test.

† Benedetti-Pichler defines the limiting concentration L.C. as the lowest concentration of a substance that always produces a positive test. Dimensions of L.C. are $M \cdot L^{-3}$.

‡ This term is used by Clarke in the same sense as "microchemical test" is used in this text.

as it does not lend itself to form the basis of an identification scheme." The real value, then, of the microchemical test is as a simple, very specific test used to confirm the results of a provisional identification by another instrumental technique of analysis. The method is equally successful in cases where the class of substance is known and the identification of the specific member of the class is needed. Stevens (1968) and Kley (1969) have assembled extensive lists of microchemical tests for organic compounds categorized either by the principal function group of the compound or by its pharmacological activity. Fulton (1969) gives extensive data tables for drugs. Since the method is essentially comparative, the range of compounds that may be identified is limited to known and studied compounds. As with inorganic microchemical tests, the sensitivity varies widely for the various functional groups.

5 Use of Optical Microscope in Textural Analysis

5.1 General

The fundamental reason for the development of the microscope was the desire of researchers to observe and measure textures of various objects with detail unresolvable by the eye. A number of important applications of the microscope have been discussed; however, it is the qualitative and quantitative analyses of textures that still occupy much of the microscopists' time.

Textural analysis as applied to particulate materials refers to the characterization of the particle, i.e., color, size, shape, surface texture, and volume fractions of solids. The microscope is used in the analysis of particulate concentrates from liquids or gases, but the success of the method depends on the collection parameters and efficiency (See Chapter 1, Volume I, and Chapter 35, Volume IV). Textural analysis of solids includes the measurement of grain sizes, the determination of grain shapes, the interrelation of phases, the measurement of grain boundary areas, and the measurement of the volume fraction of the various phases.

Textural analysis has assumed names indicative of specific fields of endeavor, e.g., metallography, ceramography, petrography, and resinography.

5.2 Sample Preparation

The particulate sample, either as received or resulting from the grinding of larger specimens, is mounted on a glass microscope slide in an immersion medium to minimize light losses and then capped with a cover slip. The immersion medium may be water or a refractive-index oil for a temporary mount. For a permanent mount, various natural or synthetic resins such as Canada balsam, Caedax,* Lakeside 70,* Arochlor,† or an unfilled epoxy can be used.

Textures of nonopaque bulk materials can be examined in a thin section; various methods of thin-section preparation are discussed by Wahlstrom (1955, p. 4), Woodbury (1967), and Richardson (1971, p. 266). The thin section is examined with a transmitted-light microscope; the polarized-light instrument usually provides the most information.

Polished and etched mounts are used for the microscopy of opaque materials. These must be examined by means of an incident-light microscope. The use of phase accessories is often necessary when the etching process does not adequately delineate the grain boundaries.

Methods for preparation of polished mounts are almost as varied as the number of materials. Details for the preparation of polished mounts are summarized by Kehl (1949), Greaves and Wrighton (1967), and Richardson (1971, pp. 287–359).

Fluorescence microscopy has been used very little in the nonbiological area. However, it is possible to stain certain phases selectively by means of appropriate fluorochrome (Section 1.6), thereby delineating the structure of the material for textural analysis.

5.3 Qualitative Textural Analysis

The optical microscope is largely used for the quantitative study of textures; however, it is also an important qualitative tool. In work with unknown materials, it is often possible to define areas for further study by means of a brief but significant qualitative examination. Features to note in a qualitative examination are given in Table 8.

Textural terms have been developed by the petrographers to describe the many and varied mineral assemblages found in the various rock types. Table 9 gives some selected textural terms suitable for most materials studies; more detailed listings can be found in Wahlstrom (1955, pp. 299–300, 331–337, 340–341, 361–371), Moorhouse (1959), and Clarke (1969).

A most interesting and useful application of the qualitative microscopy of particulates is found in various atlases of photomicrographs including those of McCrone *et al.* (1967, 1973), Jackson (1968), Konovalov, *et al.* (1962, in Russian), and Eschrich (1966, in German). The approach in each case is to compare the appearance of a particulate with various photomicrographs to effect an identification. *The Particle Atlas* of McCrone *et al.*

* Available from Wards Natural Science Establishment, Inc., Rochester, New York.

† Available from Monsanto, St. Louis, Missouri.

TABLE 8

FEATURES TO NOTE IN THE QUALITATIVE EXAMINATION OF VARIOUS MATERIAL TYPES

Feature	Crushed fragments or particulates	Thin section	Polished mounts
Bireflection			×
Cleavage	×	×	
Color	×	×	×
Coring		×	×
Crystal habit	×	×	×
Fracture	×		
Inclusions	×	×	×
Opacity	×	×	
Phase interrelations		×	×
Pleochroism	×	×	
Polarization effects	×	×	×
Preferred orientation		×	×
Relative hardnesses			×
Relief		×	
Shape	×		
Sphericity	×		
Surface character	×		
Texture		×	×
Twinning	×	×	×
Zoning	×	×	

is unique in several respects: (1) instrumental conditions, illumination type, mounting medium, and photographic film type are standardized; (2) only the highest quality photomicrographs with good resolution and illumination have been selected for reproduction; and (3) a simple six-digit binary code has been devised to permit rapid and easy comparison of the unknowns and the photomicrographs. The simplicity of the above binary code for categorizing particulates is shown in Fig. 5.

The visual identification of textures is also useful in the examination of bulk solids. In this regard, the American Society for Metals has published the *Metals Handbook,* Volume 7 (1972), an atlas of micrographs for a wide variety of industrial alloys. With this text as a guide, much can be determined concerning the thermal and physical treatment of a polished and etched specimen of a metal or alloy. Rostoker and Dvorak (1965) have written an excellent text that systematically discusses metal structures under the microscope.

TABLE 9

GENERAL TEXTURAL TERMS SUITABLE FOR COMMONLY ENCOUNTERED STRUCTURES

Anhedral	Grains possessing no crystal outlines
Subhedral	Grains possessing some crystal outlines
Euhedral	Grains fully bounded by crystal faces
Prismatic, lathlike, acicular, fibrous	Rodlike grains with increasing length-to-width ratios. May be determined with certainty on thin sections and polished specimens only if bulk material is examined in several orientations
Tabular, platelike, micaceous	Platelike grains with decreasing thickness-to-width ratios. May be determined with certainty on thin sections and polished specimens only if bulk material is examined in several orientations
Dendritic	Structure in thin sections or polished mounts resulting from skeletal crystals of one phase in a second phase. Appearance is finger-like or fernlike oriented arrangement

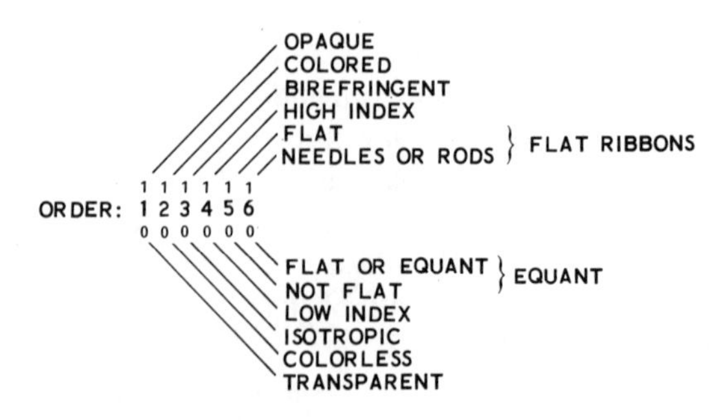

Fig. 5. Six-digit binary code used in *Particle Atlas* (McCrone *et al.*, 1967) for the classification of a particle. Depending upon the characteristics of the particle, either a "0" or "1" value is assigned for each of the six digits. The resulting binary number can then be used to locate the appropriate group of photomicrographs in the atlas for comparison with the observed particle. [Courtesy of Ann Arbor Science Publishers.]

5.4 Quantitative Textural Analysis

Quantitative analyses are generally applicable to two general types of sample, i.e., the particulate sample and the polished mount. Thin sections are suitable for quantitative work only at lower magnifications, since the exact positions of boundaries that are at angles other than 90° to the section are difficult to measure accurately at higher magnifications.

5.4.1 Quantitative Textural Analysis of Particulates

A number of methods have been described in Chapter 1 (Volume I) for the collection of particulates from liquids and gases; of these, the cascade impactor (Green and Lane, 1964) the thermal precipitator (Silver, 1951), and the membrane filter (Goetz, 1953) are ideally suited for optical microscopic examination. In each case, the particles can be quantitatively collected (at least down to the resolution limit of the microscope), and the area on which the particles are collected is of reasonable size and flatness for microscopic examination.

The optical microscope provides a number of important advantages for the examination of particulates: (1) It yields a complete size distribution; (2) it shows presence of large particles or fiber; (3) it provides an opportunity for visual identification of particulates; and (4) it provides greater sensitivity than other methods for low concentrations of particles.

The sample from a thermal precipitator, if collected on a glass microscope slide, should be used without a cover slip and immersion medium to prevent disturbing the particle distribution. It should be examined with a transmitted-light microscope. At higher magnifications ($>200\times$) an objective without cover-glass correction should be used.

If the sample is collected on a membrane filter, it is transferred to a microscope slide of the appropriate size and wetted with a quantity of immersion oil. The immersion oil should have the same refractive index as the filter; this renders the filter essentially transparent. The filters may be purchased with inked grids to simplify measurement with a transmitted-light microscope. If an incident-light microscope is used, the immersion oil should be omitted.

Procedures used in making the counts vary greatly, and literally hundreds of "in-house" procedures have been developed by individual companies. Several published reports covering a broad range of applications are listed:

1. Procedure for the Determination of Particulate Contamination of Hydraulic Fluids by the Particle Count Method, SAE ARP-589, Society of Automotive Engineers.
2. Method 3009-T Lubricants, Liquid Fuels, and Related Products; Method of Testing, Federal Test Method Standard No. 791a.

3. Procedure for the Determination of Particulate Contamination of Air in Dust Controlled Spaces by the Particle Count Method, SAE ARP-743, Society of Automotive Engineers.
4. Tentative Method for Sizing and Counting Airborne Particulate Contamination in Clean Rooms and Other Dust-Controlled Areas Designed for Electronic and Similar Applications, ASTM F25-63T.
5. Tentative Method for Measuring and Counting Particulate Contamination on Surface, ASTM F24-62T.

5.4.2 Quantitative Textural Analysis of Polished Mounts

Many techniques have been developed for acquiring quantitative information about the random-plane surfaces of polished mounts. These are generally used for materials with homogeneous structures. Lineal intercept and point counting are two such techniques.

In the lineal-intercept technique, measurements are made on lines placed randomly across the structure. These lines may be made by (1) traversing the image of the specimen by means of a micrometer eyepiece; (2) moving the specimen by means of a mechanical stage, using the cross wire in the eyepiece as a reference; or (3) drawing lines on photomicrographs of the material.

In the point-counting technique, a uniform grid is superimposed on the structure. This may be accomplished by (1) the use of a grid pattern in the eyepiece reticle for visual counting; (2) traversing the specimen in a two-dimensional stepwise pattern, with the cross wire of the eyepiece as a reference; or (3) placing a grid pattern on a photomicrograph of the structure.

One of the major applications of quantitative textural microscopy is the determination of volume fractions of the various components in a polyphase material. The fundamental equation relating those parameters, measured by lineal analysis or point counting, is

$$V_P/V_T = L_P/L_T = N_P/N_T \tag{4}$$

Thus, for lineal analysis the volume fraction V_P of the phase P in the total volume V_T is equal to the length of random line L_P in P divided by the total length of the random line L_T. For point counting the volume fraction of P is equal to the number of intersections N_P falling on P divided by the total number of points N_T.

A second important measurement, that of grain size, is most simply performed by means of lineal analysis. In this case, the number of grains per unit length N_L of a random line is a measure of the grain size. A useful application of N_L is its relation to the much used ASTM grain size (ASTM, 1968). Underwood (1961) has experimentally verified this relationship, and it is shown graphically by Richardson (1971, p. 604).

Grain boundary area may be determined by

$$S_V = 2N_L \quad (5)$$

where S_V is the grain boundary area in square millimeters per cubic millimeter.

The spacings between particles of a dispersed phase are given by Fullman (1953) as

$$\mathrm{MFD} = 1 - V_{Pt}/N_{Pt} \quad (6)$$

where V_{Pt} is the volume fraction of the particles, N_{Pt} is the number of particles per unit length, and MFD is the mean free distance between particles.

The mean center-to-center distance between particles S_{Pt} is

$$S_{Pt} = 1/N_{Pt} \quad (7)$$

For other quantitative measurements, see DeHoff and Rhines (1968) and Underwood (1970).

If a material is known or suspected to be inhomogeneous, a sufficient number of samples must be prepared in various known orientations with respect to the bulk material so that the inhomogeneity may be observed and measured.

For complicated structures or inhomogeneous materials, a great deal of time can be spent in quantitative analysis; various mechanical aids that increase the speed of both lineal analysis and point counting are given by Richardson (1971, pp. 631–639). The recent advances in electronic image analysis, however, provide for relatively straightforward and rapid analyses.

5.5 Automated Textural Analysis

The number of specimens to be examined, the complexity of the structure, or the speed of analysis often make it necessary to invest in one of the automatic-image analysis systems; these are currently available from Image Analyzing Computers, Inc. (Imanco), Bausch and Lomb, E. Leitz and Co., Carl Zeiss, and The Millipore Corporation.

These generally consist of a TV-type camera affixed to the microscope, the output of which is fed to a computer memory system. The output is visually examined on a monitor, and the quantitative information is obtained either from the monitor, or from a teletype output, punched tape, or magnetic tape. In addition, the output of the system can be used to trigger a scanning stage. In this way, different areas of the specimen can be examined in a preselected manner. The information gained provides greater precision (through the examination of more area) and can be used in the study of the homogeneity of the sample.

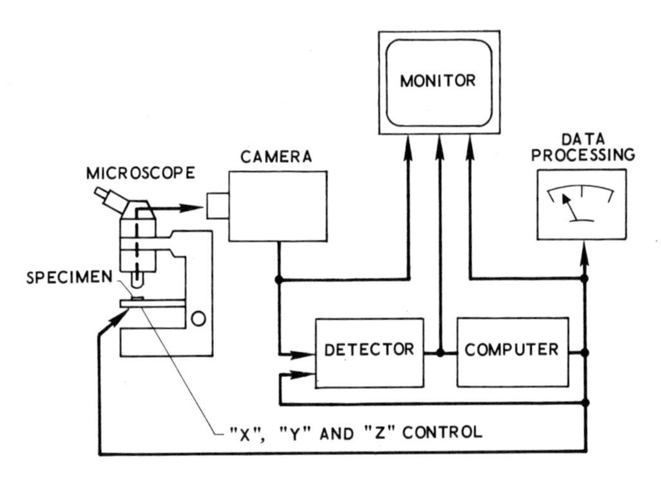

Fig. 6. Block diagram of essential parts of typical automatic-image analysis system.

The capabilities of these devices vary considerably; hence it is beyond the scope of this chapter to discuss these in detail. A typical system is shown schematically in Fig. 6.

References

Amelinckx, S. (1967). Trace Characterization (W. W. Meinke and B. F. Scribner, eds.) pp. 427–540. U.S. Nat. Bur. St. Monograph. 100, Washington, D.C.

Amelinckx, S., and Votava, E. (1953). *Nature* (*London*) **172,** 538.

American Society for Testing Materials (ASTM) (1968). ASTM Standards, Pt. 31, p. 446.

Benedetti-Pichler, A. A. (1964). "Identification of Materials." Academic Press, New York.

Brown, K. M., McCrone, W. C., Kuhn, R., and Forlini, L. (1963). *Microsc. Crystal Front* **14,** 2.

Cameron, E. N. (1961). "Ore Micrscopy." Wiley, New York.

Chamot, E. M., and Mason, C. W. (1960). "Handbook of Chemical Microscopy," Vol. 2, 2nd ed. Wiley, New York.

Clarke, E. G. C. (1969). "Isolation and Identification of Drugs," p. 135. Pharmaceutical Press, London.

Crossman, G. C. (1948). *Anal. Chem.* **20,** 976–977.

Crossman, G. C. (1949). *Science* **110,** 237–238.

DeHoff, R. T., and Rhines, F. N. (1968). "Quantitative Microscopy." McGraw-Hill, New York.

Dodge, N. B. (1948). *Amer. Mineral.* **33,** 541–549.

Dodgson, J. (1963). *Nature* (*London*) **199,** 245–247.

Durrell, C. (1949). "A Key to the Common Rock Forming Minerals in Thin Section." Freeman, San Francisco, California.

El-Hinnawi, E. E. (1966). "Methods in Chemical and Mineral Microscopy." Elsevier, Amsterdam.

Eschrich, W. (1966). "Pulver-Atlas der Drogen des Deutschen Arzneibuches." Gustav Fischer Verlag, Stuttgart.

Forlini, L. (1969). *Microscope* **17,** 1st quart., 29–54.

Forlini, L., and McCrone, W. C. (1971). *Microscope* **19,** 3rd quart., 243–254.

Fullman, R. L. (1953). *Trans. AIME* **197,** 447–452.

Fulton, C. C. (1969). "Modern Microcrystal Tests for Drugs." Wiley (Interscience), New York.

Galopin, R., and Henry, N. F. M. (1972). "Microscopic Study of Opaque Minerals." Heffer, Cambridge, England.

Goetz, A. (1953). *Amer. J. Public Health* **42** (2) 150–159.

Goodman, R. A. (1970). *Microscope* **18,** 1st quart., 41–50.

Greaves, R. H., and Wrighton, H. (1967). "Practical Microscopical Metallography," 4th ed., pp. 5–20. Chapman and Hall, London.

Green, H. L., and Lane, W. R. (1964). "Particulate Clouds: Dust, Smokes, and Mists," p. 267. Spon, London.

Hartshorne, N. H., and Stuart, A. (1969). "Practical Optical Crystallography." 2nd ed. Arnold, London.

Hartshorne, N. H., and Stuart, A. (1970). "Crystals and the Polarizing Microscope," 4th ed. Arnold, London.

Heinrich, E. W. (1965). "Microscopic Identification of Minerals." McGraw-Hill, New York.

Hess, H. H. (1949). *Amer. Min.* **35,** 624.

Indenbohm, V. L. (1961). *Dokl. Akad. Nauk SSR* **141,** 1360.

Indenbohm, V. L. (1962). *Fiz. Tverd. Tela* **4,** 231.

Jackson, B. P. (1968). "Powdered Vegetable Drugs; An Atlas of Microscopy for Use in the Identification and Authentication of Some Plant Materials Employed as Medicinal Agents." Amer. Elsevier, New York.

Johannsen, A. (1968). "Manual of Petrographic Methods." Hafner, New York.

Julian, Y., and McCrone, W. C. (1970). *Microscope* **18,** 1st quart., 1–10.

Kehl, G. L. (1949). "The Principles of Metallographic Laboratory Practice," 3rd ed., pp. 1–79. McGraw-Hill, New York.

Kley, P. D. C. (1969). "Microscopical Identification of Organic Compounds" (R. E. Stevens, Trans.). Microscope Publ., Chicago, Illinois.

Konovalov, P. F., Volkonskii, B. V., and Khashkovaskii, A. P. (1962). "Atlas of Microstructures of Cement Clinkers, Refractory Materials and Slangs." Gos. Izd. Lit. po Stroit., Arkhitekt. i materialam, Leningrad.

Larsen, E. S., and Berman, H. (1934). The Microscopic Determination of the Non-Opaque Minerals, 2nd ed., Geolog. Surv. Bull. No. 848. U.S. Dept. of Interior, Washington, D.C.

McCrone, W. C., Draftz, R. G., and Delly, J. G. (1967). "The Particle Atlas." Ann Arbor Sci. Publ., Ann Arbor, Michigan.

McCrone, W. D., Draftz, R. G., and Delly, J. G. (1973). "The Particle Atlas," 2nd ed. Ann Arbor Sci. Publ., Ann Arbor, Michigan.

"Metals Handbook" (1972). Vol. 7, 8th ed. Amer. Soc. for Metals, Metals Park, Ohio.

Moorhouse, W. W. (1959). "The Study of Rocks in Thin Section." Harper, New York.

Murdoch, J. (1916). "Microscopical Determination of Opaque Minerals." Wiley, New York.

Parker, R. B. (1961). *Amer. Min.* **46,** 892–900.

Poldervaart, A. (1950). *Amer. Min.* **35,** 1067–79.

Richardson, J. H. (1971). "Optical Microscopy for the Materials Sciences," Dekker, New York.

Rostoker, W., and Dvorak, J. R. (1965). "Interpretation of Metallographic Structures." Academic Press, New York.

Schouten, C. (1962). "Determination Tables for Ore Microscopy." Elsevier, New York.
Short, M. N. (1940). Microscopic Determination of the Ore Minerals, Geolog. Surv. Bull. No. 914. U.S. Dept. of Interior, Washington, D.C.
Shubnikov, A. V. (1960). "Principles of Optical Crystallography." Consultants Bureau, New York.
Stevens, R. E. (1968). *Microchem. J.* **13,** 42.
Tolansky, S. (1970). "Multiple-Beam Interference Microscopy of Metals," pp. 123–130. Academic Press, New York.
Underwood, E. E. (1961). *Metals Eng. Quart.* **I** (3), 62–71.
Underwood, E. E. (1970). "Quantitative Stereology." Addison-Wesley, Reading, Massachusetts.
Uytenbogaardt, W., and Burke, E. A. J. (1971). "Tables for Microscopic Identification of Ore Minerals," 2nd ed. Elsevier, New York.
Wahlstrom, E. E. (1955). "Petrographic Mineralogy." Wiley, New York.
Wilner, T. (1951). *Amer. Ind. Hyg. Ass. Quart.* **12,** 115.
Winchell, A. N. (1937). "Elements of Optical Mineralogy," Pt. I, 5th ed. Wiley, New York.
Winchell, A. N. (1957). "Elements of Optical Mineralogy," Pt. III, 2nd ed. Wiley, New York.
Winchell, H. (1964). "Optical Properties of Minerals." Academic Press, New York.
Winchell, A. N., and Winchell, H. (1951). "Elements of Optical Mineralogy," Pt. II, 4th ed. Wiley, New York.
Winchell, A. N., and Winchell, H. (1964). "The Microscopical Characters of Artificial Inorganic Solid Substances." Academic Press, New York.
Woodbury, J. L. (1967). *Metal Digest* **13** (1), 12.

CHAPTER 28

X-Ray Diffraction

G. M. Wolten

The Aerospace Corporation
El Segundo, California

Introduction

The interaction of x rays with crystalline matter produces characteristic interference patterns that are widely used for identification purposes and are the principal means of exploring the structure of solids.

1 The Crystalline State

Solids, with the exception of glasses and most high polymers, are *crystalline.* That is, the atoms of which the solid is composed constitute an ordered geometrical array.

Usually, this ordered atomic arrangement extends undisturbed only through a very small volume of material, called a *grain* if part of a larger mass or a *crystallite* if isolated. The size of a grain commonly ranges from a fraction of a micrometer to a fraction of a millimeter, but it can reach several millimeters, especially in metals. In a bulk sample of material, the grain is met at the *grain boundaries* by other grains, which have precisely the same atomic arrangement but are rotated randomly with respect to the first grain. Such a material is said to be *polycrystalline.* In general, all possible orientations of the grains are equally likely. If some orientations occur much more frequently than is expected statistically, i.e., if the orientations of the grains are correlated rather than strictly random, the material is said to possess *preferred orientation,* also called *texture* by metallurgists.

Sometimes an isolated grain, or crystallite, can be made to grow, either by man or nature, to visible dimensions, producing a bulk sample without grain boundaries, and in which the atomic arrangement pervades the entire mass without change of orientation. Such a specimen is called a *crystal,* whether it has the natural flat bounding surfaces called *faces* commonly associated with this term or is merely an irregular fragment. Crystals can range in size from a tenth of a millimeter to several feet (e.g., quartz). Gemstones are examples of large natural crystals, although their facets are usually not the natural crystal faces but are produced artificially by grinding or cutting.

The regular atomic arrangement within a grain or crystal consists of a basic grouping of atoms that is repeated over and over again (translated) parallel to itself along three noncollinear and noncoplanar directions in space, just as the motif of a patterned wallpaper is repeated in two directions. If one draws a straight-line outline around this basic repeating unit of atoms, one obtains a sort of box, or parallelepiped-shaped block of space, which is called a *unit cell.* There are conventions governing the selection of unit cells that make the choice unique for a particular arrangement (Kennard *et al.,* 1967). The entire crystal is made up of these boxes stacked together in parallel orientation. When they are stacked in this manner, space must be filled and no gaps are permitted, so that the allowable shapes are restricted.

The shape of the unit cell reflects at least some aspect of the symmetry of the underlying atomic arrangement, and six basic shapes are sufficient to cover all cases. In some instances there is an alternative to one of the shapes, so that one counts either six or seven *crystal systems*.

All unit cell shapes are parallelepipeds, so that the six edges of the box are equal in pairs. The lengths of these edges are the *lattice parameters*. In some crystal systems, the angles between three adjacent edges are variable and must also be specified as additional lattice parameters. This does not mean that the angle varies between two crystals of the same material; it does not. Rather, two different materials whose unit cells belong to the same crystal system will have different values of the angle. The two extremes are the cubic and the triclinic systems. In the cubic system, the shape of the unit cell is a cube, so that all angles are 90° and invariant, and all three edges have the same length, so that only one lattice parameter, the single edge length, is required. In the triclinic system, three edge lengths and three angles, six lattice parameters altogether, must be specified.

A corner of the unit-cell parallelepiped may be chosen as the origin of a system of coordinates, and the three adjacent cell edges that meet at that corner become unit vectors of the coordinate system and are called the *crystal axes*. This axial system is closely related to the concept of the *crystal lattice*.*

The actual arrangement of the atoms within the unit cell usually has a certain amount of symmetry that can be expressed in terms of *symmetry elements*, such as mirror planes, rotation axes, and others. One of these symmetry elements (centering) can be added to the concept of crystal systems, and this divides the 6 or 7 crystal systems into the 14 *Bravais lattices*, which embody the concepts of the axial system plus some selected considerations of basic symmetry.

If one considers all those symmetry elements of the atomic arrangement that deal with all the different ways in which one can assemble atoms about one particular point, consistent with the requirement that the assembly must be repeatable in three dimensions, then one arrives at the 32 *point groups* or *crystal classes*. Finally, the sum total of all symmetry operations includes those that move an atom from one point to another by a translation, and this yields the 230 *space groups*. The terms discussed above, crystal system, Bravais lattice, point group, and finally space group, represent a more and more complete way of describing the *symmetry* of the atomic arrangement.

* The rigorous definition of the term *lattice*, as used by the geometrical crystallographer—not necessarily by everyone else—implies an array of points such that a *symmetry operation*, for example, rotation of an interpoint vector about a point by $\pi/3$ or reflection of the array in a plane, will lead from one point to another, and all the symmetry operations possible in the lattice form a group in the mathematical sense. This means that any combination of symmetry operations is itself a symmetry operation and a member of the set. The 230 space groups mentioned below were derived with the aid of group theory, but one need not be conversant with group theory to appreciate the results.

None of these terms describes the arrangement itself, which is described by specification of the nature, number, and individual locations of all the atoms in the unit cell; this description is called the *crystal structure*.

It is not uncommon to find statements such as "The structure of diamond is face-centered cubic, with a_0 = 3.5669 Å; the space group is Fd3m." This is a very loose, not to say incorrect, use of the term *structure*. The statement gives only the lattice (face-centered cubic), the size of the unit cell (a_0), and the full symmetry of the atomic arrangement (the space group). All this must be known if the structure is to be described, but the description must state that there are eight carbon atoms in the unit cell and

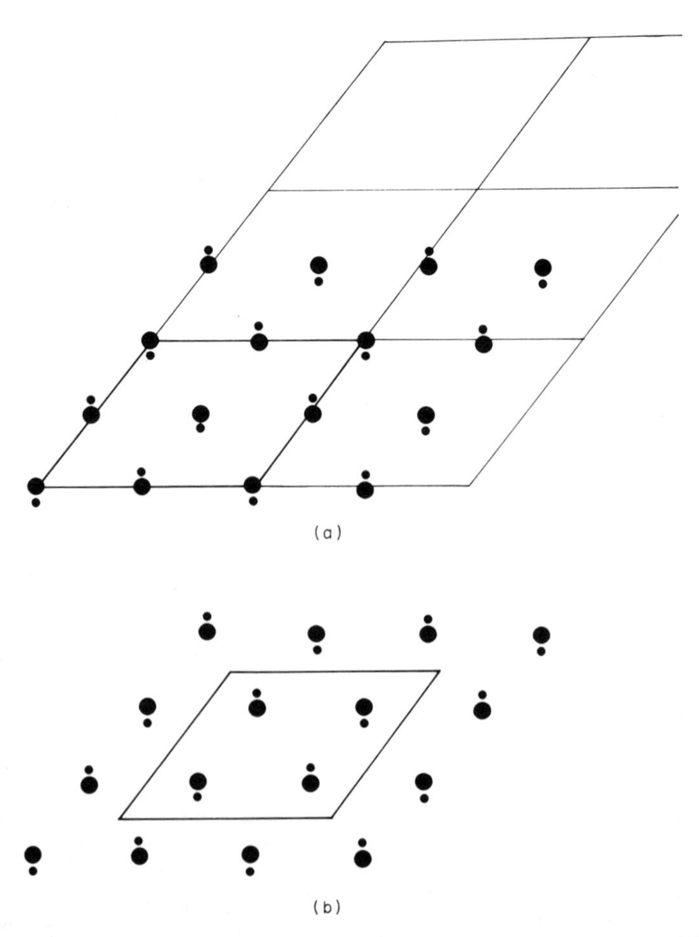

Fig. 1. (a) A structure and stacked unit cells; (b) origin of coordinate system of (a) shifted to reveal contents of unit cell more clearly.

that they are to be found at certain locations, whose coordinates must be given.

Figure 1a illustrates the choice of a unit cell for a particular structure and the stacking of the unit cells to fill space and form a lattice. In Fig. 1b the origin of the lattice (a corner of a unit cell) has been shifted to make the atomic contents of the unit cell more obvious. Since the lattice is a mathematical abstraction and is considered infinite in extent, the origin can be placed anywhere because no fundamental considerations dictate a particular choice. There are, however, conventions governing the choice of origin so that everyone may agree on the values of the atomic coordinates and the mathematical descriptions resulting therefrom.

2 X-Ray Diffraction Processes

The foregoing discussion has emphasized the repetitive nature of crystals; they are *periodic* structures. A periodic structure is any collection of identical objects that are located at constant intervals. Such a structure will *diffract* a wave of any kind if the wavelength is of an order of magnitude close to the spacing of the objects in the structure. Therefore, in the field of optics, a periodic structure is called a *diffraction grating*. The interaction of a plane parallel wave front with a diffraction grating will produce a *diffracted beam* at an angle to the incident beam that is a function only of the wavelength and the repeat distance of the grating.

Figure 2 shows some stacked unit cells from which all the contents have been removed. What is left is merely the lattice, which illustrates the repetitive nature of the arrangement. The corners of all the unit cells, or the *nodes* of the lattice, are called *lattice points*, and any straight line that goes through lattice points is a rational *direction* of the lattice. Figure 2 also shows that it is possible to construct many different *families of planes* (two are illustrated) from the lattice points and that these families of planes have a constant spacing, called the "d spacing," that gives rise to diffraction effects under the proper conditions. The d spacings of crystals are of the order of Angstrom units (10^{-8} cm). For simple compounds they are rarely larger than 20 Å, and they range down without limit to fractions of an Angstrom because one can draw more and more tightly spaced planes through fewer and fewer lattice points without limit. The wavelengths of easily produced x rays are of the same order; e.g., x rays produced from a copper target, after elimination of a "white radiation" background, have a wavelength of 1.5 Å, and those from a molybdenum target have a wavelength of about 0.7 Å. Because of this dimensional match, x rays are diffracted efficiently by crystals. It should be emphasized that, unlike the

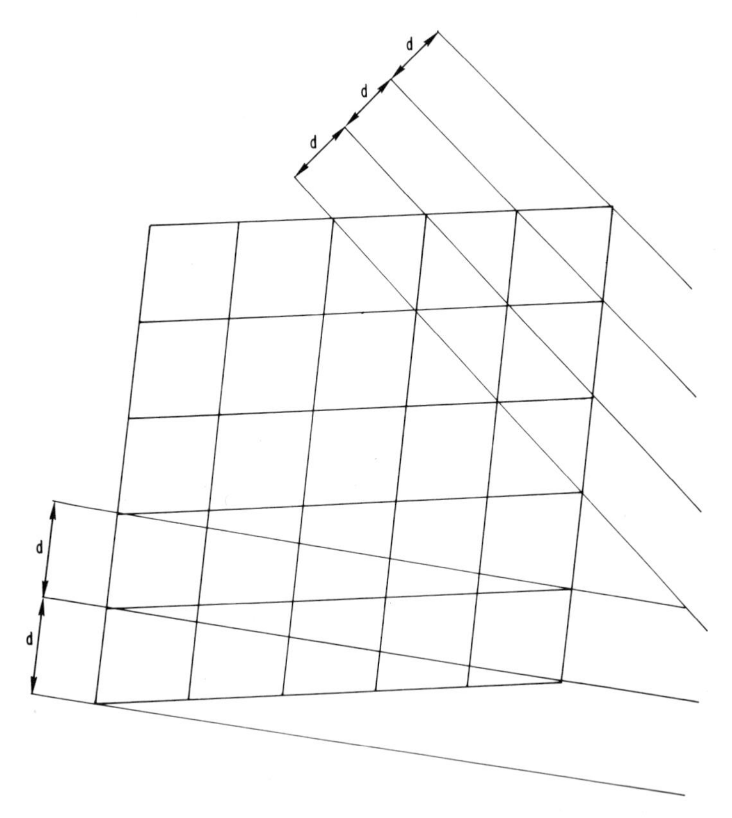

Fig. 2. A lattice, showing two families of lattice planes and their interplanar spacings (*d* spacings).

artificial linear grating used in spectroscopy, a crystal is a three-dimensional array of many gratings in one and will produce a multitude of diffracted beams in many different directions in space. Each of its "gratings" is characterized by a value of *d*, and all can be measured with x rays from the largest present in the crystal down to those approaching in size one-half the wavelength of the x rays used. For physical reasons, spacings smaller than this do not produce diffracted beams.

A simple relationship between the spacing *d*, the wavelength λ, and the diffraction angle 2θ was first derived by Sir W. Lawrence Bragg. The derivation is not rigorous, but it affords an easily visualized if somewhat naïve model, which leads to the correct angular relationship. In Fig. 3a, the planes P_1 and P_2 represent a family of planes with the spacing *d*, one of many possible families. The planes are planes of lattice points; they are not, in general, sheets of atoms. The latter might be the case, however, in a very

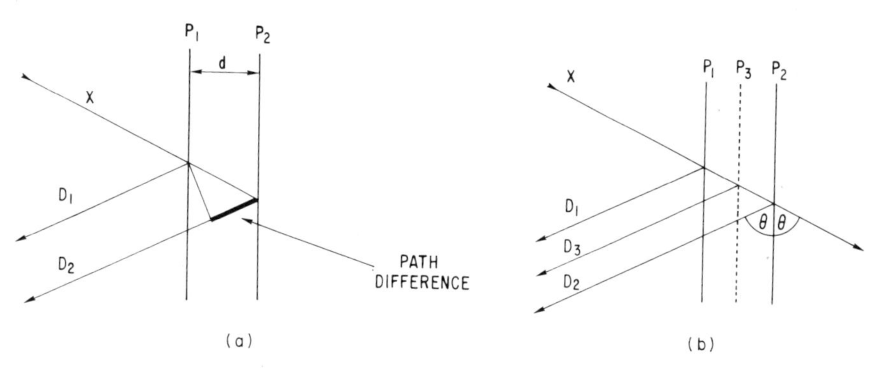

Fig. 3. (a) Bragg's law: Diffracted beam produced when path difference is integral number of wavelength—incident beam X, diffracted beams D_1, D_2, as well as the normal-to-the-lattice planes P_1, P_2, all lie in one plane. (b) Effect of added plane—drawing also shows that angle between the propagation directions of primary beam and diffracted beam is 2θ.

simple structure, and we will make such assumption here because it will allow us to explain another phenomenon later in a very simple fashion.

Let X be an incoming beam of x rays, and let us assume that a small portion of its energy is *specularly* (i.e., as if by a mirror) reflected by P_1 and some more by P_2. If the path difference between the two reflected beams D_1 and D_2 is a whole number of wavelengths, then D_1 and D_2 are in phase and a diffracted beam is observed. If they are not in phase, *destructive interference* cancels the beams. They are in phase if the diffraction angle θ satisfies the "Bragg equation"

$$n\lambda = 2d \sin \theta \qquad (1)$$

where n is the number of whole wavelengths by which the paths differ. If we gradually increase θ, we will first observe a diffracted beam when $n = 1$, again when $n = 2$, and so on, and this will occur at those values of the angle at which (1) is satisfied. Thus a single d spacing, that is, a single family of planes, will give us several diffracted beams, which are actually the several *orders* of one reflection. One can then calculate the spacings by

$$d/n = \lambda/2 \sin \theta \qquad (2)$$

In practice, one ignores n and lists $d/2$, the second order from a family of planes with spacing d, as if it were the first order from a family with half that spacing.

Let us now assume that there is an extra sheet of atoms exactly midway between P_1 and P_2, as indicated by the dotted line P_3 in Fig. 3b. If the path difference between D_1 and D_2 is still one wavelength, then it is one-half wavelength for D_1 and D_3 and also for D_2 and D_3. Destructive interference takes place, and the reflection is not observed. But for the second order, the path difference between D_1 and D_2 is two wavelengths, so that it is one wavelength for D_1 and D_3 and for D_2 and D_3. The second order is therefore still observed.

Of course, the presence of the extra plane exactly midway ordinarily implies the presence of a symmetry element, and so it is now plausible, although not rigorously proved, that many (but not all) symmetry elements make their presence known by just such *systematic absences* in certain classes of reflections.

What would happen if the extra plane were not *exactly* midway but just slightly off center? Then, of course, there would no symmetry element, but merely an extra sheet of atoms. If it were slightly off center, then destructive interference would not be quite complete and there would be a reflection, although a very weak one. Moving the sheet farther off center would lessen the destructive interference further, and the reflection would become stronger.

This enables us to make the following statements: (1) The angular positions of the diffracted beams are determined solely by the geometry of the lattice, i.e., by the size and shape of the unit cell. (2) Systematically absent reflections are caused by some of the symmetry elements. The intensities of the reflections depend very strongly upon the positions of the atoms, i.e., the content of the unit cell (among other things). (3) Actually, the intensities depend upon a number of factors of which the positions of the atoms constitute only one, called the *geometric structure factor.* Before atomic coordinates can be determined, we must reduce intensities to structure factors by accounting for the other components of the intensities.

Some further elaboration of the first of these three statements will be helpful at this point.

Figure 4 shows a set of unit cells with a superimposed coordinate system parallel to the edges of the cells. The x axis is parallel to the edge of the cell with the length a_0, the y axis is parallel to b_0, and the z axis is parallel to c_0, which is here assumed to be normal to the plane of the drawing. The lengths a, b, and c are the unit distances along their respective axes. Also shown are the edges of a set of vertical lattice planes.

Taking the plane closest to the origin, we see that it cuts the x axis at the unit distance a and the b axis at one-half the unit distance $b/2$ and that it is parallel to z. The *intercepts* of the plane on the axes are, therefore 1, 1/2, ∞. The reciprocals of these numbers, 1, 2, 0, are known as the *Miller indices h,k,l,* and when written (120) they serve to identify the set of planes. When written 120, without parentheses, they serve to identify the first-order reflection from the set of planes. The second-order reflection is written 240. The d spacing for this set of planes is often written d_{120}. Sets of planes (and possible reflections) exist for all combinations of integers *hkl.*

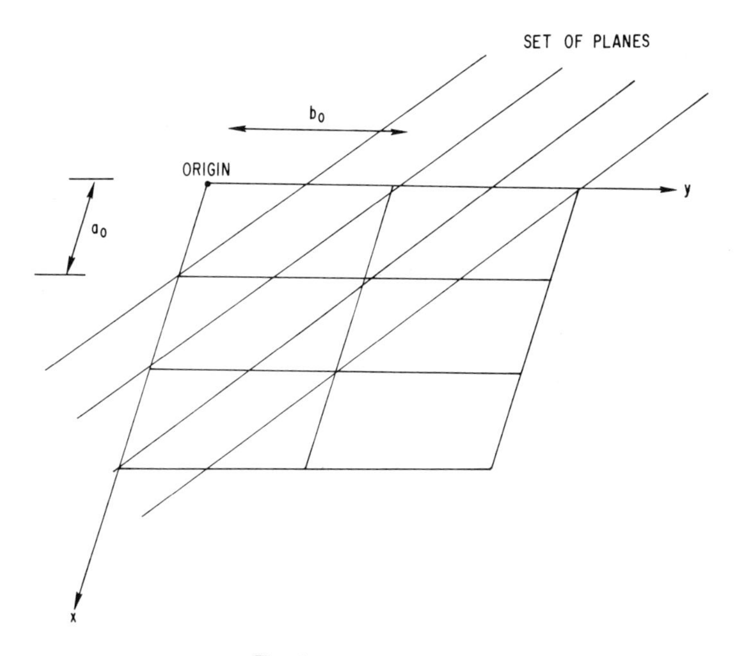

Fig. 4. Indexing of planes.

The relationship of the spacings to the latttice parameters depends on the shape of the unit cell, and there are seven different equations describing the relationship. When the angles of the unit cell are all right angles, the equations are relatively simple, as for the orthorhombic system given here as an example:

$$d_{hkl} = [(a^2/h^2) + (b^2/k^2) + (c^2/l^2)]^{1/2} \tag{3}$$

3 Preliminary Discussion of Powder and Single-Crystal Patterns, and Their Information Content

In the discussion of Fig. 3a, it is pointed out that the incident x-ray beam, also called the *primary beam;* the normal to the diffracting planes, which is called the *diffraction vector;* and the diffracted beam all lie in one plane. This, however, does not define a unique spatial relationship between the beam and the crystal. It is possible to roll the diffraction vector (and hence the crystal) around the incident beam and still maintain 2θ at the correct value, as is shown in Fig. 5. The entire surface of the solid inner cone represents locations of the diffraction vector with the same value of 2θ. The outer cone shows corresponding directions of the diffracted beam. If our specimen were polycrystalline instead of the single crystal heretofore envi-

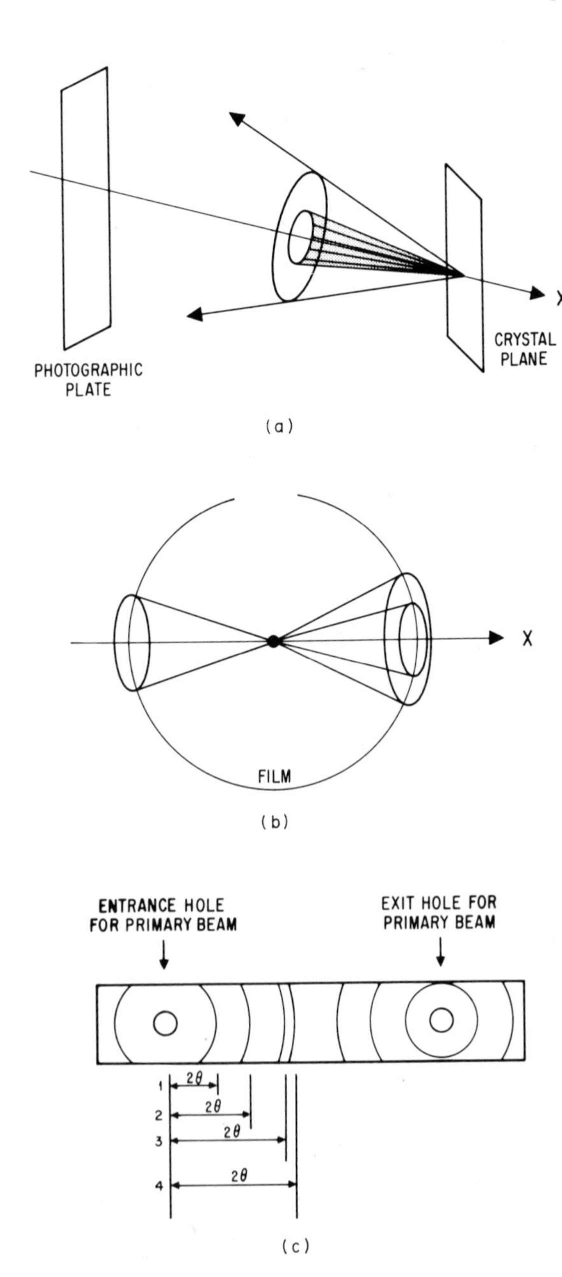

Fig. 5. (a) Production of diffraction rings; (b) principle of Debye–Scherrer camera; (c) powder diffraction film.

sioned, the various grains would simultaneously occupy all possible orientations. It follows that if we placed a photographic film to the left, as indicated in the figure, a single crystal would cause just one spot to appear where the diffracted beam penetrates the film, but a polycrystalline specimen would give a complete ring, corresponding to the opening of the outer cone. Moving the crystal in some way other than the one indicated would alter the value of 2θ, so that the Bragg condition would not be fulfilled and no diffracted beam would be produced from the particular set of planes under consideration. However, the angle might be satisfied for some other set of planes, so that the film would show not just one, but a set of several concentric rings, corresponding to all possible values of 2θ, and hence d, that are observable from this material and arise in directions that can be intercepted by the film.

Flat films like the one pictured are used in some of the techniques for single crystals, but ring patterns are rarely recorded in this manner. Instead, one of the most common arrangements, used in the Debye–Scherrer camera, is to place a narrow strip of film (35 mm) about the specimen in a circular arc. It intercepts short segments of the rings along their diameters, which results in slightly arced "lines" on the film. The arrangement is shown schematically in Fig. 5b, and the final film, again schematically, is given in Fig. 5c.

Note that there is a crucial difference between the information content of single-crystal and *powder patterns*. The latter are patterns arising from polycrystalline materials whether they are actually powders or solid chunks.

In a single-crystal pattern, each set of planes gives rise to a single diffracted beam, which is characterized not only by its value of 2θ but also by a unique position in space with respect to the incident beam and the crystal. Since the orientation of the unit cell within the crystal is known or can be determined, each reflection can be uniquely identified (indexed) with a particular set of planes.

By contrast, the reflections in a powder pattern are all produced simultaneously from a large set of randomly oriented unit cells. In this case, no correlation of the spacings with any aspect of the unit cell is obtained. If the unit cell is known from prior work, then the d spacings can be calculated from an equation for the appropriate crystal system similar to that given for the orthorhombic system [Eq. (3)]. The powder pattern can then be indexed by comparison of the list of observed d spacings (which are not indexed) with the calculated list (which is). In low-symmetry systems, the calculated list is apt to become very densely populated toward the small-spacing end, so that ambiguities remain.

If the unit cell is not known, one may try to determine whether the entire set can be reproduced by one or the other of the aforementioned formulas by appropriate choice of the lattice parameters. This is reasonably successful if the symmetry is high, because the cubic, tetragonal, and hexagonal systems have only one or two adjustable parameters (the lattice parameters). Some rather sophisticated methods exist (Ito, 1950; Zsoldos, 1968; Kharin and Davydov, 1970) to achieve the same results for low-symmetry systems, which have 3, 4, or 6 lattice parameters, when high-quality patterns are available. Although these methods quite often will yield a unit cell, the probability that this is the correct one is no higher than 50% because the powder pattern of a low-symmetry crystalline substance usually reveals only 1 to 10% of the number of reflections that would be observable with a single crystal, and the full set not infrequently tells a different story than the small selected set observed. If a single crystal is oriented to produced a reflection, the entire bulk of the crystal (typically from 0.02 to 0.3 mm in size) contributes to the reflection. Although in some powder diffraction methods the sample can be considerably larger, only a very small fraction of the grains is correctly oriented, so that the total amount of contributing material is much smaller and only the more intense reflections are observed. A low-symmetry structure has relatively few intense reflections and very many weak ones. In high-symmetry systems, on the other hand, the structures are usually simpler, so that the reflections are more intense, and the fraction of strong reflections is higher also. Furthermore, the high symmetry creates many *symmetry-related* or *equivalent reflections;* these are due to the repetition of the same set of planes in a number of different orientations, which increases the amount of contributing material.*

We have now justified the following statements: (1) The experimental data derived from a powder pattern consist mainly of a list of interplanar spacings (and the intensities of the corresponding reflections). (2) A powder pattern is not the method of choice for determining the unit cell. (3) Single-crystal patterns not only allow observation of a much larger number of reflections,† but each reflection is also unambiguously indexed in terms of a known or derivable unit cell.

In subsequent sections, we shall discuss the uses of the information contained in powder and single-crystal patterns.

* Symmetry-related reflections have the same intrinsic intensity. It is also possible for unrelated reflections to have the same *d* spacing and hence give rise to a single arc in a powder pattern. For structure determinations, intensities of the separate reflections can be measured only in single-crystal experiments.

† Powder patterns contain from 5 to 200 reflections. A single crystal allows the measurement of several hundred to a few thousand reflections for small- and medium-sized structures. For a structure as large as a small protein, the number can be of the order of 50,000.

4 The Production and Monochromatization of X-Rays for Diffraction

4.1 X-Ray Tubes

Most diffraction tubes in common use today have a heated tungsten filament at a large negative potential and a massive, water-cooled anode or *target* at ground potential. Electrons are evaporated from the filament and accelerated toward the anode, where they are stopped and their energy is converted into x rays. At fairly low voltages only a continuous spectrum of x rays (*white radiation*), called the *background*, is produced. The cutoff at short wavelengths is determined by the maximum energy available according to

$$\lambda = 12.4/\mathrm{kV} \tag{4}$$

When the voltage is increased, electronic transitions occur in the target and *characteristic* x-ray spectral lines appear, superimposed upon the background. The wavelengths of the spectral lines depend only upon the target material. The minimum voltage (the *excitation potential*) at which the K_α lines appear for copper is about 9 kV and for molybdenum 20 kV. Figures 6a and 6b are schematic representations of the output of copper and molybdenum x-ray tubes at 50 kV. The highest intensity ratio between characteristic lines and background is obtained at a voltage from 3.0 to 3.5 times the excitation potential, but this figure may be somewhat modified by the use of filters.

4.2 Beta Filters

The absorption of x rays by a pure element increases regularly with increasing wavelength but drops sharply at *absorption edges*. Thereafter it rises again (Fig. 7). Elements either one or two atomic numbers lower than the target material have an absorption edge that falls between the K_β and K_α lines, such that the K_β falls into a high-absorption region and the K_α falls into a low-absorption region, as is shown in Fig. 8. If the thickness of the absorber (called a *β filter*) is chosen so as to reduce the intensity of the K_α line to one-half its former value, then K_β should be reduced to one-one hundredth of its original intensity. Since the unfiltered β intensity is about one-fifth that of α, the final intensity ratio, after filtering, should be 500:1.* This is usually sufficient to eliminate diffraction lines due to the K_β component from most powder photographs. Thus, the lines on a diffraction

* These ratios are approximate and vary slightly for different target and filter combinations and voltages. Except for manganese filters used with iron targets, the author has found experimental values to approximate the numbers given quite well.

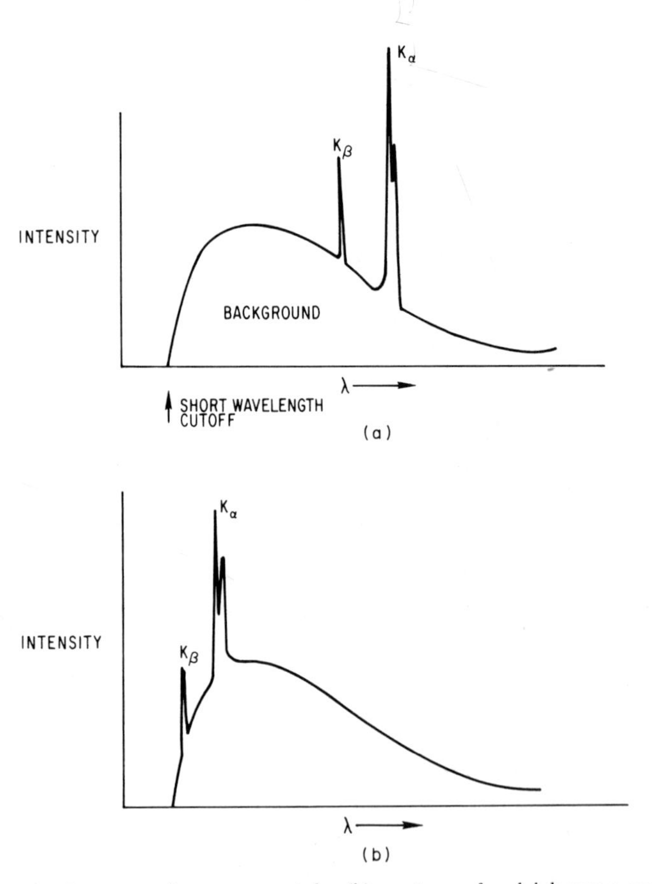

Fig. 6. (a) Spectrum of copper x-ray tube; (b) spectrum of molybdenum x-ray tube.

film taken through a β filter are due to the K_α wavelength. They are seen against a background of general darkening, which is due to the residual white radiation. Sharp changes in the background near the beginning of the film are effects produced by the absorption edges of silver and bromine in the photographic emulsion. K_α is a narrow doublet, which is resolved at larger angles.

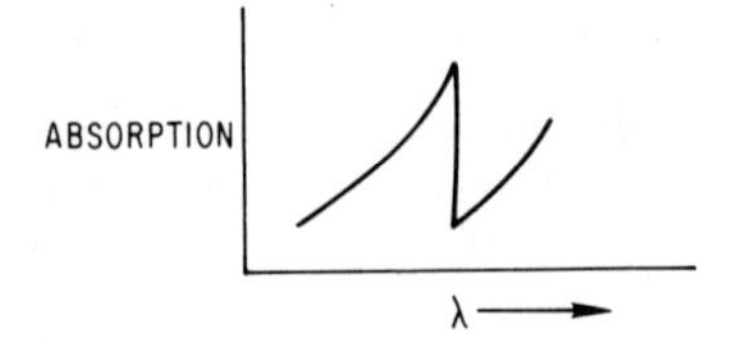

Fig. 7. Absorption edge.

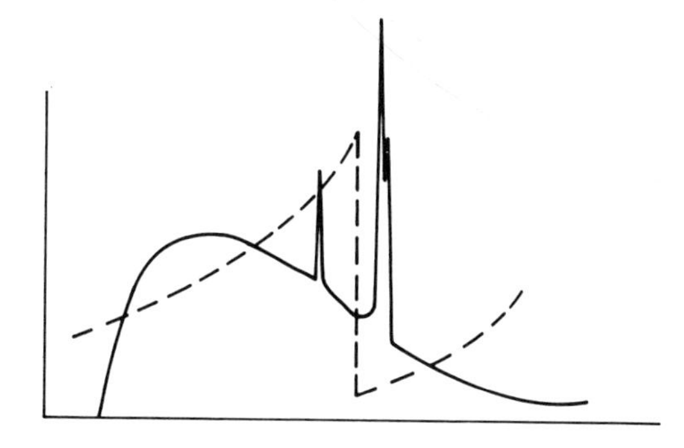

Fig. 8. Absorption curve of nickel filter superimposed on emission spectrum of copper target.

A second look at Figs. 6a and 6b shows that the β filter will also eliminate part of the short-wavelength hump in the background for copper radiation but not for molybdenum radiation.

4.3 Balanced Filters

One can record a diffraction pattern with a β filter and then record it again with another filter having an absorption edge on the other side of K_α, and the thicknesses of the two filters can be adjusted to give approximately the same attenuation in regions remote from the edges. If the intensity transmitted through the second filter is subtracted, point by point, from the intensity transmitted through the first filter, what remains is primarily the K_α and its immediate vicinity, with very little (in theory none) of the background remaining anywhere else. Since this procedure requires taking the pattern twice, keeping everything except the filter constant, and then making the subtraction, it is obviously cumbersome to carry out manually. However, it is well suited to automatic operations. The two filters are called *balanced filters* or *Ross filters*; Fig. 9 illustrates their principles of operation.

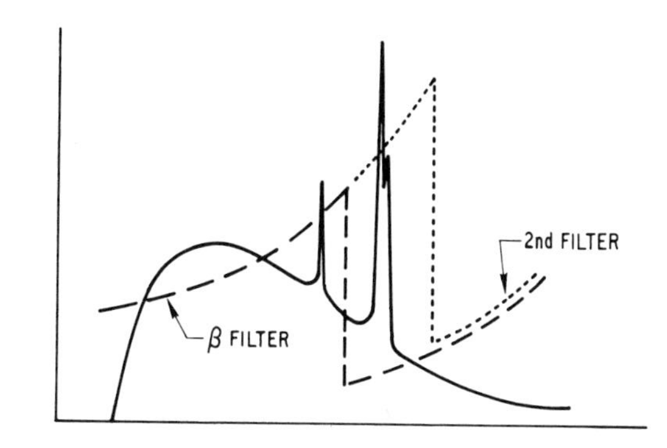

Fig. 9. Balanced filters.

4.4 Crystal Monochromators

According to the Bragg equation, 2θ depends upon the wavelength; therefore, a set of lattice planes can be considered a spectrum analyzer because, when set at the correct θ for a particular λ, it cannot diffract any other λ except harmonics, and the diffracted beam is quite monochromatic. Thus it is feasible to use a beam diffracted from a large single crystal as the incident beam for the specimen to be analyzed. This is the principle of the crystal monochromator. Crystal-monochromatized beams are, of course, substantially weaker than the primary beam from the x-ray tube, and low intensities are one of the difficulties of this method. Usually, however, the background is reduced to such an extent that the signal-to-noise ratio is much improved despite the lower overall intensity. Thus, weaker lines that merge into the background without monochromatization can now be seen, provided that the detector is sensitive enough or, if film is used, the exposure has been long enough. The narrowness of monochromatized beams also tends to improve resolution.

4.5 Focal Spots

In the type of x-ray tube described, the target is usually a rectangular block of metal that is viewed by the specimen at a shallow angle (2–6°) to its surface, almost at right angles to the path of the electron beam inside the tube. If the view is toward the long edge of the target, the latter appears in perspective as a narrow line of x-ray intensity (Fig. 10a), called a *line focus*. This type of beam illuminates a long length of specimen and results in high resolution along the narrow width of the line. It is used for diffractometer powder patterns, for the Guinier camera, and for making d spacing measurements of single crystals in the diffractometer. Single-crystal intensity measurements and the Debye camera require a *spot focus*. This is obtained by "viewing" (on the part of the specimen, not the experimenter!) the target, end-on at an angle of 5 to 6°, so that in perspective the beam has a square cross section (Fig. 10b), and—ideally—constant intensity over most of its diameter.

5 The Use of X-Ray Diffraction for the Analysis of Materials

5.1 Identification of Crystalline Unknowns from Powder Patterns by the Fingerprint Method

The powder pattern of a crystalline substance, after measurement but without any further interpretation, consists of a list of d spacings with intensities. The latter are not overly precise unless special efforts are made,

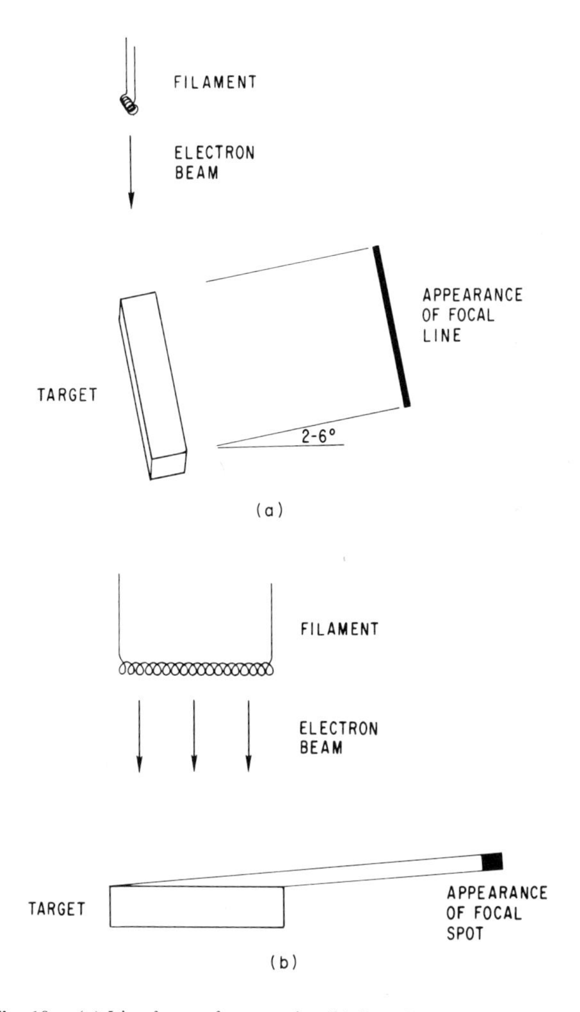

Fig. 10. (a) Line focus of x-ray tube; (b) Spot focus of x-ray tube.

and this is not warranted for the purposes discussed here. The intensities can vary somewhat with the experimental parameters, and they are usually the more simply determined *peak intensities* rather than the structurally more significant, but more difficult to determine, *integrated intensities*. Reduction of intensities to structure factors is not generally practiced for powder patterns, but it is customary to adjust them to a scale of 0 to 100. The reflections may or may not be indexed.

It should be noted that the degree of blackening of a photographic film, called the *optical density*, is proportional not to the incident x-ray energy but to its logarithm (and this only over a limited range of energies, called the linear portion of the film's response curve). If one estimates blackness of diffraction lines on a linear scale from 1 to 10, then the actual intensities on a relative scale of 100 are given by

$$I_{\mathrm{log}} = 5 \times \log I_{\mathrm{lin}}$$

where I_{lin} and I_{log} are intensities expressed on the linear and logarithmic scales, respectively.

For many years the ASTM and, more recently, the "Joint Committee on Powder Diffraction Standards" (Swarthmore, Pennsylvania) have published a collection of experimental powder patterns which as of 1973 contained more than 20,000 individual patterns and grows currently at the rate of 1000–2500/yr. In the past, the collection was printed on individual 3×5 cards and was widely referred to as the "ASTM Card Index of Diffraction Patterns." It is now available in a variety of formats and bears the official name of "Powder Diffraction File" (PDF).

Without any attempt at interpretation, structural or otherwise, this formidable collection serves as the basis for the identification of crystalline materials by matching the diffraction pattern of an "unknown" against an entry in the file. In principle, the method is the same as that of identifying a person by matching his fingerprints against a similar collection. In each case, uniqueness of the pattern or the fingerprint is a basic assumption, but for diffraction patterns, this assumption is subject to some qualification.

In the early, '40s, when there were but a few thousand entries in the file, it was generally held that there should be only one case in a few million in which two patterns would match in the d spacings and relative intensities of the four strongest lines. This contention was based on the unjustified assumption that the lines of all diffraction patterns would be randomly distributed over the range that was experimentally accessible. Actually, for any one pattern, the d spacings have to one another nonrandom relationships that are determined by the crystal system, and it is this fact that makes possible the attempts described previously to deduce the crystal system from the pattern.

For example, the d values for tetragonal crystals obey the formula

$$1/d^2 = [(h^2 + k^2)/a^2] + (l^2/c^2) \tag{5}$$

The d values for all reflections where $l = 0$ $(hk0)$ are inversely proportional to the square roots of all integers that can be formed as the sum of two squares, $h^2 + k^2$; hence they are in the ratio of the square roots of 1, 2, 4, 5, 8, 9, etc. The d values for $00l$ reflections are inversely proportional to the square roots of all the integers. The relationship of the two sets to one another and to the general set hkl are determined by the *axial ratio* c/a and are similar for crystals having similar axial ratios. Admittedly, this set of relationships is somewhat complicated even for this rather simple crystal system, but the set of *d* values is not a random collection of numbers, and the chances for coincidences are greater than originally suggested.

There are many isostructural compounds, i.e., compounds having the same crystal structure although composed of different kinds of atoms and, in some cases, even different numbers of atoms. If the lattice parameters are similar, the entire list of *d* spacings for the two compounds will be similar; if they are composed of atoms that do not differ greatly in atomic number, then the intensities may be similar also, at least within the limited precision of the usual powder pattern intensity measurements.

For example, it is virtually impossible to distinguish with certainty between the diffraction pattern of the rutile phase of TiO_2 and that of $AlTaO_4$. The unit cell of rutile contains two titanium and four oxygen atoms; that of the second compound contains one aluminum and one tantalum atom in the same positions that are occupied by Ti in rutile, and four oxygens in the same positions. Actually, the structures differ very slightly, but not enough to affect the observed powder patterns. Partial coincidences are turning up with increasing frequency as the powder diffraction file grows, and it is becoming increasingly important to make more precise measurements and to utilize supplementary chemical information. It is obviously wise to maintain a critical attitude and exercise judgment, but fortunately, cases like the above are still rare enough that the value of this method of identification is not seriously impaired. At present, it remains the most widely practiced application of x-ray diffraction. It is also the application that requires the least advanced knowledge of crystallography, and it is often adequately practiced on a nonprofessional basis.

The index to the PDF is a numerical listing of the three strongest lines of all the patterns, permuted so that each pattern is listed three times, once each under the *d* spacing of the first, second, and third most intense lines. Such permutation allows for variations in the observed intensities that might alter the ranking of the lines.

Another index lists the eight strongest lines, with the first six permuted. This was intended to be helpful with electron diffraction patterns, which can give intensities radically different from those seen in x-ray work, but this index is useful for the latter also.

Generally, the search of the PDF is begun by picking the three strongest lines of the unknown pattern and finding a match in the index. When this is found, the entire pattern must be looked up for detailed comparison with the unknown. If some lines of the unknown are not accounted for at this time, then either the identification is incorrect or a second phase is present. One then treats the unaccounted-for lines as a new pattern and repeats the procedure. It may be that one of the three strongest lines belongs to a second phase, and a variety of combinations may have to be tried.

If all the lines that remain unidentified after the first component has been found are very weak, the second component is probably present in low concentration. However, it is also possible, although not common, that one has observed weaker lines than were observable under the conditions under which the file pattern was taken. A knowledge of the unit cell would enable one to calculate whether the lines belong to the same substance.

If one has positively identified the unknown as a mixture of two phases, the concentrations can be estimated within 1 to 5% if one prepares mixtures of known composition for comparison by intensity.

The diffracting power of an element is proportional to the square of its atomic number, its x-ray absorption to the first power of the atomic number. If the elements in a mixture of compounds do not differ greatly in atomic number, one can generally see 3 to 4% of a second phase in a matrix of the first. These figures do not hold when atomic numbers (and hence both diffracting power and absorption) differ greatly. It is possible to see ½% of UO_2 in a matrix of BeO but not less than 30% BeO in a matrix of UO_2. Both of these figures refer to routinely taken diffractometer patterns. If the presence of the other ingredient is suspected, measures can be taken to improve the detection level.

When recording a diffraction pattern with a diffractometer, one adjusts the electronic amplification of the signal so that the strongest line runs as nearly as possible across the full height of the chart paper (100). Intensities of 3 or 4 will then generally merge into the background. Weaker lines can be observed by increasing the amplification so that the strong lines run off the chart. This procedure is limited by the accompanying amplification of the background itself or if the latter is quite low, by the artificial background due to electronic noise. When film methods are used, exposure times can be increased until the general background begins to blacken the film uniformly. But if a monochromator is used with film methods, then there is virtually no background, and very long exposure times can be used. Although the actual exposure increases only with the logarithm of the time, setting a practical limit to what can be done, very weak lines can be brought into view in this manner and the detection limit pushed down appreciably.

It is quite common to have unknown patterns that are due to mixtures of three, or many more, different phases. A search conducted in the indicated manner is likely to fail because the lines due to the additional phases may be few in number or weak in intensity or may overlap other lines, or merely because the method becomes very tedious. In such cases, greatly superior results can be obtained by use of computer search methods developed in recent years. A computer never gets tired, it can keep track of a large amount of information without difficulty, it can apply objective criteria for the quality of match, and it can go through the entire file from beginning to end many times in a few seconds.

The PDF is currently available from the Joint Committee in the form of magnetic tape. Several computer programs for searching the magnetic tape file have been developed; for example, one was developed at the Dow Chemical Co. by Frevel and Adams (1968). The program distributed by the Joint Committee (Johnson, 1970), which scans the entire file, was created and refined over a number of years at the Pennsylvania State University. Although the Committee's literature states that the program has been used successfully on a variety of computers, only one version is being supplied at present. This is for an IBM 360 or 370 computer with H-level FORTRAN IV capability. Users of other computers must convert both the data base and the program, or else obtain the conversion and search programs from someone who has already done so. The author has run version 12 of the search program on CDC-6600 and CDC-7600 computers.

5.2 Use of Single-Crystal Data for Identification of Materials

While the use of the PDF is the major avenue for the identification of unknown crystalline materials by x-ray diffraction, an alternate route exists.

Crystal Data (Donnay and Ondik, 1972) is a collection of unit-cell constants (lattice parameters), listed numerically within crystal systems by values of axial ratios. For example, within the orthorhombic system, crystals are listed by the ratios a/b and c/b, used as "determinative" numbers. Each entry contains the individual lattice parameters, the space group if known, the density if known, an indication whether the full structure has been determined, and literature references. Some optical properties may be listed. The third edition of *Crystal Data* contains over 24,000 entries.

If we compare the PDF with *Crystal Data,* we note that well over 25% of the entries in either are unique and not duplicated in the other. In principle, then, the combined use of both data collections extends the range of unknown substances that can be identified. In practice, other considerations frequently prevail.

If the unknown specimen is polycrystalline, one will of course use the PDF method. The difficulties of deriving reliable unit-cell from powder patterns have been discussed. Nevertheless, if the PDF does not yield an answer, the attempt could be made.

If the unknown is in the form of small single crystals, a choice exists. Although determining the unit cell of a crystalline substance from single-crystal data requires more sophistication both in knowledge and in equipment, it is nevertheless a relatively simple matter and yields a certain answer.

If the crystals are plentiful and of no special value, then the most expeditious procedure is still to grind some of them into a powder and use the PDF. It may be, however, that one has only a few crystals, prepared with difficulty, which may be valuable for further research, or perhaps a new compound. It is even possible, for example, in the case of a rare mineral or a material of biological origin, that one finds oneself in possession of the only good crystal in existence. Under such circumstances, it would be most inadvisable indeed to turn "crystalloclast" (Donnay) and smash the crystal. For investigations of properties and structure of any material suspected of being a new compound, crystals are far more valuable intact than ground up. Given this situation, one would surely choose to determine the lattice parameters by single-crystal methods and determine whether the species has been previously reported via *Crystal Data.*

Happily, it is now possible to have one's cake and eat it too. In recent years, *Gandolfi* cameras have become more readily available. This camera allows one to obtain a powder pattern from a single crystal without destroying it. It is discussed in the section on cameras.

6 Methods for Recording Diffraction Patterns of Materials

6.1 The Debye–Scherrer Camera

The simple circular geometry of the Debye–Scherrer camera is schematically indicated in Fig. 5b, and the appearance of the films taken with it in Fig. 5c. Most Debye–Scherrer cameras in use today have an inside diameter of 114.59 mm. The entrance and exit collimators for the primary beam are, of course, 180° apart on the circumference of the camera, and the quoted diameter results in a corresponding linear distance of 180 mm on the film when it is laid flat. A slight shrinkage of the film, typically from 1 to 3 parts per thousand, occurs during processing. If the film is weighted while it dries, it may expand instead. Hence for best accuracy, the actual length of the dried film should be measured and a correction applied, which will also compensate automatically for any inaccuracy of the camera diameter. It is not permissible to measure the distance between the holes to this end. Not only is the film shrinkage nonuniform in the immediate vicinity of the holes (and near the ends of the film), but the holes are punched by a relatively crude instrument, which does not locate them very precisely. Instead, one determines the 0 and 180° points by measuring the diameters of several diffraction rings that are visible on both sides of the holes (see Fig. 5b). If there are no well-defined rings about the rear center, the correction is omitted or an average value is used. Having determined the exact location of the front center (0°), one measures the linear distances of the diffraction lines from it along the midline of the film, applies the shrinkage correction if available, and obtains a list of values of 2θ along with estimated intensities. Intensities can be measured with a densitometer if desired, or the entire process can be automated. Values of 2θ are then converted to d spacings by means of the Bragg equation (1). Commercially available film-reading devices allow the film to be read to within 0.05 mm and hence 0.05° (2θ) if the 114.59-mm-diam. camera is used. Because of the nature of the films, it is impractical to try for a higher precision. A look at the Bragg equation shows that a constant precision of 2θ results in an increasing precision of d (which is proportional to $\sin \theta$) as θ increases. Thus, the low-angle lines, which are the most important ones in the initial stages of an identification, are the least precise. The change in precision with θ is much steeper than the Bragg equation suggests, because both the nature of the diffraction process and instrumental limitations introduce systematic errors, which also vary with θ, most of them in such a manner that they decline to zero as 2θ approaches 180°.

A number of trigonometric functions have been suggested to correct for various errors under a number of experimental conditions. Generally, they are used only in conjunction with least-squares procedures to refine lattice parameters from powder data.

6.2 Sample Preparation for the Debye–Scherrer Camera

An adequate specimen required for the Debye–Scherrer camera is a small grain of material, perhaps 0.25 mm in diameter, mounted on the end of a fine glass fiber and centered in the x-ray beam. More typically, the specimen is a cylindrical rod, about 1 cm long and from 0.2 to 0.5 mm in diameter. It can be prepared in three different ways:

(1) The powdered specimen is mixed with a small amount of noncrystalline glue, rolled into a rod, and allowed to harden.

(2) A thin glass fiber is very lightly coated with petroleum jelly and rolled in the powder, which will coat its surface.

(3) The specimen is sifted into a closed-end glass capillary, which is then cut to the proper length and may be sealed. Capillaries made from a special glass with low absorption for x rays are available in several diameters (e.g., 0.2, 0.3, and 0.5 mm). The wall thickness is only 0.02 mm, and the open end is flared into a funnel to facilitate introduction of the specimen.

This last technique is suitable for materials that cannot be exposed to air. The capillaries can be filled in a glove box and the funnel end sealed temporarily with a dab of silicone grease. The final seal is made outside the glove box.

6.3 The Focusing Camera

In most cases, the uncorrected data from a Debye–Scherrer camera are sufficient for identification purposes. Sometimes, however, it is necessary to distinguish between very similar phases; in structural investigations, for example, one asks whether a broad line might not be several closely spaced lines, and often one wishes to use powder data to derive accurate lattice parameters. In such cases, it is desirable to increase both the precision and the resolution of powder patterns. This is generally done by abandoning the Debye–Scherrer camera in favor of some device that employs a *focusing geometry*.

Figure 11 shows the general appearance of a focusing arrangement. If the specimen SP forms part of the periphery of a circle of proper diameter, and if the x-ray source XS is placed on another part of the same circle, then all diffracted beams with the same value of 2θ, even though coming from different parts of the extended specimen, will come to a sharp focus at DB, where one locates either a piece of photographic film or a radiation detector.

The diameter of the focusing circle must change with 2θ. In film cameras, this is not practical for mechanical reasons; hence there are no focusing cameras that can cover the range from 0 to 180°. The Guinier camera extends from 0 to 90°, with some compromise. It is difficult to use but can be a fine instrument in competent hands. Some commercial Guinier cameras are clumsy, and experimenters often build their own. Films from a Guinier camera, because of the focusing geometry and the use of a crystal monochromator, can be read profitably to a precision of 0.01 to 0.02 mm, using specially constructed film readers or a traveling microscope.

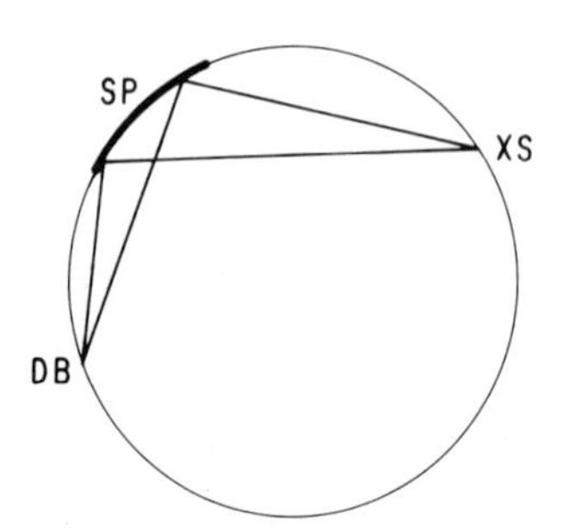

Fig. 11. Focusing geometry.

The symmetrical focusing backreflection camera is used almost exclusively to measure one or two of the highest angle lines of a high-symmetry (mostly cubic) substance in order to obtain the best possible lattice parameter from a single measurement. It plays no significant part in identification procedures.

6.4 Sample Preparation for the Focusing Camera

Figure 11 shows the specimen to be a segment of a circle. In the backreflection camera described, the specimen is spread as a thin film of powder on a strip of Mylar or Scotch tape, which is then attached to the cylindrical specimen backing plate. In other devices employing this geometry, a flat specimen is placed tangent to the focusing circle, and this works well enough for most purposes.

6.5 The Gandolfi Camera for Single-Crystal Materials

There is a way of obtaining a powder pattern experimentally from a single crystal without destroying it. The Gandolfi camera is a modified Debye–Scherrer camera in which a small crystal is given extra motions that spread the single-crystal spot pattern out into the conventional (line or ring) powder pattern.

6.6 The Diffractometer

The *diffractometer* employs photon-counting radiation detectors in place of film. Values of 2θ are indicated on instrument dials, and intensities are read on a *scaler* in the form of a total number of counts and on a *ratemeter* as counts per unit time. Angles and count rates are also combined in the output of a strip-chart recorder as a plot of intensity versus 2θ. The diffractometer continuously varies the diameter of the focusing circle and is thus usable over the entire angular range. This makes it a general-purpose instrument similar to the Debye–Scherrer camera but with superior precision and resolution. Only its ability to detect very weak lines is somewhat inferior. However, sensitivity, resolution, and precision are adjustable through manipulation of scanning speeds, time constant, and slit widths.

Since the diffractometer is a complicated mechanical device, it requires far more elaborate alignment procedures than does a camera. Exacting standards for diffractometer alignment (primarily for single-crystal work) have been described by Samson and Schuelke (1967). In powder work, the accuracy is particularly sensitive to the displacement of the specimen normal to its surface. The resulting error, like many others, extrapolates to zero as 2θ approaches 180°. In addition to these mechanical aspects, the diffractometer also requires the proper functioning of several pieces of electronic equipment concerned with the direct measurement of radiation. The gain realized from this greater complexity is not only a higher precision of the measured intensities but also the capability of direct measurement techniques to handle intensities over a range of 10^5, while a piece of photographic film will respond linearly only over a range of about 20. Azaroff (1960), Azaroff and Buerger (1958), and Wilson (1963) provide more extended treatments of some of these topics.

With the diffractometer, one must accept the fact that a radiation-measuring device makes an instantaneous measurement only. No matter how sensitive the detector, it can hardly compensate for the integrating effect that results when film is exposed to radiation for a prolonged period of time.

The diffractometer sample must be considerably larger than the one required for the Debye–Scherrer camera, typically in excess of 100 mg. It may be a solid piece of material with a flat surface, but commonly it is a powder that is packed into a shallow depression, about 2.5×2.0 cm, in a flat slide to a depth of 0.15 to 0.25 cm, giving a volume slightly under 0.5 cm^3. It is possible, however, with a slight sacrifice of accuracy, to spread a much smaller amount of material thinly on the surface of a slide. The specimen holder can be replaced with a large variety of special-purpose gadgets, making the diffractometer a very versatile instrument.

7 Additional Uses of X-Ray Diffraction Data

It is frequently necessary to obtain lattice parameters to a precision and accuracy greater than those obtained from routine powder patterns. The objective may be the characterization of a new material; an investigation of the influence of dissolved impurities; the analysis of an alloy, the lattice parameters of which change continuously with composition; or the determination of a precise value for the volume of the unit cell so that the theoretical density can be calculated.

Usually, precise lattice parameters are obtained by (1) careful measurement of the *d* spacings from a Guinier camera or from a well-aligned diffractometer, (2) verification of correct indexing (Wolten, 1970) (3) least-squares refinement, and (4) corrections for instrumental aberrations and absorption errors, usually done by means of various extrapolation techniques. One of the best known and highly efficient computer programs for this purpose can be obtained from Dr. M. H. Mueller of the Argonne National Laboratory.

With proper care, precisions and accuracies of one or two parts in 10^4 for cubic materials and 10^3 for monoclinic materials are not too difficult to obtain. A rather large effort is required to exceed these precisions and accuracies, and thermal expansion must be taken into account. It is illuminating to read the report on an extensive study of this subject (Parrish, 1960) sponsored by the International Union of Crystallography.

Thermal expansion can be studied by measurement of lattice parameters at different temperatures. This is highly effective for the limited range of temperatures over which the equipment can be thermostated. For more extended temperature ranges, the precision is modest because of the difficulties associated with the use of heated specimen holders. The method has the great advantage, however, that any change with temperature of a particular d spacing is a direct measure of the expansion of the material in the *particular direction* normal to the set of lattice planes identified with that spacing. The equipment used for measuring thermal expansion over large temperature ranges (some experimenters have approached 3000°C) actually is used much more frequently for the investigation of phase transformations.

If a bulk specimen is polycrystalline, then all reflections are observable independent of the orientation of the specimen. In the presence of preferred orientation, the intensity of the same reflection observed in different orientations varies and furnishes a measure of the preference. A plot of intensity versus orientation, somewhat like a contour map, is called a *pole figure.*

Many kinds of treatment can leave a specimen with a retained uniform *residual stress,* in which case lattice planes normal to the direction of the applied stress have had their spacings slightly altered, and the changes in d can serve as a measure of the residual stress. It is difficult to make the required measurements with high precision, and the results are usually considered to be valid within 10%.

In plastic deformation, different grains usually have suffered differing degrees of deformation, and we speak of *nonuniform stress or microstress.* The specimen *as a whole* exhibits a small range of d spacings instead of a single value. The resulting *line broadening* is a measure of microstress.

Line broadening can also result from small particle sizes, and it may be necessary to separate the two effects or distinguish between them. In the derivation of Bragg's law, we considered only two or three lattice planes and saw how, depending on angle and spacing, the "reflected" beams either reinforced one another (in-phase condition, *constructive interference*) or canceled one another (out-of-phase condition, *destructive interference*). For this two-plane case, if the intensity were high enough to be observable, it would vary smoothly (as a cosine function) from one condition to the other, as shown by curve 1 of Fig. 12. If more planes are added to the stack, both constructive and destructive interference are enhanced, and curve 2 is the

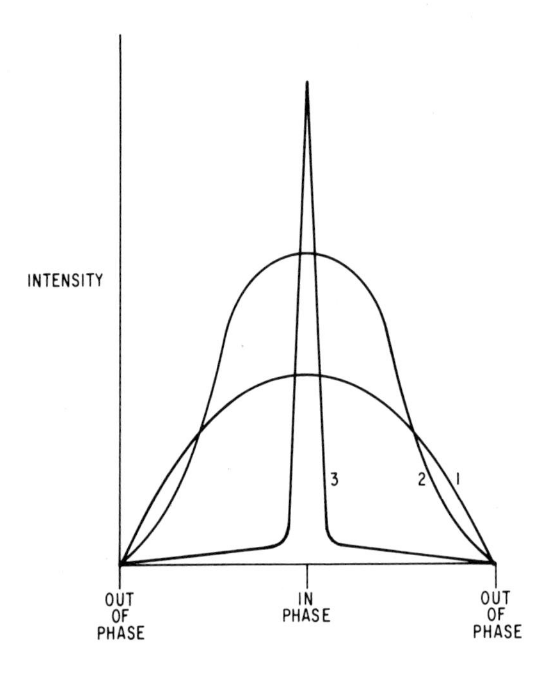

Fig. 12. Effect of number of cooperating planes on sharpness of Bragg peak (line broadening).

result. In the limit, when a few hundred planes are involved, the sharp peak of curve 3 is obtained.

It follows that diffraction peaks become shallower and broader when the number of cooperating planes decreases. This type of line broadening, which becomes observable when the particle or grain size drops below approximately 1 μm, has been used as a measure of particle size. Further information on these techniques is available in the volumes by Guinier (1952), Barrett (1952), and Cullity (1956). *X-ray microscopy* and *x-ray topography* can be applied to large plates of rather perfect single crystals, such as are available for some semiconductors. The crystal is scanned by a narrow beam, and either the transmitted beam or a diffracted beam is recorded. The photographic image produced by either beam shows contrasts produced by the variation in perfection of the areas scanned. These and other methods are described in detail by Newkirk and Wernick (1961).

The single crystals that are used for their optical, electrical, or mechanoelectrical (piezoelectric) properties are anisotropic in these properties. When such crystals are cut and mounted for their intended applications, they must therefore be oriented properly. Consequently, one must determine the orientation of the unit cell or a specific lattice direction with respect to the external form of the crystal. Sometimes the properties themselves can be used for this purpose. It is always possible to use x rays, and this is frequently done.

Holders that accommodate large specimens are available for diffractometers. These allow the crystal to be positioned and rotated about several axes, so that the spatial orientation of characteristic reflections can be found.

The *Laue camera* uses a flat photographic film perpendicular to the x-ray beam, which in this case is polychromatic (being *white radiation*). Because large specimens do not transmit the x-ray beam, the camera is generally used in the *backreflection* geometry, in which the film is between the specimen and the source of the x-ray, which passes through a hole in the film before it strikes the specimen. The film records the diffracted beams that are returned toward it.The symmetry of the resulting pattern is related to the symmetry of the crystal direction along the x-ray beam so that this direction can be identified.

8 Structure Determination

It has been indicated above that the information concerning the positions of the atoms is contained in the intensities of the reflections and that structure determination requires measurement of a very great number of these intensities.

These measurements are made on a crystal, 0.02 to 0.3 mm in size, mounted on a *goniometer head,* a rather small gadget with four of five independent motions. There are two lateral translations at right angles, and usually a height adjustment, for centering the crystal, and two arcs, segments of two orthogonal circles, for orienting it. The goniometer head, with the crystal, is inserted either into a single-crystal camera or into an *orienter* or *goniostat,* which occupies the specimen position on a diffractometer. In the latter case, the orienter allows the many reflected beams to be brought into the detector one by one for measurement. There are a number of types of single-crystal camera, the most popular being the *Weissenberg* camera, but they all accomplish their task by imparting such motions to both the goniometer head and the film that each reflection records as a separate spot on the film in a pattern that permits identification of the reflection by its position on the film.

The determination of crystal structures using single crystals is the central problem of crystallography. It requires the greatest amount of computer time of all the procedures based on x-ray diffraction. Even a birdseye view of the procedures involved would require a separate chapter. Since the objective of this text is to inform the reader of what a particular discipline can do for him, we discuss here only the "why," rather than the "how," of crystal structure determination. For the latter, the reader is referred to the texts by Stout and Jensen (1968), Lipson and Cochran (1966), Buerger (1942) and Buerger (1960).

The properties of a solid are determined by a number of parameters; these parameters have a hierarchical relationship to one another, as shown in Fig. 13, with the chemical composition at the upper level and crystal or molecular structure at the second level. Purity, microstructure, and (where applicable) crystal perfection constitute a third level.

In the preceding section, brief mention has been made of various crystallographic studies that supply information on the perfection and microstructure of crystalline materials. The influence of these factors on properties is marked only in rather simple structures, but in such case it can be of very major importance indeed. Point defects (*interstitial* atoms and *vacancies*), line defects (*dislocations*), area defects (grain boundaries), and grain size, as well as deformation, all are studied assiduously, particularly by metallurgists, because most metals have the very simple structures in which such factors profoundly influence the mechanical and sometimes the optical properties of the material. The electrical properties of semiconductors, which by most standards are extremely pure materials, depend crucially upon the nature and distribution of their low-level impurities.

On the second level of the hierarchy, crystal structure is quite obviously the exclusive domain of crystallography. By way of illustration, consider the startling differences between the properties of graphite and those of diamond. Both are pure carbon; the differences lie simply in the crystal structures (which implies a major difference in the nature of the bonding). Rutile and anatase are two different crystal structures of titanium dioxide, chemically identical. The former has a high refractive index, good gloss,

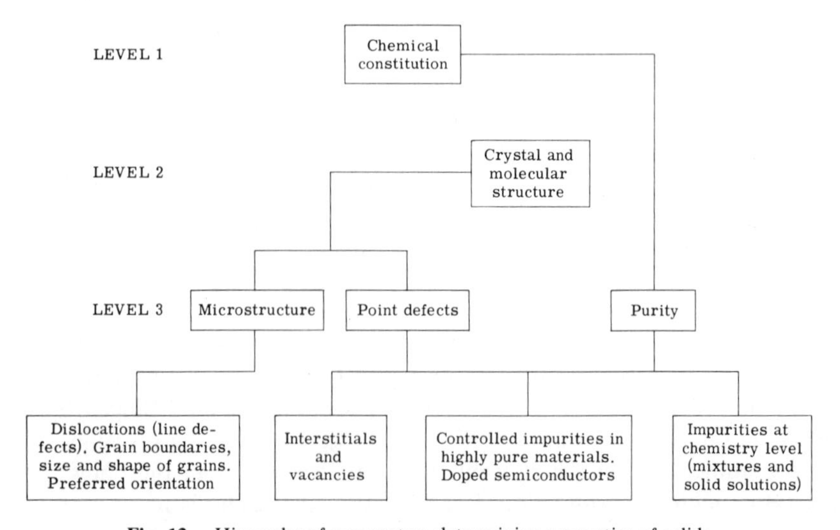

Fig. 13. Hierarchy of parameters determining properties of solids.

and excellent hiding power when used as a pigment in paints; the latter has none of these and can be used only as a filler.

In compounds that are predominatly bonded covalently, specifically in large organic molecules, it is the molecular structure, rather than the crystal structure, that is important for an understanding of the properties. But since molecules remain intact when dissolved, the results are not limited to the solid state. Traditionally, organic chemists have spent much of their time determining molecular structures by taking the molecules apart bit by bit to learn how they are put together. This activity is being supplanted more and more by crystallography.

If a compound can be crystallized, then a crystal structure determination will give a full three-dimensional picture of the molecular arrangement, showing not only the sequence in which the various atoms are joined but also all interatomic distances and angles, the nature of the thermal vibrations, and the distribution of electron density. Thus this approach yields more information than can be obtained by other methods.

The biological action and the chemical reactions of a complex molecule such as a protein or DNA depend on the molecule's overall shape and on the shape, size, location, and orientation of all its component parts. The ability of the crystallographer to analyze compounds of this size and complexity is a quite recent achievement and is still in a state of development.

The examples of the last two paragraphs illustrate another truth about crystallography, namely, that its subject matter is the structure of matter, and that crystals are often incidental, merely providing a suitable specimen.

9 Crystallography by Diffraction of Radiation Other Than X-Ray

A beam of electrons interacts with the electron shells of an atom in a manner similar to that of x rays and produces quite similar diffraction effects. At accelerating voltages of 50–100 kV (high-energy electron diffraction or HEED), the equivalent wavelength is 0.05–0.04 Å, which is about one-thirtieth to one-fifteenth (for copper and molybdenum, respectively) of the range of x-ray wavelengths used. Insertion of this value into the Bragg equation shows diffraction angles 2θ to be less than $1°$. However, the diffracted beams are passed through an electron lens, which spreads them so that the pattern recorded on film has dimensions similar to those of x-ray patterns. Electrons in this energy range penetrate to a depth about one-tenth that of x rays. When used in a reflection geometry, electrons can thus be used to investigate thin surface layers that would be largely transparent to x rays. Not only are materials much more opaque to electrons, but the depth of material in the path of the beam required to produce diffraction is very much less. In the transmission mode, electron diffraction can thus be used for films much too thin and for particles or crystallites much too small for x-ray work. This is the principal application of HEED. It is not preferable for specimens suitable for x-ray diffraction, because samples cannot be manipulated as well, the precision of d spacings is less, and the interpretation of intensities is much more difficult.

At substantially lower voltages (*low-energy electron diffraction* or LEED), the penetration is so reduced that essentially only the top layer of atoms is involved; the study of this layer is called two-dimensional or *surface crystallography*. LEED is used to study absorbed monolayers on surfaces and the structures of surfaces themselves. The latter can differ from the bulk structure, because the outer layer of atoms is subject to less symmetrical force than are atoms in the interior.

HEED and LEED are discussed in Chapters 4 and 5 (Volume I), respectively.

Thermal neutrons have wavelengths comparable to those of x rays used in diffraction and also yield diffraction patterns with crystalline materials. Neutrons are very penetrating and can be used only with large specimens. Beams of adequate intensity are available only from nuclear reactors. Despite these drawbacks, neutron diffraction is very important in crystallography because of two major differences from x rays.

First, neutrons are scattered by atomic nuclei rather than by their electron shells. As a consequence, neutron scattering factors do not vary with atomic number in the same way that x-ray scattering factors do, and they do not vary as much. Therefore, the ratios of light-atom to heavy-atom scattering factors are generally more favorable, which makes it possible to determine hydrogen positions in hydrides and in organic compounds and oxygen positions in the oxides of heavy metals with much better precision than with x rays.

Second, some atoms have magnetic moments, either from unpaired electrons or from orbital motions. X rays do not respond to magnetic moments, but the neutron, which has a magnetic moment of its own, will interact with them. Neutron diffraction with polarized neutrons can thus be used to investigate the *magnetic structure* of materials. Symmetry elements beyond those observable by x rays come into play, and the number of magnetic space groups greatly exceeds that of the 230 conventional space groups.

Neutron diffraction is discussed in detail in Chapter 34, Volume IV.

References

Azaroff, L. V. (1960). "Introduction to Solids." McGraw-Hill, New York.

Azaroff, L. V., and Buerger, M. J. (1958). "The Powder Method in X-Ray Crystallography." McGraw-Hill, New York.

Barrett, C. S. (1952). "Structure of Metals." McGraw-Hill, New York.
Buerger, M. J. (1942). "X-Ray Crystallography." Wiley, New York.
Buerger, M. J. (1960). "Crystal Structure Analysis." Wiley, New York.
Cullity, B. D. (1956). "Elements of X-Ray Diffraction." Addison-Wesley, Reading, Massachusetts.
Donnay, J. D. H., and Ondik, H. M. (1972). "Crystal Data, Determinative Tables," 3rd ed. (two volumes). Nat. Bur. St. and Joint Committee on Powder Diffraction Standards.
Frevel, L. K., and Adams, C. E. (1968). *Analy. Chem.* **40,** 1335.
Guinier, A. (1952). "X-Ray Crystallographic Technology." Hilger and Watts, London.
Ito, T. (1950). "X-Ray Studies of Polymorphism." Maruzen Co., Tokyo. (Method described in many textbooks, see Azaroff and Buerger, 1958.)
Johnson, G. G., Jr., (1970). "Fortran IV Programs, Version XII, for the Identification of Multiphase Powder Diffraction Patterns." Joint Committee on Powder Diffraction Standards, Swarthmore, Pennsylvania.
Kennard, O., Speakman, J. C., and Donnay, J. D. H., (1967). *Acta Cryst.* **22,** 445.
Kharin, V. M., and Davydov, G. V. (1970). *J. Appl. Cryst.* **3,** 522.
Lipson, H., and Cochran, W. (1966). "The Determination of Crystal Structures." Cornell Univ. Press, Ithaca, New York.
Newkirk, J. B., and Wernick, J. H. (eds.) (1961). "Direct Observation of Imperfections in Crystals." Wiley (Interscience), New York.
Parrish, W. (1960). *Acta Cryst.* **13,** 838.
Samson, S., and Schuelke, W. W. (1967). *Rev. Sci. Instrum.* **38,** 1273.
Stout, G. H., and Jensen, L. H. (1968). "X-Ray Structure Determination." Macmillan, New York.
Wilson, A. J. C. (1963). "Mathematical Theory of X-Ray Powder Diffractometry." Philips Tech. Library, Eindhoven.
Wolten, G. M. (1970). *Metallography* **3,** 219.
Zsoldos, L. (1968). *Acta Cryst.* **11,** 835.

CHAPTER 29

X-Ray Fluorescence and Absorption Spectrometry

N. Spielberg

Department of Physics
Kent State University
Kent, Ohio

Introduction

X-ray fluorescence spectrometry is used for qualitative and quantitative elemental analyses of materials by means of measurements of the intensity and wavelength (or quantum energy) of characteristic x radiation emitted by the individual atoms in a sample. Usually the sample is caused to emit x rays by irradiation with other x rays, but gamma-ray, electron, or other charged-particle irradiation may also be used. The technique features simple nondestructive (with very few exceptions) excitation, complete coverage of the periodic table (except for H, He, and Li) with slowly and smoothly varying sensitivity, and ease of verification of reproducibility. While the optimum concentration ranges for analysis are for major and minor constituents, applications to the study of trace and ppm analysis are quite common. The specimen to be analyzed may be in almost any form in a condensed state. Major constituents of gases are analyzable with special techniques.

X-ray absorption spectrometry is used for qualitative and quantitative elemental analysis of materials by means of measurement of the absorption of x rays by a sample. The technique is applicable from about atomic number 13 upward. Applicability is usually limited to major concentrations of similar atomic-number elements or to minor concentrations of heavy elements in light-element matrices. The specimens should be reasonably x-ray-transparent specimens, but may be solid, liquid, or gaseous, provided they have fairly uniform thickness and homogeneity.

1 Essentials of X-Ray Fluorescence Spectrometry

1.1 X-Ray Spectra

There are two types of spectra of interest for chemical analysis:

1.1.1 Continuous Spectrum

When a beam of electrons strikes a target, as in an x-ray tube, there is generated a continuous x-ray spectrum, whose wavelength distribution is shown schematically in Fig. 1a. The spectrum rises smoothly from a short wavelength limit λ_0 ($=12.4/V$ Å, where V is the kilovoltage applied to the x-ray tube) to a maximum at 1.5 λ_0, and then decreases smoothly with increasing wavelength. The shape of the spectrum depends only on V and is almost entirely independent of the target material. The overall intensity is proportional to V, to Z, the atomic number of the target, and to the electron-beam current. To a rough approximation, the x rays are emitted hemispherically from the point of impact of the electrons on the target and are partially polarized parallel to the electron-beam direction. The degree of polarization is greatest for wavelengths close to λ_0.

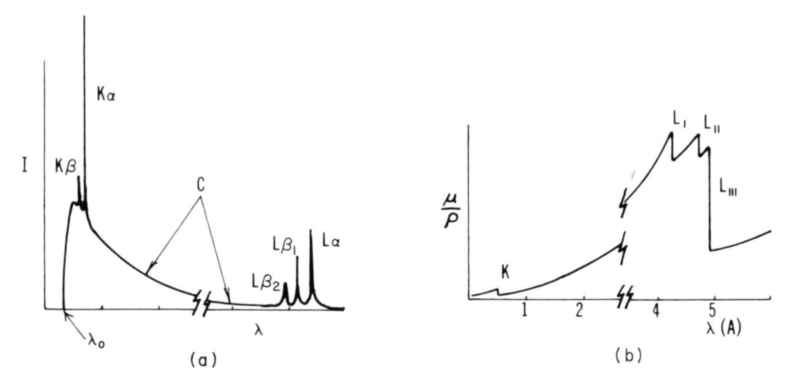

Fig. 1. Schematic x-ray emission and absorption spectra. (a) Emission spectrum from x-ray tube: *I*, intensity; λ, wavelengths; *C*, continuous spectrum, λ_0, short wavelength limit; Kα, Kβ, Lα, $L\beta_1$, $L\beta_2$, lines from characteristic spectrum. (b) Absorption spectrum: (μ/ρ), mass absorption coefficient, proportional to logarithm of ratio of incident to transmitted intensity; λ, wavelength in Angstroms, K, L_I, L_{II}, L_{III} , absorption edges.

1.1.2 Characteristic Spectra

The characteristic x-ray emission spectrum, which is the basis of fluorescence spectrometry, consists essentially of a set of discrete narrow lines arising from radiative transitions between atomic energy levels. The wavelengths of the lines are characteristic of the emitting atoms. Any means of removing inner shell electrons from an atom will excite characteristic x rays; such means include electron bombardment, as in an x-ray tube, or irradiation with other x rays. When excited by electron bombardment, the characteristic spectrum is superposed on the continuous spectrum with comparable overall intensity (Fig. 1a).

Characteristic x rays are unpolarized and emitted isotropically in all directions from the emitting atom. Other atoms, however, absorb x rays; the intensity and angular distribution of the observed x rays depend upon the spatial distribution of the excited atoms relative to other atoms.

Excitation of characteristic x rays requires the creation of a vacancy in an inner electron shell, i.e., the complete removal of the inner shell electron from the atom. An electron transition which will fill this vacancy is most likely to take place from some other inner shell of the atom, thereby creating a vacancy in that other shell. The levels in the x-ray energy-level diagram correspond, therefore, to the ionization energies of the various electron shells and subshells. A schematic x-ray energy-level diagram for a heavy atom is shown in Fig. 2.

Because of various quantum mechanical selection rules, only certain radiative transitions are allowed between the energy levels; these are labeled in Fig. 2. All transitions originating from the K level give rise to lines

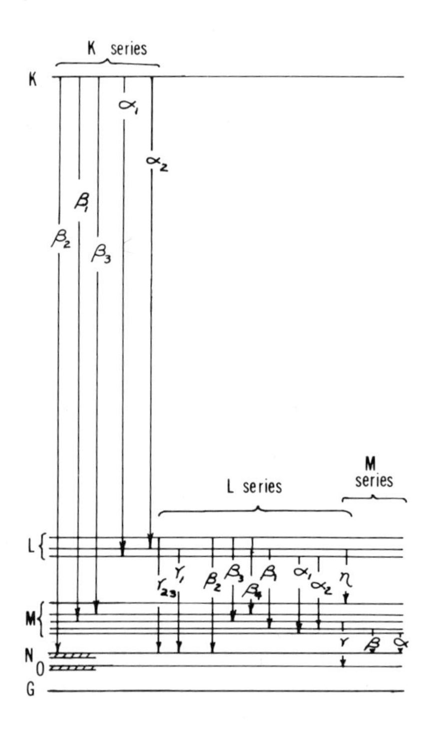

Fig. 2. Schematic one-electron x-ray energy-level diagram for a heavy element. K, L, M, N, O, energy levels corresponding to ionization energies of inner shell electrons; G, ground state. Not all allowed radiative transitions shown. Lengths of the arrows indicating transitions roughly proportional to energies of emitted photons.

in the K series; transitions originating from one of the L levels give rise to L-series lines; etc. The average wavelengths of lines in different series of a given element differ by very roughly an order of magnitude.

Absorption spectra are continuous spectra, marked by discontinuities (called absorption edges) which are characteristic of the absorbing atoms. The wavelength or energy of a given absorption edge corresponds to the energy difference between the ground state and one of the levels of the x-ray energy-level diagram. Absorption edges appear because x-ray photons of wavelength just longer than the K edge, for example, no longer have sufficient energy to eject a K-shell electron from the atom, and hence they are not absorbed as strongly as x rays of slightly shorter wavelength. There are no absorption lines corresponding to a transition between x-ray levels at normal x-ray fluxes, because any atoms excited to an x-ray state return to the ground state long before they absorb a second x-ray photon.

The energy (expressed in electron volts) required to ionize the atom in its inner shell numerically equals the minimum or excitation voltage at which an x-ray tube must be operated before the characteristic spectrum will be generated in the tube or in an irradiated specimen. The excitation voltage in kilovolts equals $12.4/\lambda$, where λ is the wavelength of the corresponding absorption edge.

TABLE 1

RELATIVE INTENSITIES OF SELECTED SPECTRAL LINES FOR A HEAVY ELEMENT[a,b]

K series		L series		M series	
Designation	Relative intensity	Designation	Relative intensity	Designation	Relative intensity
α_1	100	α_1	100	α	100
α_1	50	α_2	10		
α	150	α	110		
β_1	35	β_1	55	β	55
β_2	10	β_2	20		
		β_3	7		
		β_4	5		
		γ_1	10		
		$\gamma_{2,3}$	4		

[a] Based on Compton and Allison (1935, pp. 640–646).
[b] α_1 line of each series taken to have intensity of 100. No comparison between series.

The fluorescence yield, or probability that an excited atom will emit x rays, depends upon both the atomic number of the atom and the particular electron shell which is ionized. The K-shell fluorescence yield is less than 0.5 for atomic number less than 35, and decreases markedly with decreasing atomic number. Fluorescence yields for L and M shells are one-half to two-thirds of those for the same energy K shells. The relative intensities of the various lines comprising a particular series are quite different. Table 1 presents the values of relative intensities for a typical heavy element.

Usually x-ray spectra are independent of the state of chemical combination of the emitting atoms because the inner electron shells involved in the transitions are affected relatively little by chemical bonding. However, for wavelengths greater than about 4 Å, effects on both the wavelength and intensity of some spectral lines are observable with medium- or high-resolution instruments. These wavelengths correspond to the K series of chlorine or lighter elements.

1.2 INTERACTION OF X RAYS WITH MATTER

1.2.1 Absorption

X-ray absorption obeys Beer's law: The intensity I of a well-collimated pencil of x-rays, after passing through a thickness t of matter, is given by $I = I_0 e^{-\mu t}$, where I_0 is the initial intensity of the x-ray beam and μ is the linear absorption coefficient of the particular material. The magnitude of the absorption coefficient is a function of the wavelength of the x rays.

The linear absorption coefficient is proportional to the density ρ of the absorbing material. The mass absorption coefficient, $\mu_m = \mu/\rho$, is independent of the physical state (gaseous, liquid, or solid) of the material. The mass absorption coefficient of any substance is the weighted sum of the mass absorption coefficients of the chemical elements of which that substance is composed, the weights being the fractional densities of the pure elements composing the substance. For example, for Fe_2O_3 the fractional densities of Fe and O are 111.6/159.6 and 48/159.6. Thus μ/ρ for Fe_2O_3 equals $(111.6/159.6)(\mu/\rho)_{Fe} + (48/159.6)(\mu/\rho)_O$. Mass absorption coefficients in square centimeters per gram for the various pure elements as a function of wavelength are available in various tables (see texts cited in Section 4).

X-ray absorption involves three major mechanisms: the photoelectric effect, Thomson or coherent scattering without change of wavelength, and Compton or incoherent scattering with change of wavelength. These three mechanisms contribute additively to the absorption coefficient. In photoelectric absorption, the energy of an absorbed photon is used to eject an electron from the atom with kinetic energy equal to the energy of the photon less the binding energy of the electron. The atom is left in an excited state. Thus photoelectric absorption is the excitation mechanism in fluorescence spectrometry. The photoelectric contribution to the mass absorption coefficient varies roughly as $Z^3\lambda^3$, and is the largest portion of the absorption coefficient except when Z is significantly lower than ~ 25 and/or λ is less than ~ 0.5 Å. The shape of absorption spectra largely reflects changes in the photoelectric absorption coefficient.

1.2.2 Scattering

The contribution of Thomson or coherent scattering to the absorption coefficient varies more slowly than $Z\lambda$. Scattering is the mechanism underlying the diffraction effects by which x rays are dispersed into a spectrum. Thomson scattering may most easily be understood as electromagnetic radiation from oscillating atomic electrons, which are driven by the electromagnetic field of the incident x rays. The scattered radiation is necessarily of the same frequency as the incident x rays. The directional dependence of the phase and amplitude of the scattered radiation results from the polarization of the incident x rays, the details of the atomic and molecular structure of the scattering substance, and the "oscillator strength" of the atomic electrons, as calculated from quantum theory.

1.2.3 Compton Effect

Compton or incoherent scattering, which contributes a component to the absorption coefficient with relatively slight wavelength dependence, is most easily explained in terms of collisions between photons and electrons. Because momentum and energy are conserved, the scattered photons have increased wavelengths by amounts which depend only on the scattering direction. The effect is largest for light elements having loosely bound electrons and for very short wavelengths.

The actual observation of both coherent and incoherent scattering requires that the specimen both be thick and have sufficiently small photoelectric absorption that the radiation from within the specimen shall escape. In fluorescence spectrometry, scattering effects result in the addition of unwanted spectral components to the emitted characteristic spectra.

1.3 Excitation Techniques

1.3.1 Photon Excitation

Irradiation of the specimen with photons from a high-output, high-power x-ray tube makes possible the versatility and easy applicability of the method. The irradiating wavelengths should be short enough to ionize the atoms of interest in the K or L shells, but long enough that photoelectric absorption, and hence excitation of the sample, takes place close enough to the surface of the sample to minimize self-absorption of the characteristic radiation by the sample itself. Thus extremely high-voltage x-ray tubes (100 kV) or gamma-ray sources are not very efficient for photon excitation. On the other hand, optimum excitation voltages are not very sharply defined, and it is usually adequate to operate the x-ray tube at maximum power.

1.3.2 Electron Excitation

The use of high-voltage electrons directly incident on the sample gives a factor of 100 to 1000 higher x-ray excitation per unit of electrical energy consumed, but most specimens cannot tolerate the necessary operating conditions. Electron excitation of x rays is employed in electron microprobe analysis (see Chapter 6, Volume I).

1.3.3 Proton and Heavy-Particle Excitation

Low-power proton and alpha-particle excitation of a sample can be obtained by use of radioactive sources and has the advantage that there is essentially no continuous spectrum excited and no scattered x rays. The x-ray excitation efficiency per particle for long wavelength x rays is also much higher than for electrons. Because of the very short range of these particles in matter, self-absorption effects are minimized. Particle accelerators have also been used to give high-intensity beams of exciting particles, but the specimens must be capable of withstanding the bombardment conditions.

1.4 Detection Techniques

1.4.1 Photographic Techniques

Photographic techniques lend themselves well to the determination of geometrical distributions of radiation but do not have the precision and accuracy of other techniques. The technique is made quantitative by the use of relationships between the relative blackening of a photographic plate or film, as measured in various ways, and the total exposure of the film to x rays.

1.4.2 Gaseous Ionization

Photoelectric absorption of x rays by a gas results in the creation of a number of ionized molecules and free electrons. If all the absorbed energy is dissipated within the gas, then the amount of ionization produced will be proportional to the energy of the photons absorbed, and a measurement of the ionization determines the intensity and/or photon energy of the x rays impinging on the gas. If an electric field is applied to the gas volume where ionization has taken place, an electrical discharge is triggered, and thus the absorption of individual photons of x rays may be detected and counted as electrical pulses. If the discharges are not permitted to approach saturation sizes, then the magnitude of the electrical pulses will, on the average, be proportional to the energy of the photons being detected. Such an x-ray detector is called a proportional counter. Actually, even if monochromatic x rays are detected, the pulses are not uniform in size, but rather vary according to a pulse-height distribution as determined by the statistics of the ionization and multiplication processes in the gas and are proportional to the square root of the energy of the radiation. These proportional counters are generally of the side-window type with an active area up to 5 cm^2, since these usually have the best resolution. The depth of the detecting volume ranges from 2 to 5 cm. The gas filling may be argon, krypton, or xenon, plus a few percent organic quench gas, at a total pressure ranging from ½ to 1 atm.

If the discharges in a gaseous ionization detector are permitted to reach saturation, then all pulses will be of the same size regardless of the energy of the x-ray photons being detected. Such an x-ray detector is a Geiger counter.

1.4.3 Solid Ionization

It is also possible to measure x-ray-induced ionization in certain semiconductors. At present, this is most effectively done in "lithium-drifted" silicon or germanium crystals, cooled to liquid nitrogen temperature. The ionization, measured as an electrical pulse, is again proportional to the energy of the x-ray photons absorbed in the semiconductor. The resulting pulse-height distribution is considerably narrower than that for the proportional counter, with the width determined mostly by fluctuations in the "dark current" of the detector and first amplification stages. At present, this very high resolution is obtained only with semiconductor detectors of rather small dimensions (roughly 10- to 25-mm^2 area $\times$ 3-mm thickness).

1.4.4 Scintillation Detectors

In certain liquid and solid materials, the ionization produced as a result of the absorption of radiation is partially converted into a flash of visible or ultraviolet light, the amount of which is proportional to the energy of the x-ray photon absorbed. Such a scintillation of light may be detected with a highly sensitive photomultiplier tube, yielding an electrical pulse which is proportional to the energy of the x-ray photon. The pulse-height distribution obtained has a width, determined by the statistics of the scintillation, photodetection, and multiplication processes, which is significantly wider than that of the proportional counter for x-ray photons.

For the present application the scintillating element is a thallium-activated sodium iodide crystal with an active area of up to 5 cm^2 with a thickness ranging from 0.5 to 3 mm.

1.5 Dispersion Techniques

A number of dispersion techniques, individually or in combination, may be used to measure an x-ray spectrum.

1.5.1 Wavelength Measurement. Bragg's Law

Wavelength measurement is the most widely used dispersion technique. Wavelength dispersion of an x-ray spectrum depends upon the diffraction or cooperative coherent scattering of x rays by the atomic electrons of atoms arranged in a periodic or crystalline lattice. The result may be pictured as constructive interference of "reflections" from successive parallel planes of atoms. If the angle between the incident (and reflected) x rays and the surface of the atomic planes (*not* the normal to the planes) is defined as θ, and d as the spacing between successive planes, then Bragg's law states that $n\lambda = 2d \sin \theta$, where λ is the wavelength of the x rays, and n is an integer specifying the order of reflection. The use of large single crystals, appropriate slits and apertures, and suitable gearing devices then makes it possible to disperse the incident beam into an x-ray spectrum spread over an angular range, with wavelength increasing with angle. The factor n may cause overlapping successive orders of reflection of various intensities, but these can be distinguished or suppressed by use of the other dispersion techniques discussed below.

Although diffraction gratings have been used to disperse x rays, their applicability is limited except for long wavelengths.

1.5.2 Photon Energy Measurement

As indicated in the discussion of detection techniques (Section 1.4), the electrical impulses resulting from the detection of x-ray photons by gas-proportional counters, semiconductor ionization detectors, and scintillation counters have average magnitudes proportional to the energy of the detected photons. A plot of number of photons N versus the size of the resulting electrical impulses V can be converted into a spectral distribution by virtue of the relationships $V = kE = khf = khc/\lambda$, where k is an instrumental constant, E is the energy of the photon and f its frequency, h is Planck's constant, c is the velocity of light, and λ is the wavelength of the photon. Such a spectrum is referred to as an energy-dispersive spectrum.

1.5.3 Absorption and Filtration

Since x-ray absorption coefficients are wavelength dependent, measurement of the intensity of an x-ray beam as affected by various known absorbers placed in the beam gives an indication of the spectral composition of the beam. The existence of absorption discontinuities makes it possible to use two or more judiciously chosen absorbers to isolate various portions of the spectrum by subtracting measurements made with the absorbers individually and sequentially inserted into the beam. This is called the Ross balanced-filter technique. Similarly, various detectors having different spectral sensitivities because of the different absorption properties of their active media or other component parts may be used to infer a spectral distribution.

1.5.4 Critical Excitation

If the excitation conditions of the spectrum of interest can be varied, then it is possible to measure the x-ray intensity as the exciting voltage is varied. In particular, as the voltage is increased above the excitation voltage for a given line series, sudden increases in x-ray intensity should be noted.

2 Essentials of X-Ray Absorption Analysis

X-ray absorption analysis depends on Beer's law and the fact that the equivalent x-ray mass absorption coefficient for a homogeneous composite sample depends linearly on the relative concentrations of the elements making up the sample.

2.1 Absorption Edge Spectrometry

The magnitude of the change in the effective mass absorption coefficient at the K absorption discontinuity, for example, of a given element in a specimen is proportional to the concentration of that element. If the density

ρ and the thickness t of the sample are known, then it is possible to calculate directly, without calibration curves, the concentration C_x of the element from the formula $\ln(I_2/I_1) = C_x \rho t \Delta(\mu/\rho)$, where I_1 and I_2 are the intensities transmitted by the sample on the short and long wavelength sides of the discontinuity, respectively, and $\Delta(\mu/\rho)$ is the corresponding change in the tabulated values of the mass absorption coefficient for the pure element.

2.2 Monochromatic Absorptiometry

The transmitted intensity for a given wavelength is measured. If the element under analysis is a major contributor to the absorption coefficient at this wavelength, then an appropriate graph of transmitted intensity versus concentration can be developed. If measurements are made at more than one monochromatic wavelength, particularly if two of the wavelengths closely straddle the absorption edge of the desired element, then more accurate and specific analyses are possible.

2.3 Polychromatic Absorptiometry

If an unmonochromatized beam is used, then variations in overall mass absorption coefficient or specimen density can be determined and can reflect changes in average atomic number of the specimen. This is basically the technique used in x radiography.

3 Instrumentation

3.1 General

For general-purpose elemental analysis only moderate spectral resolution is required, but at present this is attained only by crystal-dispersion instruments. The so-called "nondispersive" instruments, which include energy-dispersive techniques based on semiconductor radiation detectors, do not have sufficiently high dispersion for general-purpose analysis; however, they are quite useful for many problems, and their number and applications are growing.

3.2 Crystal-Dispersion Instrumentation

3.2.1 Basic Components

A schematic representation of the basic components of a photon-excited crystal-dispersion x-ray spectrometer is shown in Fig. 3.

a. X-Ray Tube and Generator. The x-ray tube used in the usual commercial instrument typically operates at a voltage of 30 to 100 kV and a

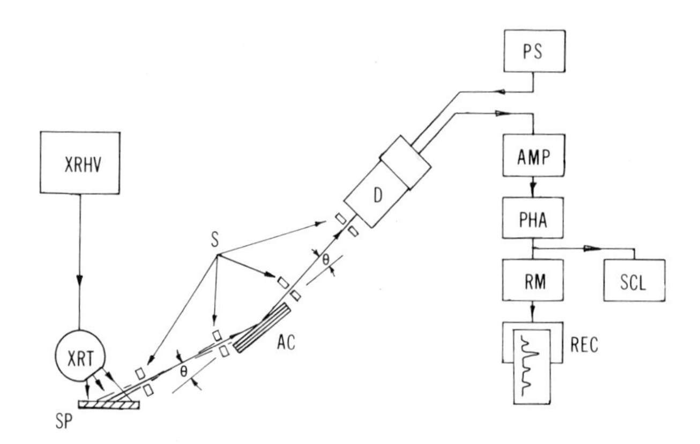

Fig. 3. Schematic photon-excited, crystal-dispersion spectrometer: XRT, x-ray tube; SP, specimen; S, slits; AC, analyzing crystal; D, detector; θ, Bragg angle; XRHV, x-ray high-voltage generator; PS, power supply for detector; AMP, amplifier, PHA, pulse-height analyzer; RM, ratemeter; SCL, scaler; REC, strip-chart recorder.

power of 1 to 5 kW. The tube has a thin beryllium window (0.3–1.5 mm thick) permitting the use of suitably long wavelength x rays for a more efficient excitation of the specimen. The x-ray high-voltage generator includes a high-voltage transformer, x-ray-tube filament transformer, rectifiers, and usually a filter capacitor to smooth the rectified high voltage. Appropriate voltage and current stabilization circuits, and controls for varying the high voltage and current applied to the x-ray tube are provided.

b. Specimen Holder. The specimen is positioned in the instrument in such a manner that it is as close as possible to the x-ray tube consistent with the requirement that its irradiated surface be "visible" to the geometrical optics of the instrument. In order to obtain full utilization of the photon flux from the x-ray tube, the holder should be able to accommodate specimens of at least 5-cm^2 area. The specimen holder often provides facility for spinning the specimen in its own plane so as to average out inhomogeneities in the specimen and the irradiating x-ray source.

c. Analyzing Crystal. Geometrical Optics. The resolution and intensity of the observed x-ray spectrum is determined by the analyzing crystal and geometrical optics. Through Bragg's law the particular crystal and diffracting planes used in the instrument determine the angular dispersion of the x-ray spectrum and the accessible wavelength range. The quality or perfection of the crystal, as well as various inherent diffraction effects, determines how narrow is the interference pattern associated with a particular reflection angle. Typical analyzing crystals are LiF (200 planes), topaz (303), and ethylenediamine-*d*-tartrate (020).

Geometrical or ray optics, as determined by appropriately dimensioned and located slits, make it possible to detect and measure the intensity and wavelength of x rays diffracted at a particular angle. The angular apertures determined by the width and separation of the slits, together with the crystal interference pattern, determine the spectral resolution and intensity of the observed radiation. Overall effective angular apertures are of the order of 0.1°. There is also substantial absorption by the air in the ray-path length (20–30 cm) from specimen to detector for wavelengths beyond 2.5 Å. Such absorption is eliminated by operating the ray path in a helium or vacuum atmosphere, which adds somewhat to the mechanical complexity of the instrument. The crystal, geometrical optics, and detector are mounted on an appropriate mechanical device, such as a goniometer, so that the dispersed spectrum can be observed by variation of the Bragg angle θ. Instruments of this type, of adequate resolution for almost all analytical applications, are available from several manufacturers.

d. DETECTOR. RECORDING SYSTEM. The three most common choices of detector are the Geiger counter, the gas-proportional counter, and the scintillation counter, no one of which covers the entire spectral region of interest. The Geiger counter, which is the least expensive and simplest to operate, gives satisfactory performance only over a limited wavelength region for low counting rates. The gas-proportional counter, particularly when used with continuously flowing gas and very thin windows, has good sensitivity from medium to very long wavelengths. The scintillation counter has good sensitivity from medium to very short wavelengths. The proportional and scintillation counters can handle very high counting rates and offer the possibility of discrimination against higher order reflections but require the use of stable high-gain pulse amplifiers and power supplies. The output pulses from the detector and associated electronic circuits are registered on appropriate scalers or ratemeters to determine the x-ray intensity as measured by the number of pulses per second detected. The ratemeter output is then connected to a strip-chart recorder driven in synchronism with the goniometer to record the x-ray spectrum emitted from the specimen. (A typical recording is shown in Fig. 7.)

e. OTHER COMPONENTS. It is possible and often desirable to include some of the features of a nondispersive instrument, particularly energy-dispersion and filtration techniques. Components for automation are discussed below in Section 7.2.

3.2.2 Flat-Crystal Geometry

An instrument using the slit arrangement shown in Fig. 3 would "see" only a few percent of the total irradiated specimen area. The entire irradiated specimen area can be "seen", with corresponding increase in x-ray

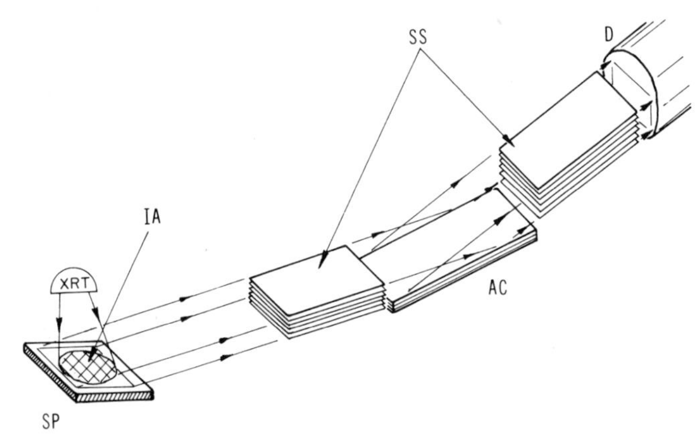

Fig. 4. Flat-crystal geometry: SP, specimen, IA, irradiated area of specimen; SS, Soller slit collimators; AC, analyzing crystal; D, detector.

flux, if the slits shown are replaced with Soller slit collimators, which are sets of long parallel metal foils, as shown in Fig. 4. Any two adjacent foils are equivalent to a pair of slits in Fig. 3. The foils are about 10 cm long, with spacings of 125 to 500 μm depending upon the resolution desired. The cross-section dimensions of a collimator assembly are of the order of 1.5 cm^2. A large flat analyzing crystal, 5 to 7.5 cm long, is then needed in order to intercept at small values of θ all the radiation passing through the collimators.

3.2.3 Focusing-Crystal Geometry

An alternative to the use of Soller slit collimators is to increase the acceptance angular aperture of the geometrical optics by the use of a focusing diffraction geometry, analogous to the Rowland focusing grating spectrometer for visible wavelengths. The crystal is bent to twice the radius of the focusing circle, and the radiation entrance slit, center point of the crystal surface, and radiation exit slit lie on the focusing circle. The focusing obtained is only approximate, unless the surface of the crystal is ground to the radius of the focusing circle. Suitable mechanical devices are required to maintain proper ray geometry while varying the Bragg angle θ (see Fig. 5).

3.2.4 Multichannel Instruments

The normal spectrometer geometry is such that a given set of diffracting planes reflects only one wavelength at a time to the detector. For many applications, however, it is desirable to measure two or more wavelengths simultaneously. There are two principal means to accomplish this purpose. The most commonly used technique is to deploy several independent spectrometers, each with its own crystal, collimators or slits, and detector, and

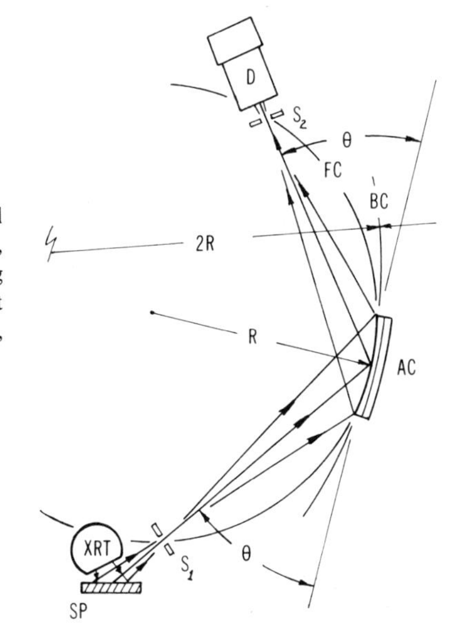

Fig. 5. Schematic of focusing-crystal geometry: XRT, x-ray tube; SP, specimen; S_1, radiation entrance slit; R, radius of focusing circle FC; 2R, radius of bending circle BC of bent analyzing crystal AC; S_2, radiation exit slit; D, detector; θ, Bragg angle.

each of which can "see" the irradiated specimen area. Another techique involves taking advantage of the fact that diffraction of different wavelengths can take place simultaneously from differently oriented atomic planes within the same crystal, by placing appropriate collimators or slits and detectors at the proper angles to receive the diffracted radiation.

3.3 Nondispersive Instrumentation

3.3.1 Basic Components

"Nondispersive" instrumentation is the generic term for instruments which do not use crystal dispersion to study the x-ray spectrum (see Fig. 6).

a. Excitation Sources and Specimen Geometry. For the same strength excitation source and the same source-to-specimen distance, nondispersive techniques give detection signals which are several orders of magnitude larger than those obtained from crystal-dispersion techniques. This is true because it is possible to place the detector very close to the emitting specimen and also because the large intensity losses inherent in the diffracting crystal are eliminated. Thus the various types of radiation from radioisotope sources or miniaturized low-power x-ray tubes may be advantageously employed to excite the sample. Indeed, high-powered x-ray tubes often cannot be directly used because the resulting fluorescence and scattered radiation spectrum would overload the x-ray detector and/or associated electronic circuits.

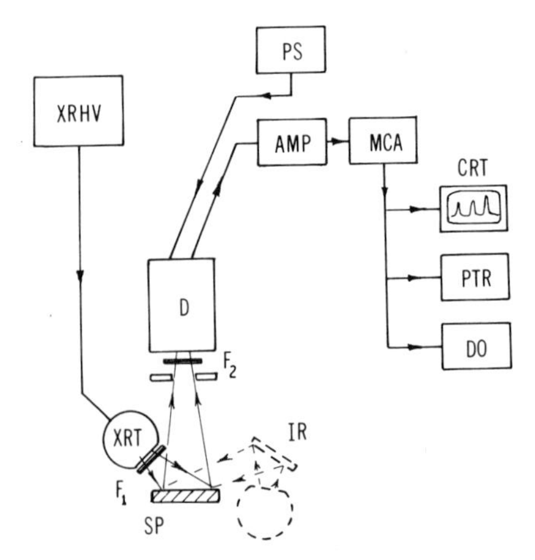

Fig. 6. Schematic "nondispersive" spectrometer elements: XRT, x-ray tube; SP, specimen; D, detector; PS, power supply for detector; AMP, amplifier; MCA, multichannel pulse-height analyzer; CRT, cathode-ray-tube display of MCA memory; PTR, printer; DO, other digital outputs; F_1, F_2, absorber filters; IR, alternative intermediate radiator; XRHV, x-ray high-voltage generator.

b. ENERGY-DISPERSIVE DETECTORS. The utility of the nondispersive techniques depends primarily upon the spectral response and the width of the pulse-height distributions of the detectors used. In nondispersive systems, the resolution of the cooled semiconductor detectors should make possible a few orders of magnitude decrease in minimum detectible quantities of analysis under proper conditions. This resolution is sufficient to resolve the Kα from the Kβ lines of a given element. Unfortunately, the achievement of this decrease is frustrated by the inability of the semiconductor detectors and associated electronics to handle high pulse-counting rates without serious deterioration of resolution. The escape-peak phenomenon mentioned below can also be troublesome. The resolution attainable with a proportional detector is sufficient to resolve the characteristic K emissions of adjacent atomic number elements from each other, when they are present in comparable amounts. The resolution for a scintillation counter is about half as good as that for a proportional counter in this portion of the x-ray spectrum.

In all energy-proportional detectors, the detection of monoenergetic x rays does not always result in a single pulse-height distribution. Sometimes a second distribution, called an "escape peak," is observed, whose average pulse height is smaller than that of the main distribution by an amount proportional to the energy of the K radiation of some constituent of the detecting medium. The "escape peak" is due to the "escape" from the active medium of K x rays excited by the radiation being detected. The effect is more noticeable with increasing atomic number of the constituents of the detecting medium because of the higher fluorescence yield and low absorption. It can complicate the interpretation of an energy-dispersed spectrum.

c. Electronic Analysis Instrumentation. The measurement of the spectral distribution of the radiation detected in an energy-proportional detector is carried out most conveniently with electronic pulse-height analyzers, which may be of the single- or multichannel type. By very rapid measurement of the size of the pulses from the detector, the multichannel pulse-height analyzer can sort all the pulses detected into an energy-dispersive spectrum, as described in Section 1.5.2 above.

d. Filtration and Critical Excitation Components. These accessories modify the spectral response of the detector and/or the exciting photon spectrum incident on the specimen. Thus filters may be placed between the specimen and detector and/or between the x-ray tube and specimen. Critical excitation instruments usually have a smoothly variable but highly stable voltage control to permit settings just below the critical excitation voltage of an element, thereby suppressing its presence in a spectrum. An alternative technique is to use an intermediate radiator, whose fluorescence radiation is excited by a high-powered x-ray tube. The characteristic radiation from the intermediate radiator in turn excites the specimen. By judicious choice of the atomic number of the intermediate radiator, semiselective excitation of the specimen can be achieved.

3.4 Instrumentation for Absorption Measurements

Instruments for absorption measurements make provision for the insertion of a specimen in the x-ray beam and measuring the transmitted intensity. Absorption edge spectrometry requires the use of a crystal-dispersive instrument directly analyzing the continuous spectrum emitted from an x-ray tube. This can be done by inserting a single-crystal plate in the specimen holder of an x-ray diffractometer or by suitably modifying a fluorescence spectrometer. A low-intensity continuous spectrum can be obtained from a normal fluorescence spectrometer by inserting a low atomic number scatterer such as a block of Plexiglas in the specimen position. In any crystal-dispersive instrument, it is highly desirable to use an energy-proportional detector with appropriate pulse-height selection techniques to suppress higher order reflections from the crystal.

Monochromatic absorptiometry is easily done with a crystal-dispersion instrument. It may also be done using any of the nondispersive techniques described above, either singly or in combination, and with appropriate definition of an x-ray beam by slits or collimators. It is preferable that the absorbing specimen be placed between the source and analyzing crystal, or as far as possible from the detector in order to minimize the effects of scattering and fluorescence in the specimen.

Polychromatic absorptiometry requires only definition of the x-ray beam.

4 Literature

4.1 Reference Texts for X-Ray Fluorescence Spectrometry

Several books specifically on x-ray fluorescence spectrometry have been published in English during the past ten years. Bertin (1970) has written a particularly comprehensive and detailed book with an extensive bibliography. Jenkins and deVries (1969) have also written a rather complete book of about one-third the length. Two other recent books are by Adler (1966) and Birks (1969). A book originally written in 1967 in German has recently been translated into English (Müller, 1972). Liebhafsky *et al.* (1972) have written a book on a variety of x-ray analytical techniques. An older book by Liebhafsky *et al.* (1960) is still useful. The classic text by von Hevesy (1932) is considerably out of date.

4.2 Reference Texts on X-Ray Physics

The most recent book-length publications on general x-ray physics are a translation of a Russian text by Blokhin (1961) and an elementary book by Brown (1966). Volume 30 of the *Handbuch der Physik*, edited by S. Flügge (1957) is devoted to x rays, with all but the first article in English. Older books are by Worsnop and Chalklin (1950) and the classic, but still useful, work by Compton and Allison (1935). A handbook edited by Kaelble (1967) is also useful, as is the book by Clark (1955).

4.3 Periodicals and Serials

The proceedings of the annual Denver conferences on the application of x-ray analysis, published under the title *Advances in X-Ray Analysis* are a source of much useful information. The greatest number of individual papers on x-ray spectrochemical analysis appear in *Analytical Chemistry, American Mineralogist, Spectrochimica Acta, Applied Spectroscopy,* and *Analytica Chimica Acta.* A new journal specifically devoted to x-ray fluorescence spectrometry, *X-Ray Spectrometry*, published by Heyden & Son, Ltd., London, has recently been announced. An abstracts service has also been established (Masek *et al.*, 1970). Extensive bibliographies have been published in the Analytical Reviews numbers of *Analytical Chemistry* (Birks, 1972; Campbell and Gilfrich, 1970; Campbell and Brown, 1968), and by the American Society for Testing Materials, the U.S. Bureau of Mines, and the Norelco Reporter. Papers on applicable physics and instrumentation have appeared in various journals devoted to applied physics or scientific instrumentation.

5 Applications

5.1 Qualitative Uses

5.1.1 Selectivity

In almost all cases only sufficient resolution to partially separate the components of the Kα doublet is required. In a very large percentage of applications, even less resolution is needed and the instrument need only be capable of separating the Kα line of element Z from the Kβ line of element $Z + 1$. Crystal-dispersion instrumentation gives adequate resolution to yield qualitative and semiquantitative analyses on any sample for individual specific elements within the range of sensitivity of the particular instrument employed. This is particularly true when use is made of the nondispersive techniques and devices which are usually part of such instrumentation. Commercial instruments which cover the range of elements from atomic number 11 on up are available, as well as special instruments which extend the range downward to atomic number 5 or 4. The selectivity of the energy-dispersive techniques alone often depends upon the number and concentration of elements and their proximity to each other in the periodic table. Thus the semiconductor ionization detectors may be extremely useful for elements having atomic number 12, 13, or higher in small or dilute samples, when concentrations of adjacent elements in the periodic table are not too disparate from each other, and when it is not necessary to deal with both K- and L-series spectra in a small spectral range. On the other hand, nondispersive techniques based on the gas-proportional and scintillation counters are not very specific for adjacent elements in the periodic table unless they are present in comparable amounts. If the determination of a specific atomic number Z is not required, but rather $Z \pm 2$, then energy-dispersive techniques alone are quite useful. Energy-dispersive instruments based on the gas-proportional counter can in principle have an elemental range from $Z = 5$ or 6 on up, while scintillation counter instruments range from about $Z = 20$ on up.

The use of critical excitation and judiciously applied absorption techniques may enhance the selectivity of nondispersive or energy-dispersive instrumentation for specific elements.

5.1.2 Sensitivity

Sensitivity is proportional to the slope of the calibration curve relating observed intensity of the element of interest to the concentration of the element in the specimen matrix. The larger the slope, the greater the sensitivity. The slope depends upon the inherent spectral line intensity per atom of the pure element, and the effect of the specimen matrix on the apparent intensity finally detected. The inherent intensity per atom depends upon the atomic number and spectral line excited, and upon the excitation conditions. The matrix always absorbs some of the exciting radiation as well as some of the emitted intensity. It may also enhance the emitted intensity if other elements in the matrix emit x rays which can excite the element of interest. Thus the sensitivity for aluminum in a matrix of average atomic number 29 is one-third the sensitivity for aluminum in a matrix of average atomic number 7.

At very low concentrations, the sensitivity determines the ultimate minimum detectible quantity of the element. Statistical considerations show that the ultimate minimum detectible quantity is inversely proportional to a figure of merit (f.m.) given by the ratio of the slope of the calibration curve P to the square root of the background intensity B observed from the specimen matrix at the spectral line when the element is absent; i.e., f.m. $= P(B)^{-1/2}$.

Two of the major factors contributing to the background intensity observed from the specimen matrix are the intensity of a partially overlapping adjoining spectral line and the scattered radiation registered in the x-ray detector. Both of these can be affected by resolution techniques. Even the crudest energy-dispersive or filtration techniques introduced into a wavelength-dispersive instrument can have a marked effect on the background radiation.

The minimum detectible limit for wavelength-dispersive instruments varies smoothly from about 100 ppm at atomic number 11 or 12, to 1 ppm at atomic number 30, to 10 ppm at atomic number 50, to 20 ppm at atomic number 96. In energy-dispersive instruments using cooled semiconductor detectors, minimum detectible limits approaching those for wavelength-dispersive instruments can be obtained if the specimens are not too complex.

For analysis by x-ray absorption, qualitative and quantitative analyses can be estimated from the change in apparent absorption coefficients. The minimum detectible limit is of the order of 1% by weight for elements similar in atomic number to the average atomic number of the matrix. If the element of interest is much heavier than the matrix, the sensitivity and minimum detectible limit may improve tenfold.

5.1.3 Size and Kind of Specimens

There is no inherent limit to the physical dimensions of specimens which can be analyzed, other than restrictions which may be imposed by the actual configuration of the instrument. The maximum size specimen for most instruments is of the order of 7- or 8-cm^2 area, and 2- or 3-cm thickness. For qualitative analysis, shape is not significant. Suitable adapters or fixtures are furnished or readily fabricated to handle small specimens or odd-shaped specimens. In order to obtain maximum sensitivity for small quantities of specimen, if the choice is possible, it is preferable that the specimen be spread over the useful irradiated area (usually of the order of 5 cm^2 or less) and be correspondingly thin, rather than concentrated into a thick lump.

Quantities as small as a few micrograms can be analyzed if the element of interest is a major constituent; in particular, thin evaporated films down to 50- or 100-Å thickness can be analyzed. In absorption, the quantity of specimen analyzed depends very much on the element, matrix, and technique used. In cell tissue, analyses for calcium and sulfur detect quantities as small as 10^{-12}g (Liebhafsky *et al.*, 1960, p. 300).

For fluorescence spectrometry the specimen may be in the form of amorphous or crystalline solid, powder, grease, or fluid. It is even possible to devise a cell for analyzing gases. For qualitative analysis the specimen surface need not be particularly smooth or clean, except for detection of light elements less than about atomic number 20. If a vacuum spectrometer is used for the light elements, there may be some concern about specimens which volatilize under vacuum. In such cases a thin cover may be required over the surface of the specimen, and this will significantly reduce or obliterate sensitivity for the lightest elements. Often this difficulty can be overcome by the use of a helium atmosphere rather than a vacuum atmosphere in the instrument. For absorption measurements, the specimen may be solid, liquid, or gas but should be reasonably homogeneous and uniform in thickness.

5.1.4. *Nonconsumptive Nature of Measurement*

The fluorescence spectrometry technique, like all x-ray techniques, is essentially nonconsumptive of the specimen. Thus repeated tests may be made on the same sample. There is some possibility of apparent radiation damage to the sample, which, however, is often reparable by infrared bleaching techniques, for example. Occasionally, it is also necessary to allow for radiation-induced precipitation of components of the specimen from solution, or for an increased rate of oxidation of the specimen surface due to heating of the specimen by the infrared radiation from the target of high-powered x-ray tubes. This is much less of a problem for x-ray absorption measurements, particularly for monochromatic absorptiometry and absorption edge spectrometry.

5.1.5 *Time Required for Analysis*

a. SINGLE SPECIFIED ELEMENT. After insertion of the specimen into an operating instrument, it takes less than a minute to set the instrument to the wavelength of any desired spectral line and determine, using any convenient method of intensity measurement, whether this line is significantly present or absent; thus an indication that the corresponding element is present as a major or minor constituent, or as trace or less, is obtained.

b. Range of Unspecified Elements. The time required to examine an unknown sample for a range of elements is the time required to produce a ratemeter recording of the spectral range appropriate to the elemental range. It takes from one-half to two hours, depending upon the scanning speed chosen, to scan over the entire spectral range accessible to a given analyzing crystal used in a wavelength-dispersive instrument. For a LiF crystal, this would include all elements from atomic number 20 on upward. A more restricted range of elements is scanned more rapidly. Identification of all the spectral lines in a given record may take from five minutes to an hour, depending upon the complexity of the specimen. However, it is also possible to index or identify the various spectral lines in real time, as the spectrum is being recorded, in which case essentially complete identification of spectral lines takes only a few minutes longer than the actual recording of the spectrum.

Since the sample is not consumed by the analysis, one technique used for qualitative analysis is to make several rapid scans through the spectral range, appropriately varying instrumental parameters and/or using auxiliary energy-dispersive techniques, such as pulse-height discrimination, critical excitation, and absorption.

In some energy-dispersive instruments, such as the cooled semiconductor detectors coupled with multichannel pulse-height analyzers, the time for multielement qualitative analysis is essentially no longer than that for single-element qualitative analysis of the minor constituents of interest because all parts of the spectrum are recorded simultaneously.

5.1.6 X-Ray Absorption Spectrometry

Measurements in x-ray absorption spectrometry are carried out more rapidly than for x-ray fluorescence spectrometry.

5.2 Quantitative Uses

5.2.1 Selectivity

The selectivity of the method for quantitative analysis is generally the same as that for qualitative analysis, as discussed in Section 5.1.1.

5.2.2 Sensitivity

The minimum detectible limit for quantitative analysis is also the same as that for qualitative analysis (Section 5.1.2), but modified in terms of statistical considerations as to the confidence level with which such a limit can be expressed. The ultimate limiting factor, which is often reached in practice, is the statistical fluctuation in the number of photons detected in a reasonable measuring interval. There are many cases, however, where random or systematic errors in sample presentation prevent the achievement of this ultimate limiting factor.

Since the sensitivity for changes in the concentration of an element in a sample is proportional to the slope of the curve relating concentration and intensity of the spectral line, it too is ultimately dependent on statistical fluctuations in the number of photons detected. The smaller the slope of the curve, the greater the relative precision that is required of the intensity measurement for a desired sensitivity. For minor and major constituents, it is relatively easy to accumulate a sufficient number of counts in the detection system for any reasonably desired precision. The precision then becomes limited by fluctuations arising from the nature of the actual sample or its preparation, or by instrumental instabilities.

5.2.3 Accuracy

Assuming a properly functioning instrument, accuracy of analysis is determined by the accuracy of calibration. Because of the specimen preparation problems discussed below and the absorption-enhancement effects of the matrix on the observed intensity of a given concentration of element, as already mentioned in Section 5.1.2, there is no universally useful instrument calibration procedure available, but rather specimen calibration curves must be established for each type of specimen to be analyzed. In the external standards technique, x-ray fluorescence measurements are made on a sufficient number of independently analyzed or synthesized specimens to establish the shape of the plot of concentration versus x-ray line intensity. It is assumed that the unknown samples are sufficiently like the standard samples in overall composition and preparation that they will fall on the same plot, and that the only significant difference between and among the unknowns and the standards is the concentration of the single element of interest. Often, this is not the case, and the resulting deviations must be allowed for by corrections based on other measurements or information, or else different calibration curves must be constructed to allow for significant variations in the matrix.

In terms of basic physics, one of the major factors causing inaccuracy and imprecision in specimen preparation is the substantial amount of absorption of x-rays by the specimen. Because of absorption there is an effective sampling depth, below which specimen material does not contribute significantly to the observed spectrum. This depth is roughly equal to the reciprocal of the linear absorption coefficient of the specimen for the emitted wavelength or energy of interest. For pure Cu the sampling depth is 20 μm for the Cu Kα line, while for 10% Cu in Fe it is about 4 μm for the Cu Kα line. For a few percent of Al in Fe, the sampling depth is only about 0.5 μm for the Al Kα line.

For accurate analysis, therefore, it is necessary that the uppermost layer of the specimen of the thickness of the sampling depth be both representative of the bulk of the specimen and homogeneous. Thus, direct measurements on powders or minerals which may have their constituents significantly segregated among different grains or particles are subject to large inaccuracies unless calibration standards are prepared from similarly segregated specimens. These problems become particularly severe when specimens must be ground to powders which are then packed in a holder or briquetted. Similarly, specimen surface finish requirements are affected by this sampling depth (a few scratches, of course, are not significant). Polishing of specimens must be carried out in such a way as to avoid "smearing" or gouging out soft constituents over the sampling depth. It may be necessary to fuse or dissolve the specimen in order to obtain sufficient homogeneity.

Sometimes it is not possible or convenient to obtain suitable external standards for the preparation of calibration curves. It is sometimes advantageous in such circumstances to make use of an internal standard, which may already be present in the sample or which may be added in the specimen preparation process. The internal standard may even be some particular wavelength emitted by the x-ray tube, and which is scattered by the specimen. The analysis is then made in terms of the relationship between the concentration of the element of interest and the ratio of its line intensity to the line intensity of the internal standard. The method assumes that the relative effects of matrix and other variations on both the spectral line of interest and the internal standard are either identical or known. Known amounts of the element of interest may be added to the specimen, in which case it is assumed that the general shape of the calibration curve is known and is not changed by the addition. The added amounts serve to particularize the general shape. The accuracy of the added internal standard technique is of course affected by the various specimen preparation factors already mentioned.

5.2.4 Precision

a. Counting Statistics. As already indicated, the ultimate limit on the precision of the x-ray technique is set by fluctuations in the number of photons detected over a given time interval. It is assumed that the arrival of photons at the x-ray detector is governed by a Poisson distribution law. Thus if N x-ray photons are detected in a given time interval, then the relative standard deviation of N is $N^{-1/2}$. To obtain a 1% relative standard deviation requires that $N = 10{,}000$. If it is necessary to correct the observed intensity for background radiation, then the relative standard deviation of the net intensity will be substantially greater than 1%. For example, if a particular spectral line has an intensity of 50 counts/sec, then 200 sec are required to accumulate 10,000 counts. If the background intensity is 40 counts/sec, 250 sec are required to determine it to a relative standard deviation of 1%. The net intensity (10 counts/sec), however, will have a relative standard deviation of about 6% (Jenkins and deVries, 1969, Chap. 5). To achieve a relative standard deviation of 1% in the determination of net intensity, a total measurement time of 36 times (200 + 250) = 16,200 sec = 4.5 hr is required.

b. REPRODUCIBILITY AND STABILITY OF INSTRUMENTATION. In order to attain the limiting precision imposed by counting statistics, it is necessary to minimize fluctuations due to instrumental instability and irreproducibility. The sources of these fluctuations are of two types: (1) instabilities in the actual construction (electrical, mechanical, etc.) of the instrument and (2) changes in external parameters, such as electrical line voltage, atmospheric pressure, room temperature, and vibrations. The effect of many of these changes on x-ray output or on x-ray detection is often greatly magnified. For example, a 1% change in the high voltage applied to an x-ray tube may lead to a 2% change in the intensity of a spectral line; a 1% change in the voltage applied to a flow-proportional counter may cause a 10% to 20% change in the average pulse size of a particular pulse-height distribution; a 1% change in flow counter gas density, which may arise from either a temperature change of 3°C or an atmospheric pressure change of 8 Torr, may cause a change in average pulse height of 3 to 8%. Thermal expansion of analyzing crystals causes a shift of expected reflection angles. In an instrument using a LiF crystal, at standard resolution, a 3°C temperature change would cause a decrease in observed intensity of about 4% for Ti $K\alpha$ (2.75 Å) and 13% for Sn $L\alpha$ (3.6 Å). These effects may be minimized in appropriately designed and operated instruments and measuring techniques. Internal standards, ratio measurements, or monitoring techniques are sometimes useful in this regard. Nevertheless, over long periods of time, it is difficult to achieve instrumental stabilities or day-to-day reproducibilities much better than 0.5 to 1%.

c. REPRODUCIBILITY OF SPECIMEN PREPARATION. The considerations involved in reproducibility of specimen preparation are similar to those involved with accuracy of specimen preparation (Section 5.2.3). Quite often the resulting imprecision is significantly greater than that due to counting statistics and the instrument. Sometimes, inhomogeneities over the area of the specimen can be averaged out by continuously rotating the specimen in its own plane while in the irradiating beam.

5.2.5 Size and Kind of Specimens

In general the same size and kind of specimens may be used for quantitative as for qualitative analysis. Considerably more care must be taken, of course, with regard to shape, surface finish, and positioning of the specimen in the instrument in order to assure appropriate precision and accuracy for comparison with standards, although the use of internal standards or ratio measurement sometimes helps to minimize the problems.

5.2.6 Nonconsumptive Nature of Measurement

The nonconsumptive nature of x-ray spectrometry techniques is particularly advantageous for quantitative analysis, because it makes possible repetitive measurements on the same sample over a long period of time, according to a variety of strategies and techniques. Quantitative measurements are somewhat limited, however, in those few specimens where radiation-induced precipitations take place. Radiation-induced changes which do not alter the overall elemental composition of the specimen within the sampling depth will have only slight effect on the x-ray analysis.

5.2.7 Time Required for Analysis

a. EFFECT DUE TO AMOUNT OF SPECIMEN. The instrumental time required for quantitative analysis is determined by the time required to detect the requisite number of photons for achieving the desired counting statistical precision. If the specimen is small, only a few milligrams or a few square millimeters in area, the intensities of the various fluorescence spectral lines will necessarily be low; and it will take a longer time to measure the intensities. Typical measurement times vary from 1 to 5 min, allowing for the time taken to push various buttons on the instrument, to read and record data, etc. More significant, however, will be the time required to prepare the specimen. If extensive grinding, sifting, mixing, surfacing, or chemical treatment is required, preparation times range from 5 to 30 min or more.

b. EFFECT DUE TO CONCENTRATION AND NUMBER OF COMPONENTS. If the concentration of element to be analyzed is small, particularly approaching trace quantities, then it may take a fairly long time to carry out a measurement to the desired precision. In most applications, with the majority of instruments for which measurements are made sequentially, 15 to 30 min is the maximum time that should be devoted to a single intensity measurement. If several elements are to be determined, then the total measuring time in sequential instruments is the sum of the measuring times for each element. In a multichannel instrument, on the other hand, the measuring time is determined by the time for that element which individually requires the longest time.

c. Effect of Counting Statistics. As indicated in Section 5.2.4a, the measuring time is determined by the statistical precision desired, and by the absolute and relative values of the spectral line intensity and background. Thus, the determination of a net line intensity of 10 counts/sec above a background of 40 counts/sec to a relative standard deviation of 6% requires 7.5 min; whereas a relative standard deviation of 1% requires 4.5 hr. On the other hand, a relative standard deviation of 12% would require less than 2-min measuring time.

d. Effect Due to Nature of Instrumentation. The particular instrumentation used, of course, determines the time required for an analysis. Multichannel instruments can make several or all spectral measurements simultaneously, although in some cases their speed per channel is reduced. Some instruments and instrumental techniques are more luminous than others. The problems of instrumental stability and drift mentioned in Section 5.2.4b may limit the amount of time that can be devoted usefully to intensity measurements. Some instabilities are more serious for sequential measurements than simultaneous measurements, whereas others are of greater consequence for simultaneous measurements. For example, an energy-dispersive instrument, using a cooled semiconductor radiation detector and multichannel pulse-height analyzer, with appropriate data treatment techniques can be used for measurements on a single sample which require several hours.

For comparing various instrumental arrangements, it is sometimes useful to consider a f.m. based on the ratio of net peak intensity to the square root of the background intensity. While this was introduced in Section 5.1.2 as being related to the ultimate minimum detectible limit of a particular element in a specimen, it may also be used as a basis for predicting or comparing required measurement times. The measurement time is inversely proportional to the f.m.

5.3 Literature Examples of Applications

The scientific literature contains reports of applications of x-ray fluorescence spectrometry to virtually every element of the periodic table with the exception of hydrogen, helium, and the noble gases (Campbell and Gilfrich, 1970; Campbell and Brown, 1968). Similarly, applications to problems involving alloys, glasses, powders, ores, plastics, slags, soils, brines, liquids, fibers, wood, etc., are thoroughly documented. The book by Müller (1972) contains a number of abstracts of papers (through 1964) describing applications to problems in a number of these areas, as well as in the additional areas listed below. While it is naturally expected that the technique is widely used in physical and materials sciences, the literature also describes applications to archaeology, air and water pollution studies, biology, clinical chemistry, criminalistics, lunar soils analysis, pharmaceuticals, and photography. Increasingly elaborate and complex applications to on-line process control have been described and predicted (Campbell and Gilfrich, 1970).

6 Data Form and Interpretation

6.1 Strip-Chart Records for Wavelength-Dispersive Instruments

6.1.1 *Form of Records*

As indicated in Section 3.2.1d, the detection of photons in an x-ray detector results in the development of a voltage that is proportional to the intensity of detected x rays. This voltage is used to deflect the pen of a strip-chart recorder, whose chart drive is driven in synchronism with the mechanism that varies the Bragg angle θ and hence the wavelength λ being detected. On most instruments, a recording is developed of intensity versus 2θ, as a spectrum is scanned. Figure 7 shows a chart recorded for a steel alloy sample.

6.1.2 *Indexing the Chart*

a. Tables of Elements versus 2θ. If the d spacing of the analyzing crystal is known, Bragg's law may be used to calculate the wavelength λ (or $n\lambda$) of a particular spectral line, which is then identifiable by means of a table of wavelengths and elements. It is much more convenient and simple, however, to use a table of elemental spectral lines versus 2θ for the

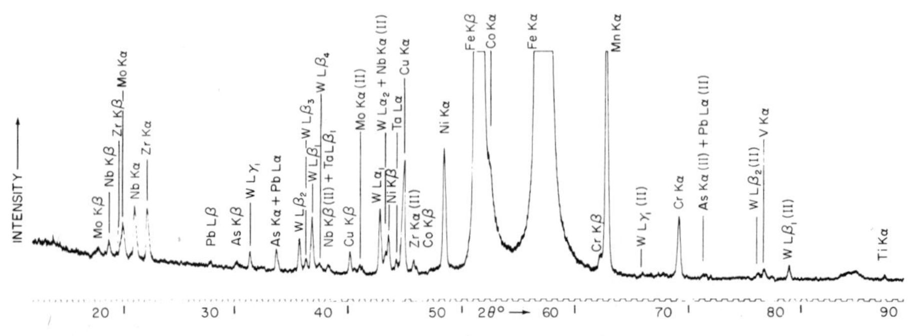

Fig. 7. Strip-chart recording of fluorescence x-ray spectrum of low-alloy steel, using air-path spectrometer, LiF(200) analyzing crystal, and scintillation detector without pulse-height discrimination. [From Spielberg, 1963.]

particular crystal to determine the spectrum directly. Because of the basic simplicity of the x-ray characteristic spectrum, it has been relatively easy to calculate by computer and publish such tables for all the analyzing crystals commonly used (Amsbury *et al.*, 1964; Bearden, 1964; General Electric, 1969; White *et al.*, 1965).

b. CRITERIA FOR SPECTRAL ASSIGNMENTS. In order to assign an observed spectral line to a given element, it is not sufficient to use the mere correspondence within experimental error of the measured and tabulated values of 2θ, particularly in view of the fact that usually more than one correspondence may be determined within experimental error. It is necessary also to verify the presence of other appropriate spectral lines for the element in the observed spectrum, as well as the presence of other orders of reflection of the same lines (if the instrument parameters and chart range permit their registration). The relative intensities of the various spectral lines and orders of reflection should also be consistent with each other, after matrix, resolution, and instrumental effects are allowed for. For example, in Fig. 7, the assignment of the spectral line at $2\theta = 20.3°$ to Mo $K\alpha$ requires that the $K\beta$ line be observed at $2\theta = 18.1°$, at one-fifth to one-eighth the intensity of the $K\alpha$ line. Similarly, it is expected that with the analyzing crystal used, the second-order reflection (i.e., $n = 2$ in Bragg's law) of the Mo $K\alpha$ line should be observed at an angle of $2\theta = 41.2°$ and considerably less intensity. The second-order reflection of Mo $K\beta$ is not expected to be observed because its intensity would be too low on the scale of Fig. 7.

c. INSTRUMENTAL EFFECTS ON OBSERVED SPECTRA. Not all the spectral features observed in a recorded spectrum can be attributed to simple dispersion of the spectrum excited in the specimen. Figure 7 also shows lines from the tungsten L spectrum. These lines are generated in the tungsten target of the exciting x-ray tube and are coherently and incoherently scattered by the specimen; thus they are detected along with the x rays emitted by the sample (see Sections 1.2.2 and 1.2.3). Similarly, the broad hump at about 84.5° 2θ is ascribed to scattering or diffraction by the specimen of a portion of the continuous or white spectrum of the x-ray tube. Effects of this type are particularly marked in specimens whose mean atomic number is less than about 20 or for thin specimens on a thick substrate. It is also possible, with certain analyzing crystals, to observe diffraction effects from crystal planes other than those normally used.

6.1.3 Determination of Spectral Intensities

Spectral intensities are obtained by measuring the height of the various spectral lines recorded on the chart. It may be necessary to make allowance for systematic recording errors introduced at a given rate of scan of the spectrum because of the response time of the complete detection system and electronics. Similarly, the recorded intensities will contain errors due to counting statistics.

6.2 Multichannel Pulse-Height-Analyzer Display for Energy-Dispersive Instruments

The multichannel pulse-height-analyzer display for energy-dispersive instruments is much simpler than that for the wavelength-dispersive instruments. The spectrum is displayed directly on an energy scale, and one needs only a table of energies of spectral lines for the elements. If the instrument has sufficient resolution, identification of a given spectral line requires the presence of other lines of the same spectral series and of the proper relative intensities, just as for the wavelength-dispersive instruments. Higher order reflections are not observed since no analyzing crystal is used; however, diffraction and scattering effects by the specimen may be expected, particularly if x-ray-tube excitation is used. Spurious peaks due to the escape-peak phenomenon discussed in Section 3.3.lb may be observed, particularly in cooled germanium detectors, xenon- and krypton-filled proportional counters, and scintillation detectors.

6.3 Digital Readouts. Computer Applications

6.3.1 Digitization of Spectral Information

a. X-Ray Intensity. Since almost all x-ray detectors used for spectrometry detect photons as discrete electrical pulses, appropriate counting and interface circuits make it possible to preserve and read out the intensity in digital form. The actual readout may be in the form of digital display, printed tape or sheet, magnetic or punched tape, or card. The time interval over which the pulses are counted may also be read out.

b. Bragg Angle or Wavelength. By use of shaft angle encoders, rotation counters, or the accumulated count of incrementing pulses from a stepper motor, the angles, θ, 2θ, or wavelength λ can be digitized. These too can be read out in any form, interspersed or not, as desired, with the corresponding intensity information.

c. Scanning Modes. The goniometer or wavelength-scanning device may be operated in either a continuous or step-scanning mode. In the continuous mode, electrical pulses from the detection system are accumulated for a short time interval and read out digitally; then pulses are accumulated again, read out, etc., while the goniometer scans continuously. The digital intensity information represents an average of the intensity over the small spectral region scanned during each accumulation period. In the step-scan mode, the instrument is stationary at a given spectral position while the intensity data are accumulated. At the end of each accumulation period, the angle or wavelength is incremented and another accumulation period is begun. Digital data obtained in either of these two modes can be plotted to yield a histogram of the spectrum.

d. MULTICHANNEL ANALYZER FOR ENERGY-DISPERSIVE INSTRUMENTATION. By the very nature of the multichannel analyzer, the spectral information is obtainable in digital form, just as the corresponding display is a histogram.

6.3.2 Quantitative Analysis

a. CALIBRATION CURVES. The digital output of x-ray intensities makes it very convenient to use analytical rather than graphical calibration curves to obtain final numerical chemical analyses of a specimen. Many instruments incorporate simple calculating devices, so that it is possible to use measurements of the spectral standards to set up the parameters for the analytical curves directly on the instrument, and to obtain directly from the instrument printouts of the chemical analysis in terms of percent concentration. Alternatively, digital intensity information may be taken to a digital computer, either directly or by means of tape or cards, for calculation of the results.

b. COMPLETE SPECTRAL ANALYSIS. The availability of high-speed computers opens up the possibility of complete spectral analysis based upon analysis of all the x-ray spectral information and any other external information, such as grain size, the presence of constituents not included in the x-ray analysis, and heterogeneity. In principle, the computer can be used to index a spectral chart or digital spectral intensity data and to assign each spectral line to an element in the sample to make a qualitative analysis. The computer could then carry out the necessary calculations to determine a quantitative analysis. There are, however, a number of problems to be solved beforehand, not the least of which is the development of a general procedure for carrying out calculations. Current efforts in this direction are confined to samples which show no grain size or heterogeneity effects (Birks, 1972).

7 Accessories and User Modifications

7.1 SPECIMEN PREPARATION AND PRESENTATION

7.1.1 Specimen Preparation Devices

Appropriate equipment is required for specimen preparation, particularly if quantitative analyses are to be carried out. Almost all of the necessary devices have been developed for use with other analytical instrumental techniques and are readily available. For solid specimens, appropriate cutoff and cutting tools, sanders, and grinders are necessary. For powders, ores, slags, etc., mills, shakers, ovens, and briquetting devices are required. Various scales, microscopes, stock supplies, and reagents normally found in analytical chemistry laboratories are also needed. An extensive discussion with detailed references is given in the book by Bertin (1970, Chap. 16).

7.1.2 Specimen Holders and Fixtures

A large number of specimen holders and fixtures have been developed for various purposes (Bertin, 1970, Chaps. 16, 17, 19). It is often convenient to modify the sample presentation components of the apparatus in order to handle special shapes of specimens, to maintain the specimen at a particular temperature, or to handle a large number of specimens. Devices have also been developed to mask all but a small part of a specimen and then to translate the specimen beneath the mask, making possible a "point-by-point" (the point is a few square millimeters in size) analysis of a specimen surface. Many of these devices and modifications are available from equipment manufacturers.

7.2 Automation. Computer Control

7.2.1 Degrees of Automation

a. General Considerations. Because techniques and devices for automation and control of x-ray instrumentation are in a state of rapid development, generalizations are difficult. A number of interrelated decisions must be made before carrying out the design and construction of an automated or computer-controlled instrument. If the kind of automation desired is rather simple, and a fixed repetitive program is to be used, then a tape- or internal-hardware-controlled machine may be satisfactory. If maximum flexibility and reaction of the control system to measurements on the specimen are desired, then computer control is needed. For computer control, it must be decided whether the control is by a computer dedicated solely to the operation of the x-ray instrument, or by a large computer involved with a number of other instruments or applications on a time-sharing basis. Among the considerations involved in this decision is the question of overall management and economics of the entire analytical system.

b. Automation of Simple Functions. Many devices and methods for automating simple functions have been developed over a number of years. These include automatic initiation and termination of intensity accumulation times, based on a preset time or preset total count condition; automatic printout of intensity data; automatic setting of the instrument to predetermined reflection angles; automatic sample changing; automatic change of resolution conditions, including selection of analyzing crystal and of Soller slit collimators; automatic choice of detector and pulse-height-selection parameters; automatic calculations of chemical concentrations based on intensity measurements.

c. AUTOMATION OF COMPLEX FUNCTIONS. More complex functions susceptible to automation include selection of excitation conditions through choice of x-ray-tube target, kilovoltage, and current; selection of filters on x-ray tube and detector; and selection of scanning speeds.

d. PROGRAMMED PARAMETERS. Programmed control of the various automated functions can be carried out with various cams, contour followers, timing devices, automatic switches, or tapes. Alterations in the program are accomplished by changing cam shapes, switch settings, patch boards, or tapes. Alternatively, digital control can be programmed more powerfully, flexibly, and complexly with a digital computer, which because of its decision-making capability can use experimental measurements to determine and modify the program to be followed for a number of automated functions simultaneously.

7.2.2 Equipment Considerations

a. BASIC INSTRUMENT DESIGN. Automation places severe demands upon instrumental precision, reproducibility, and stability, and great care must be taken to insure that the various mechanical and electromechanical devices introduced to accomplish this purpose do not interfere with the satisfactory operation of the basic instrument.

b. INTERFACING DEVICES. The choice and design of various readout devices, transducers, and interfacing devices must be carried out with the ultimate control mechanism in mind. Such considerations as the number and kind of "priority interrupts" available from a particular computer, and the effect of a power failure on the memory of the computer will determine the number and kind of interfaces required.

7.2.3 Programming Considerations

a. ANALYSIS OF OPERATING CONDITIONS. It is often as expensive, or even more expensive, to develop the necessary "software" (programs) for a computer-controlled instrument as it is to develop the hardware. Programming for decision making above the simplest level can be quite intricate. A program to direct an instrument to seek out the peak of a particular spectral line involves detailed consideration of counting statistics and spectral line shapes. Analysis and listing of all aspects of machine operation and data interpretation and an understanding of interrelationships between hardware, software, and basic instrument factors can be quite significant. For example, the mechanical design of many instruments requires that spectrum scanning be carried out in the direction of decreasing 2θ in order to take up backlash in the various components. On the other hand, spectral analysis is considerably simplified if scanning is done in the direction of increasing 2θ because identification of higher order reflections of a given spectral line can be anticipated if the lower order reflections are known to be present. Under these circumstances an appropriate program for scanning an entire spectrum would drive the instrument upward from low values of 2θ, determine the presence of a spectral line, overshoot it, and return to it from the proper direction in order to determine its peak position. Alternatively, modification of the basic instrument, if possible, may simplify the programming demands.

b. TYPES OF PROGRAMS. The number and variety of programs that may be required or possible in a computer-controlled instrument can be quite extensive. These are over and above the programs required to operate the computer itself.

Master programs give overall control and direct the system through various operations and subroutines. For example, such a program might call for the quantitative analysis of six particular elements in high-alloy steels.

Parameter setting and initialization programs are basically check lists which require the operator to supply information and make certain basic settings in order to initialize a programmed activity. This might include sample identification and specification, choice of type of analysis and information desired, instrumental parameter settings, etc. These are often written in conversational format.

"Notebook" function programs use the computer and its memory as a "notebook" to preserve information.

Control function programs actually set and check various instrumental parameters, scanning speeds, detector accumulation times, etc., according to information supplied by other programs.

Calculational programs calculate useful information required by the other programs, such as necessary pulse-height setting as a function of spectral position, intensities in terms of counts per second, statistical factors, chemical compositions according to various calibration and correction functions, angles and wavelengths according to Bragg's law, atomic number and wavelengths according to Moseley's law, or some other relationship.

Indexing and scanning programs direct the instrument through a variety of wavelength positions, either continuously as in a scan of a spectral region at a variety of speeds, or by indexing to certain spectral positions.

Also included would be programs for comparing observations from a particular specimen, with spectra already stored in the computer or in the "notebook."

Publishing programs output the desired information in printed or other form.

References

Adler, I. (1966). "X-Ray Emission Spectrography in Geology." Elsevier, Amsterdam.

Amsbury, W. P., Lee, W. W., Rowman, J. H., and Walden, G. E. (1964). X-Ray Fluorescence Tables. USAEC Rep. Y-1470.

Bearden, J. A. (1964). X-Ray Wavelengths. USAEC Rep. NYO-10586.

Bertin, E. P. (1970). "Principles and Practice of X-Ray Spectrometric Analysis." Plenum Press, New York.

Birks, L. S. (1969). "X-Ray Spectrochemical Analysis," 2nd ed. Wiley (Interscience), New York.

Birks, L. S. (1972). *Anal. Chem.* **44,** 557R-562R.

Blokhin, M. A. (1961, 1957). The Physics of X-Rays (translated from Russian), 2nd ed., AEC-tr-4502. Office of Technical Services, Dept. Commerce, Washington, D.C.

Brown, J. G. (1966). "X-Rays and Their Applications." Plenum Press, New York.

Campbell, W. J., and Brown, J. D. (1968). *Anal. Chem.* **40,** 346R–375R.

Campbell, W. J., and Gilfrich, J. V. (1970). *Anal. Chem.* **42,** 248R–256R.

Clark, G. L. (1955). "Applied X-Rays," 4th ed. McGraw-Hill, New York.

Compton, A. H., and Allison, S. K. (1935). "X-Rays in Theory and Experiment." Van Nostrand Reinhold, Princeton, New Jersey.

Flügge, S. (ed.) (1957). "Handbuch der Physik," Vol. 30. Springer-Verlag, Berlin and New York.

General Electric Co., X-Ray Dept. (1969). X-Ray Wavelengths for Spectrometer, 5th ed.

von Hevesy, G. (1932). "Chemical Analysis by X-Rays and Its Applications." McGraw-Hill, New York.

Jenkins, R., and deVries, J. L. (1969). "Practical X-Ray Spectrometry," 2nd ed. Springer-Verlag, Berlin and New York.

Kaelble, E. F. (ed.) (1967). "Handbook of X-Rays for Diffraction Emission, Absorption, and Microscopy." McGraw-Hill, New York.

Liebhafsky, H. A., Pfeiffer, H. G., Winslow, E. H., and Zemany, P. D. (1960) "X-Ray Absorption and Emission in Analytical Chemistry." Wiley, New York.

Liebhafsky, H. A., Pfeiffer, H. G., Winslow, E. H., and Zemany, P. D. (1972). "X-Rays, Electrons, and Analytical Chemistry. Spectrochemical Analysis with X-Rays." Wiley, New New York.

Masek, P. R., Sutherland, I., and Grivel, S. (eds.) (1970). X-Ray Fluorescence Spectrometry Abstracts. Science and Technology Agency, London.

Müller, R. O. (1972). "Spectrochemical Analysis by X-Ray Fluorescence" (translated from German by Klaus Keil). Plenum Press, New York.

Spielberg, N. (1963). *Appl. Spectrosc.* **17,** 6–9.

White, E. W., Gibbs, G. V., Johnson, G. G., Jr., and Zechman, G. R., Jr. (1965). X-Ray Emission Line Wavelength and Two-Theta Tables. ASTM Data Ser. 37.

Worsnop, B. L., and Chalklin, F. C. (1950). "X-Rays." Methuen, London.

Author Index

Numbers in italics refer to the pages on which the complete references are listed.

C

N

O

P

R

Subject Index

Z

A 4
B 5
C 6
D 7
E 8
F 9
G 0
H 1
I 2
J 3